GEOGRAPHIC INFORMATION SYSTEMS:

A Management Perspective

Stan Aronoff

WDL Publications
Ottawa, Canada

Canadian Cataloguing in Publication Data

Aronoff, Stanley
  Geographic information systems: a management perspective

Includes index.
Bibliography: p.
ISBN 0-921804-91-1

1. Geography — Data processing. 2. Information storage and retrieval systems — Geography.
I. Title.

G70.2.A79 1991   910'.285   C89-090266-6

Published by:

WDL Publications
P.O. Box 8457, Station T
Ottawa, Ontario K1G 3H8
Canada

Printed and bound in the United States.

Geographic Information Systems: A Management Perspective
Copyright © 1989, 1991, 1993 by Stanley Aronoff
Second printing, 1991
Third printing, 1993
Fourth printing, 1995

All illustrations designed by the author except where otherwise noted.

To Michael Schacter (1952–1989),
my oldest and closest friend, who was
always a source of encouragement and counsel.

PREFACE

This book presents the principles of geographic information systems (GIS). It is designed to provide a complete introduction to the subject, addressing both the technical and organizational issues.

Perhaps more importantly, the book as a whole illustrates an approach to using geographic information. The many examples of practical GIS applications provide the context within which a GIS can be usefully applied. This is not a technical book describing how to operate a particular system or comparing specific product offerings. The principles discussed here apply to all GISes.

The book is addressed to the wide range of practitioners and students who are now using or learning to use GIS technology in their work. The book's approach is a management perspective, in the sense that the GIS is presented as a system for making management decisions. Whether it is planning the construction of a roadway, the harvesting of timber, or the location of a nature preserve, these activities can be better assessed and managed using GIS methods.

Unusual for a book of this type is the inclusion of a section on remote sensing. Remote sensing provides a powerful data source that is often poorly understood by those who could benefit most from its use. In my experience, the information obtainable from remotely sensed data is not only more effective when used within a GIS, but the GIS can also be used to improve the accuracy of the remote sensing analysis.

I would like to acknowledge the friends and colleagues who reviewed chapters of the manuscript. They are: Tom Alfoldii, Ron Brown, Russ Congalton, Jack Dangermond, Earl Epstein, Andrea Fabbri, Ernest Hardy, Roy Mead, Paul Pearl, Donna Peuquet, Michael Schacter, Randy Thomas, Dana Tomlin, and Roger Tomlinson.

I would like to thank the many individuals who reviewed sections of the manuscript, produced graphics, or helped to obtain existing illustrations for use in the book. They are: Larry Amos, Bill Brooks, Jean Chartrand, Malgosia Chelkowska, Richard Dobbins, Neil Grant, Don Hemenway, Karl Humphreys, Mike Kirby, Ken Korporal, Vlad Kratky, Bob LaMacchia, André Leclerc, Mike Manor, Blair Moxon, David Nystrom, Elizabeth Ottaway, Simsek Pala, Rick Pierce, Jerry Porter, Brian Rizzo, Gary Roberts, Vincent Robinson, Grafton Ross, Robert Ryerson, Pamela Sallaway, Barry Schneider, Régin Simard, Roy Slaney, Roger Slothower, Susan Smith, Susan Till, Marguerite Trindade, Dave White, Elaine Wilson, Brian Wright, and Joel Yan.

Many individuals and organizations generously contributed illustrations, in some cases producing them specially for this book. They are individually acknowledged in the figure captions. However, I would like to make special mention of the Ontario Centre for Remote Sensing and the Environmental Systems Research Institute.

The Ontario Centre for Remote Sensing produced custom-enhanced Landsat images and provided financial assistance for the production of colour plates.

The Environmental Systems Research Institute provided many illustrations from their reports and map books. I would like to thank Karen Hurlbut at ESRI in Redlands, California, and Corien Greenwood at ESRI Canada in Toronto, Ontario, for their help in assembling and reproducing those graphics.

Tom Lillesand generously provided several illustrations from the book *Remote Sensing Image Interpretation* (published by John Wiley and Sons Ltd.) that he co-authored with Ralph Keifer.

In researching this book I made extensive use of the library facilities and the RESORS information retrieval service of the Canada Centre for Remote Sensing, as well as the CISTI Library of the National Research Council of Canada. I would like to thank the staff of these facilities for their assistance.

I would also like to acknowledge those who were involved in the production of the book: copy editing was done by Corien Greenwood and Audrey Kaplan, Hal Shuster and Patrick White provided valuable advice on production methods. Jerry Tarasofsky first suggested that I write this book. His sound advice and encouragement were greatly appreciated.

The writing and production of a book is a long, hard project for which large doses of moral support are regularly needed. My closest friend, Michael Schacter, was one of my most ardent supporters. He reviewed several chapters, and provided much encouragement and good counsel throughout the project. Tragically, he was not able to see its completion. He was a remarkable individual. He had a solid scientific background, a sensitive appreciation of the arts, and the wisdom to understand the interrelationships between the two. Michael was keenly interested in the book and despite his illness he invested a great deal of time in its progress. I deeply appreciate his efforts and feel that some of his thoughts and approach live on in the book.

I would especially like to thank my wife, Audrey, whose enthusiasm carried through all phases of this project. She provided quality control for the graphics, and critical review of many aspects of the book production.

Stan Aronoff
Ottawa, April 1989

FOREWORD

by Jack Dangermond, President
Environmental Systems Research Institute
Redlands, California

I believe that Geographic Information System (GIS) technology is important today because it offers an important — perhaps even a critically important — means of understanding and dealing with some of the most pressing problems of our time; problems like tropical deforestation, the future of the global climate, the need for the ecologically sensitive development of global natural resources, acid rain, and rapid urbanization, to name but a few. As Stan Aronoff describes in more detail in this book, GIS technology helps us organize the data about such problems and understand their spatial associations, and provides a powerful means for analyzing and synthesizing information about them. I look forward to the day when GIS technology will be part of the decision support environment of everyone who makes decisions that affect either the natural world or the world we have built — from decisions about local areas to decisions about the earth as a whole.

For that day to come, GIS technology must be more widely known, more widely understood, and more widely appreciated. This book can play an important role in that process.

The book is aimed at both users and managers of GIS technology, and assumes that the reader has no previous background in the field. The book is designed to provide a novice user or manager with a logically complete introduction to what she or he needs to know in order to deal effectively with GIS technology. For such persons it should provide a real service.

Managers will find in the book a broad introduction to the field that they can either read through completely, use for quick reference, or use as a way of getting into the current technical literature since the book has a good selection of citations to recent writing in the field. The organization of the chapters follows the sequence of steps that most organizations would go through in considering, selecting, and implementing a GIS. The chapter on "Implementing a GIS" will be especially valuable for those who are in the process of acquiring GIS technology. It provides good advice and identifies many of the pitfalls that are hidden in that process.

The chapter on remote sensing is an appropriate and necessary part of this book, given the essential role that remote sensing is now playing in data gathering, rectification, and updating in many GIS applications. This is especially so since more and more remote sensing data are being used chiefly within the context of GISes. The chapter provides a very good introduction to many facets of remote sensing and places remote sensing clearly in the context of Geographic Information Systems technology as a whole.

The book is up-to-date, discusses concerns critical to managers, and doesn't hesitate to deal with controversial topics. A large selection of very useful illustrations, which readers may want to return to again and again, supplement and illuminate the text. Actual examples of GIS projects, which also indicate the range of GIS applications, are cited extensively throughout the book.

The book has a clear, crisp writing style that manages to be comprehensive and thorough without being cumbersome or overwhelming to persons new to the field. The well-balanced accounts of topics, especially controversial topics, may be read with profit even by experienced professionals in the field. In its treatment of many topics, the book provides wise advice offered from the perspective of experience. Several chapters also convey some of the history of GIS technology.

Stan Aronoff has written a very good book about an important topic. If it is widely read, I believe it can make a difference in the quality of life on this planet.

I hope that will happen.

LIST OF COLOUR PLATES

TABLE OF CONTENTS

1. AN INTRODUCTION TO GEOGRAPHIC INFORMATION SYSTEMS

INTRODUCTION

Geographic information systems (hereafter abbreviated to GIS) are computer-based systems that are used to store and manipulate geographic information. This technology has developed so rapidly over the past two decades that it is now accepted as an essential tool for the effective use of geographic information.

The recent and widespread introduction of the GIS has created a sudden need for users of geographic information to become knowledgeable about this technology. Managers within public and private organizations are being called upon to make decisions about the introduction of GIS technology and to establish policies for its use. Politicians are being asked to support expensive programs to convert mapped data into digital form suitable for use with a GIS. Students and educators who use geographic information are gaining access to GIS technology that can be used to increase the depth and breadth of their analyses.

The technology has provided an exciting potential for geographic information to be used more systematically and by a greater diversity of disciplines than ever before. However, the ease with which a GIS can manipulate geographic information has also created a major difficulty. Users unfamiliar with GIS techniques or the nature of geographic information can just as easily conduct invalid analyses as valid ones. Valid or not, the results have the air of precision associated with sophisticated computer graphics and volumes of numerical tabulations. A better understanding of GIS technology by users, managers, and decision-makers is crucial to the appropriate use of the technology.

This book is addressed to those users who may have no previous experience with computer-based geographic information handling but who need to use or direct the use of GIS technology. It provides a concise introduction to the fundamentals of GIS, the capabilities of these systems, and some of the issues that arise when a GIS is implemented.

This first chapter introduces geographic information systems by describing representative applications for which a GIS can be successfully used. The examples span many disciplines, including such widely-accepted applications as municipal facilities management, agriculture, and forestry. Several more unusual applications are also presented, such as predicting the location of archaeological sites or mapping plant distributions from museum records. These sample applications are not detailed project descriptions; rather they present ideas and concepts to give a sense of the range of GIS applications. These examples do, though, represent applications that have been implemented operationally or demonstrated and described in the literature. More detailed discussions of these topics and others can be found throughout this book and in the references listed at the end of the chapters.

GIS OVERVIEW

A GIS is designed for the collection, storage, and analysis of objects and phenomena where geographic location is an important characteristic or critical to the analysis. For example, the location of a fire station or the locations where soil erosion is most severe are key considerations in using this information. In each case, what it is and where it is must be taken into account.

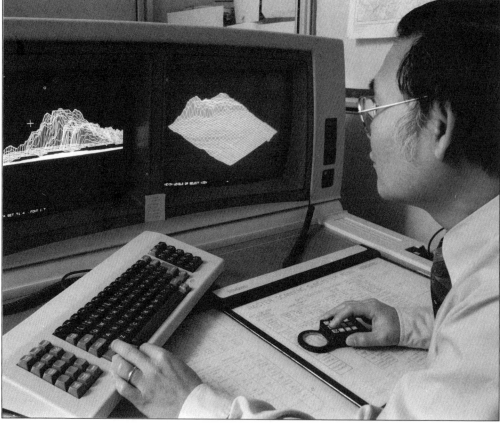

Figure 1.1 Interactive Analysis Using a GIS. (Courtesy of Energy, Mines, and Resources Canada. Ottawa, Ontario.)

While handling and analyzing data that are referenced to a geographic location are key capabilities of a GIS, the power of the system is most apparent when the quantity of data involved is too large to be handled manually. There may be hundreds or thousands of features to be considered, or there may be hundreds of factors associated with each feature or location. These data may exist as maps, tables of data, or even as lists of names and addresses. Such large volumes of data are not efficiently handled using manual methods. However, when those data have been input to a GIS, they can be easily manipulated and analyzed in ways that would be too costly, too time-consuming, or practically impossible to do using manual methods.

The applications are diverse, for example:

- finding the coincidence of factors, such as the areas with a certain combination of soil type and vegetation, or the areas in a city with a high crime rate and low income level;

- updating geographic information, such as forest cover maps to show recent logging, or updating land use maps to show recent conversion of agricultural land to residential development;

- managing municipal services, such as scheduling maintenance activities, notifying local residents of re-zoning applications, or assigning police patrol areas.

The number and type of applications and analyses that can be performed by a GIS are as large and diverse as the available geographic data sets.

Despite the analytical power of this technology, a GIS, like any other system, does not and cannot exist on its own; it must exist in a context. There must be an organization of people, facilities, and equipment responsible for implementing and maintaining the GIS. Moreover, that organization, like any organization, must have a mandate — a reason to exist — and the resources to satisfy that mandate. Without the organizational context, it becomes unclear why the considerable expense of implementing a GIS has been made, who should control the facility, and how its success or failure should be judged.

Ultimately, a GIS is used to produce information that is needed by a user, a client. That client may be a person or a group of people. They may be members of the public or representatives of an organization within government or private industry. Most importantly, the information required by the client provides the fundamental context in which the GIS should function. To be useful to the client, information must be of the right kind and quality, presented in an appropriate format for the client to use, and be available at the right time. The information in a GIS is presented in two basic forms: as maps and as tables. For example, a map can show where particular types of land use or activities occur. On the other hand, information on how much of a resource exists can be given in tabular form. For example, the quantity and types of timber in a forest can be shown as a table of quantities by tree species. In the end, the performance of a GIS is judged by those who will use the information it produces — the client.

As a result of the context in which a GIS operates, introducing a GIS is a much larger task than adding a new office machine. The GIS will fundamentally change the way information flows within the organization and between organizations. This change is organizational more than it is technical. A GIS can produce information much more quickly, achieve higher mapping standards, and keep data more current than was previously done. But, far more fundamental to the organization are the issues of who has access to the information, and what power those persons exercise in its analysis and distribution.

For example, a municipal public works department may maintain the maps of the city's water and sewage facilities. They would be responsible for the quality of the data and would also control access to them. Any other department wanting to use those maps would have to first consult the public works department. As a result, the public works department would be made aware of activities conducted by other departments. The process of the engineering department requesting maps could be an informal way for the public works department to be aware of any construction activities in the city. However, if a GIS were implemented and the map data became part of the on-line data base, the department could lose control over access to the data and its use. Any department could use the maps without the knowledge of the public works department. The informal flow of information about construction activities would cease and control of the information would no longer be in the hands of the public works department.

In themselves, these organizational changes are neither good nor bad. If the changes are anticipated, then suitable information management controls can be put in place. Therein lies the challenge. For a GIS to meet the needs of an organization, the information flows within the organization must be explicitly defined. Many of the most important information flows are through informal networks. Implementing a GIS can disrupt these informal networks, changing who has control of the information, and in so doing changing who has power.

Issues related to the flow and control of information are largely management issues and must be dealt with as such. They are addressed in Chapter 8, *Implementing a GIS*. The following section provides an overview of the range of applications to which GIS methods are being applied.

EXAMPLES OF GIS APPLICATIONS

AGRICULTURE AND LAND USE PLANNING

Agriculture, the production of food, is of such national and economic importance that it is usually better inventoried and monitored than other natural resources. Industrialized nations have well-developed national statistical reporting services that survey farmers, monitor growing conditions, monitor annual production, and predict national and international agricultural market demand and supply.

Private companies are also involved in direct crop inventory and monitoring. The CROPCAST reporting service offered by Earth Satellite Corporation is one example (Merrit et. al. 1984). Many food product companies that are dependent on large volumes of agricultural products from foreign nations have well developed in-house agricultural monitoring groups. They use a wide range of data sources including field reports, information on past production, meteorological satellite data, and the data from earth resources satellites such as Landsat and SPOT (discussed in Chapter 3). These data may be used in analyses of relative production levels by comparing current with past years' crop condition. More sophisticated analyses may involve the use of computer models to simulate the growth of a crop using weather data (e.g. temperature, precipitation, and hours of direct sun) and land use information (e.g. farming practices and previous production levels). Much of this information can be obtained from satellite observations.

At the national level, the monitoring of agricultural production has generally developed as a statistical reporting activity rather than a mapping one. The assessment of major agricultural areas is a well-developed science carried out by national agencies using procedures that were developed before powerful GIS software became available. Mapping activities are generally focussed on soils and agricultural suitability mapping. They are important mandates assigned to national and regional government organizations.

Many of the organizations responsible for monitoring agricultural land use have now adopted GIS methods. Land use and weather data provided from satellites and field measurements as well as production information from previous years are analyzed together to predict for each region the expected yield of one or more crops. Often the procedure is repeated at regular intervals throughout the growing season as conditions change. Such crops as rice, wheat, canola, potato, cocoa, and coffee are monitored using these techniques.

In addition to crop production, GIS procedures have been used to evaluate management practices for grazing lands. By tracking grazing type and intensity over time, and collecting data on vegetation condition and weather data, grazing practices can be assessed and improved (Graetz et. al. 1986).

The first national scale geographic information system was the Canada Geographic Information System (CGIS). This system has been operating continuously since the late 1960s. It evolved from a project to develop a national land capability classification to compile an inventory of all potentially productive land in Canada. In addition to agricultural capability information, other land use information categories were included such as forestry, recreation, and wildlife.

Over the twenty years since its inception, the CGIS has been extensively modified and now operates as one component of an integrated group of computer-based geographic information systems, termed the Canada Land Data Systems (CLDS). This facility operates on a cost-recovery basis, providing analysis services to national, provincial, and municipal agencies throughout Canada. It is also used by international organizations such as the United Nations Food and Agriculture Organization (FAO) (Crain 1987, Tomlinson et. al. 1976).

In Europe the CORINE system, a GIS being developed by the European Community, will be used extensively for the comparative analysis of agriculture and other land uses of the member countries.

At the local level, agriculture-related planning has usually been done in the context of more general land use planning activities. For example, the Oklahoma Conservation Commission and Oklahoma State University's Center for Applications of Remote Sensing used a GIS for soil conservation planning. The integrated analysis of soil type, slope, farming practices, and crop type were used to predict soil erosion so that erosion control programs could be targeted to the highest risk areas. This type of application is illustrated in Figure 1.2 (see Walsh, 1985).

A similar GIS application was demonstrated in the Dane County Land Records Project. Begun in 1982, the project brought together local, state, and federal agencies with the University of Wisconsin to demonstrate the use of GIS techniques for rural land use planning in Dane County, a major agricultural county in Wisconsin. County conservation officials used a GIS to analyze data on land-ownership, soils, and land use to identify farms that were in compliance with conservation standards. Using the GIS, they were able to screen the 100 to 200 farms in a township in about 4 hours. The few farms that did not appear to be complying with the conservation standards were then field checked. The method used previously required some 2 to 4 hours to assess each farm.

Economic feasibility was demonstrated by the careful recording of all costs. Substantial savings were realized by the use of scanning techniques for digitizing, the automated classification of satellite data to produce land use maps, and the use of satellites to precisely identify ground control points instead of using traditional surveying techniques. The project demonstrated that the

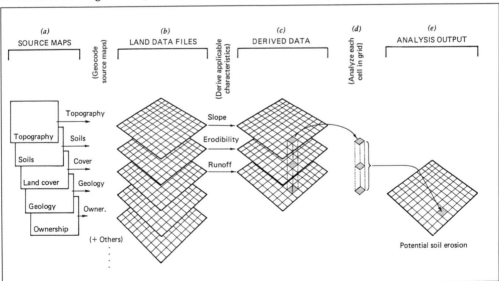

Figure 1.2 Analysis Procedure for Soil Erosion Planning. (From *Remote Sensing and Image Interpretation* by Lillesand and Keifer 1987, published by John Wiley and Sons.)

participating agencies could use automated methods to collect the same information more quickly and at lower cost than by using manual techniques. It was also demonstrated that there were substantial additional benefits. The GIS was used to perform analyses that were previously too costly or not feasible. (For more information see Ventura et. al. 1988 a,b; and Moyer et. al. 1988).

It is perhaps at the local level that GIS technology will have the greatest affect on the way agricultural land is monitored. Relatively simple GIS techniques can provide a level of information processing that allows alternative scenarios to be evaluated, amended, and re-evaluated at a reasonable cost. The production of maps showing the coincidence of conditions important for planning and management, such as erosion control, groundwater pollution control, and changes in land use, becomes relatively straight forward using a GIS.

Land use management and policy decisions are almost always based on the analysis of the interplay of factors pertaining to an issue. For example, the preservation of prime agricultural land involves political, institutional, and economic factors. A GIS can provide better information to support this type of complex decision-making. Such basic information as the distribution of current land use activities, the relationship of these activites to agricultural capability, land prices, and urban demand for land can be developed quickly using GIS tools. GIS techniques can foster better decision-making by allowing managers to conduct analyses that would be impractical or infeasible otherwise.

A major issue in the adoption of GIS techniques for land use planning at the local level has been how to put the technology in the hands of the decision-makers. One approach has been to organize cooperative projects in which a potential user agency makes use of a facility operated by GIS experts, as was done in the Dane County project. Another approach has been to bring the technology to the decision-makers. The University of Kansas Space Technology Center used this approach when it provided GIS technology to the water resource managers in Harvey County, Kansas. The GIS was used to support planning decisions for groundwater and surface water quality, irrigation, and municipal water supply. The personal computer-based GIS was operated by the land management professionals themselves. They specified the data inputs, outputs, and analytical capabilities needed, structured the planning and decision-making analyses, executed the analyses on the GIS, and evaluated the results (Merchant and Caron, 1986).

Perhaps one of the most original efforts to make GIS tools widely available is the Doomsday Project, led by the British Broadcasting Corporation in the United Kingdom. Based on a microcomputer and video disk technology, some 21,000 files of maps, photos, satellite imagery, tabular, and text information can be viewed and manipulated (Figure 1.3). The data include geology, soils, geochemistry, population, employment, agricultural production, land use/land cover, and other categories.

The complete system, hardware and software, was made available to schools for £3,000 sterling in November 1986. The general public could purchase the system for £4,600 sterling. These prices included the hardware and the digital database (containing coverage of the entire United Kingdom and valued at over £250,000 sterling).

The GIS software was developed to be easily used by school children. One of the major effects of the system was to make large amounts of geographic data, which were previously controlled by the data gatherers, accessible to the public. Perhaps this is the ultimate extension of putting the geographic data and data handling tools in the hands of the user (Rhind et. al. 1988, Rhind and Openshaw 1987).

Figure 1.3 The Doomsday GIS Developed by the BBC. The microcomputer-based GIS uses a video disk player to store some 21,000 files of maps, photos, satellite imagery, tabular, and text information. An easily learned interface enables the user to view and manipulate the data sets together. (Courtesy of The Doomsday Project and Birkbeck College. London, U.K.)

FORESTRY AND WILDLIFE MANAGEMENT

Forestry encompasses the management of a wide range of natural resources that occur in forest areas. In addition to timber, forests provide such resources as grazing land for livestock, recreation areas, wildlife habitat, and water supply sources. As a result, the public agencies responsible for forestry management typically have broad mandates. The responsibilities of the US Forest Service include management of forest harvesting, wildlife habitat, grazing leases, recreational areas, mining activities, and the protection of endangered species and archaeological sites. To satisfy these diverse responsibilities, competing resource conservation and resource use activities must be accommodated.

Assessing the compatibility of multiple uses and trading-off competing values are difficult planning processes that are greatly aided by using GIS techniques. For example, the GIS for Flathead National Forest in Montana includes digital terrain data, vegetation associations derived from Landsat satellite data, timber compartments, timber harvest history, land types, land ownership, administrative districts, precipitation, and the drainage network. The GIS has been used for such analyses as timber harvest planning, critical wildlife habitat protection, and planning the route location for scenic roadways (Hart et. al. 1985).

Over the past 5 years GIS technology has been widely accepted by public forestry agencies and private forestry companies alike. In large part, this has been a result of the clear benefits of more current forest inventory maps. The forest inventory is the primary management tool for timber production in North America. It is used to assess the existing forest resource and to develop harvest schedules and treatment programs, to project future timber supplies, and for other operational planning activities. Forest inventory data are collected using remote sensing techniques supported by field data collection. The basic forest unit is the **stand**, a forested area with relatively uniform species composition, tree size, and density. Skilled interpreters identify stand boundaries on large scale aerial photographs (scales of 1:10,000 to 1:20,000 are commonly used). Information like species composition, age, height, structure and condition is derived from the airphotos and supporting field data. This information is entered into the GIS. Other data sets commonly included in forestry GIS data bases are soils mapping, legal subdivisions, the road network, and drainage systems. Historical information pertaining to each stand, such as harvesting, regeneration, road building, or forest damage, can also be included in the GIS.

The conventional forest inventory was done progressively, a portion of the forest area being inventoried each year. However, to update the thousands of forest cover maps managed by a forest agency could take 20 years or more and was an expensive manual drafting operation. During the years between inventories, changes to the forest cover maps might be hand-drawn onto the existing map base by local personnel, but re-drafting of the maps on an annual basis was prohibitively expensive. With a GIS, forest cover maps can be updated on a continual basis and then output from the GIS as needed. In this way forest managers are provided with information more current than previously available. With GIS technology, the average age of information in the forest data base could be reduced from years to weeks. This factor alone has led to wide acceptance and demand for GIS technology in the forestry sector.

In itself the use of a GIS to update forest inventory maps is little more than automated cartography, using computer technology for an existing manual process. It is the analytical power of the GIS that sets it apart. The GIS can be used to store and analyze the forest information in ways that could not previously be done. It can be used to calculate the harvestable timber in an area, model the spread of forest fires, or develop and evaluate alternative harvest plans. The processing power of a GIS allows several alternatives to be evaluated relatively quickly. This has led to a qualitative change in the way many analyses can be approached. Plans can be progressively refined and re-evaluated to optimize a solution, a procedure that would be prohibitively expensive using manual techniques.

Virtually all government forest management agencies in North America have acquired or are considering acquiring a GIS. GISes have been widely used by the US Forest Service. Until recently they were used primarily on a project basis, such as for the development of environmental impact assessments. When the project was completed the data often were not maintained and became out-dated.

In the mid-1980s three national forests were selected as GIS Evaluation Sites. They were the George Washington National Forest in Virginia, the Tongass National Forest in Alaska, and the Siuslaw National Forest in Oregon. At each of these sites a GIS was installed and comprehensive data bases for these national forests were implemented. Data were collected on the various costs and benefits associated with implementing these systems. After a three year evaluation period, the US Forest Service has decided to implement a standardized GIS data base for the entire forest service. A $150 million procurement has been initiated for GIS hardware and software to be installed in 600 locations beginning in 1991. This expenditure is expected to represent only 15% of the total implementation budget. Seventy-five percent of the budget will be devoted to data base creation, 5% for workload analysis and data base design, and 5% for training (Mead 1989).

In Region 8 of the Forest Service (which covers the south-eastern portion of the United States), standardized GIS data bases are being completed at the rate of 2 ranger districts per month. It takes about 3 years to complete the forest data base for each district, which in this region range in size from 150,000 to 250,000 acres. Twenty-eight of the 105 districts have been completed, and half are expected to be complete by the end of 1990. GIS software and hardware are being provided by contractors until the final system is acquired. However, the data base collection effort and system specifications have been designed to ensure compatibility with whatever GIS is selected. As the data bases are completed, they are immediately put into operational use and are maintained. A set of standard analysis products has been defined that can be ordered routinely from the contractor. Plate 1 is an example of a standard Forest Cover Type map produced using the GIS. Special purpose analyses are handled on an *ad hoc* basis.

Among the major benefits that Region 8 has realized from this GIS implementation has been the improved analysis of allowable sale quantities (ASQ) for planning forest harvest operations and more accurate habitat assessment for threatened and endangered species (e.g. see the study on Gopher Tortoise habitat in Mead et. al. 1988).

In Canada, every provincial forestry agency has either implemented a GIS to manage their forest data base or is in the process of automating their forestry maps. The British Columbia Ministry of Forests and Lands has been one of the leaders in developing operational GIS and remote sensing applications for forestry. Forestry is one of the most important industries in the province. British Columbia's forests cover some 50 million hectares and contain almost 8 billion cubic metres of merchantable timber, about 40% of Canada's timber supply. The Ministry of Forests and Lands is responsible for maintaining an inventory of the forest resource. In 1978 a new provincial Forest Act required the Ministry to provide more detailed forest information and to update the forest inventory at more frequent intervals. Also, the conversion to metric map scales required that all of the 7,000 forest cover maps be re-drawn.

To satisfy these new requirements a decision was made to automate the forest inventory maps and to introduce or develop state-of-the-art GIS and remote sensing methods. Traditionally, the forest inventory data base had been developed and updated using 1:20,000 scale aerial photography and field sampling methods. Forest cover maps were updated by manual drafting and forest statistics were maintained separately from the map base.

Under the GIS implementation program, more than half of the forest cover maps have now been digitized and entered into the forestry GIS. Conversion of all of the 7,000 maps is to be completed by 1991. Each digital map sheet consists of as many as 19 layers of georeferenced information, including forest cover data, roads, drainage, land ownership data, and several levels of administrative and regulatory boundaries. Digital elevation data has also been entered into the GIS to allow elevation, slope, and aspect information to be used in analyses. All descriptive stand information is linked to the map data within the GIS, providing a very flexible information system.

The detailed forest inventory is being updated on a 10 year cycle using aerial photography. In addition, Landsat satellite imagery is being used to update individual maps on an annual basis. Depletions from forest harvest operations or fire are digitally mapped directly from the digital satellite imagery. The image data and map information are presented to the interpreter on a colour monitor. Changes are mapped interactively on the display and are then used to update the forest data base (see Plate 2). In this way the forest information for an area remains current throughout the 10 year inventory cycle.

In addition to providing data storage and retrieval functions, the B.C. Forest GIS generates a wide range of information used in forest harvest planning, regeneration surveys and monitoring, environmental sensitivity assessment, recreation planning, and watershed management (Hegyi and Sallaway 1986). The province-wide data base is also being made available at the District level. By 1991 each forest District will be provided with a GIS and the corresponding geographic data for the region.

In many sub-tropical and tropical countries, forests are poorly mapped and inaccessible. Deforestation is a serious problem, contributing to soil degradation, soil erosion, flooding, and water quality problems. The ability of an area to support forest harvesting depends on such environmental factors as the topography (steep slopes erode quickly when forest is removed), the soil type, the method used to cut the trees (e.g. clear cutting or selective cutting), as well as economic factors (e.g. market price, production and transportation costs). Yet these regions often do not have the funds to collect forest information using conventional methods. Remote sensing and GIS techniques can provide cost-effective alternatives to generate basic resource planning information. Reconnaissance level information can be used to initiate resource

management programs quickly and to identify urgent problems such as areas exposed to serious erosion. Once a resource data base system is in place, more detailed information can be added progressively as funds and man-power become available. (See for example Hutachareon 1988).

GIS methods have been used to take advantage of unusual sources of information. Mapping the distribution of non-tree plant species is a common forest resource management activity, especially for protection of endangered species. Herbariums are "plant museums" that preserve and catalogue plant specimens used for scientific research and education. In addition to the scientific name and classification of each specimen, records are kept of such information as the date and location of collection. At the University of Hawaii, a GIS was used to analyze the herbarium records.

The specimen name, date, and place of collection were entered into a GIS. The system was then used to select and map the location of specimens according to their species, genus, or collection date. The search area could be selected by latitude and longitude coordinates or by island. In this way the distribution of plant species among the different Hawaiian islands could be mapped and compared (McGranaghan and Wester 1988). Figure 1.4 illustrates the results of one analysis from this study. It is a map of the locations throughout the Hawaiian islands where specimens of *Cyrtandra paludosa* had been collected. The map provides an indication of how widespread is the distribution of the species. By combining this type of species mapping with the boundary data for existing preserves, habitat areas already protected can be identified. This relatively straightforward GIS application illustrates the diverse applications to which the technology is easily applied.

An important component of wildlife management is the prediction of the effects of human activities and natural events on the

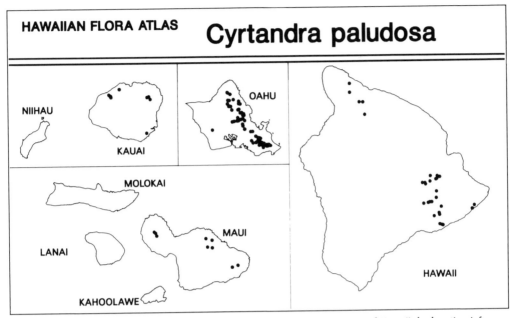

HAWAIIAN FLORA ATLAS   **Cyrtandra paludosa**

Figure 1.4 Geographic Distribution of *Cyrtandra paludosa*. At the University of Hawaii the location information from herbarium specimens was used to develop a micro-computer based system to map the distribution of plant species. The map shows the locations where specimens of *Cyrtandra paludosa* have been collected. (Courtesy of M. McGranaghan and L. Wester. University of Hawaii. Honolulu, Hawaii.)

abundance and quality of wildlife populations. This information is used in making such decisions as choosing a right-of-way for road construction or closing critical wildlife areas to public access. Wildlife depends on the presence of an appropriate mix of resources within a geographically defined area. A GIS can be used to analyze such factors as the availability of food and cover, protection from predators, and the suitability of areas for nesting or denning sites. GIS techniques have been used to analyze the habitat of a wide range of animal species, including grizzly bear, elk, deer, caribou, and various bird species. Data layers for wildlife habitat suitability and critical wildlife areas are commonly included in forestry data bases in North America.

Figure 1.5 is a plot of Gopher Tortoise habitat suitability in the Ocala National Forest in Florida. (The Gopher Tortoise is an endangered species.) A GIS was used to produce this map from four data layers; soil type, forest cover type, forest condition class, and understory cover type. The GIS was also used to predict the tortoise population and habitat availability forward and backward in time. Figure 1.6 is a histogram of the quantity of available Gopher Tortoise habitat over several years. This type of analysis was used to evaluate the trend of habitat availability and to develop a management plan for this species. (See also Mead et. al. 1988.) Similar GIS techniques are also being used to predict and control forest pest outbreaks such as spruce budworm in New Brunswick (Jordan and Vietinghoff 1987) and gypsy moth in Michigan (Montgomery 1987).

One of the most innovative uses of GIS and remote sensing technology has been the tracking of wildlife by satellite telemetry. The U.S. Fish and Wildlife Service has been monitoring caribou in Alaska since 1985. A satellite transmitter is incoporated into a collar on the animal and the signals are received by one of the NOAA meteorological

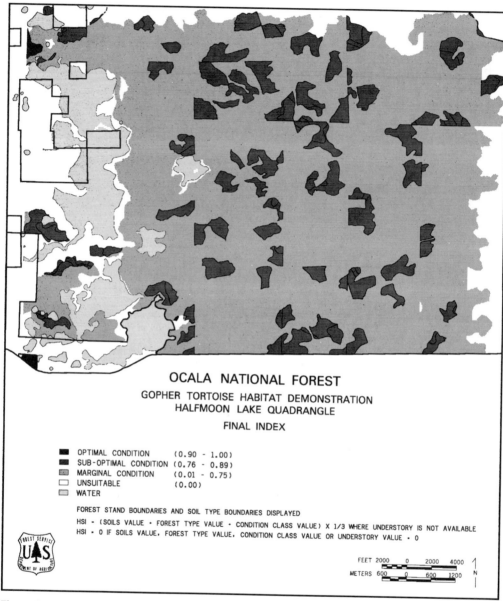

OCALA NATIONAL FOREST
GOPHER TORTOISE HABITAT DEMONSTRATION
HALFMOON LAKE QUADRANGLE

FINAL INDEX

	OPTIMAL CONDITION	(0.90 - 1.00)
	SUB-OPTIMAL CONDITION	(0.76 - 0.89)
	MARGINAL CONDITION	(0.01 - 0.75)
	UNSUITABLE	(0.00)
	WATER	

FOREST STAND BOUNDARIES AND SOIL TYPE BOUNDARIES DISPLAYED

HSI = (SOILS VALUE · FOREST TYPE VALUE · CONDITION CLASS VALUE) X 1/3 WHERE UNDERSTORY IS NOT AVAILABLE
HSI · 0 IF SOILS VALUE, FOREST TYPE VALUE, CONDITION CLASS VALUE OR UNDERSTORY VALUE · 0

FEET 2000   0   2000   4000
METERS 600   0   600   1200

Figure 1.5 Map of Gopher Tortoise Habitat Suitability in the Ocala National Forest, Florida. (Courtesy of the US Forest Service. Atlanta, Georgia.)

satellites. Initially, 10 females from the Porcupine herd, and 10 females from the Central Arctic herd were tracked. The program has since been expanded so that more animals can be monitored. Data, collected five or more times a day, give the geographic location of each animal (with an average positional accuracy of 500 m), an index of air temperature, and a measure of activity level.

A GIS was used to analyze the animal location data to assess the potential impact of pipeline, infrastructure, and other development on caribou that use the Alaska National

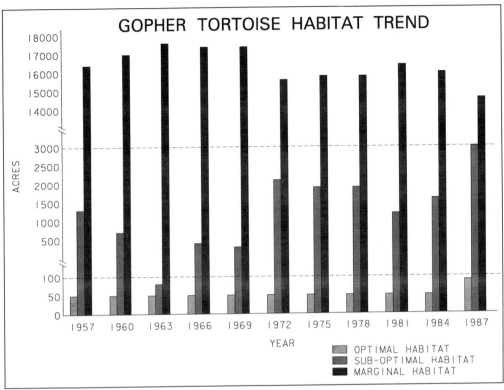

Figure 1.6 Histogram of Gopher Tortoise Habitat Availability Over Time. The histogram shows the area of available tortoise habitat in the Ocala National Forest in each year. Land cover changes such as natural plant succession, fire, and forest harvesting can change the amount of habitat available to this species over time. (Courtesy of the US Forest Service. Atlanta, Georgia.)

Wildlife Refuge. Plate 3 is a plot from the GIS of the movements of the animals during five different time periods between April 1985 and December 1986. This study has shown that some caribou migrate as much as 2000 km in a single year and that individuals do not consistently use the same migration routes, wintering areas, or post-calving areas each year. Similar studies are being conducted to track musk-ox, walrus, wolf, mountain sheep, deer, elk, and polar bear using satellite telemetry for data acquisition, and GIS analysis methods (Pank 1989). These techniques have enabled wildlife biologists to track animals over larger distances more accurately and more efficiently than could be done previously. Animals can also be tracked across international boundaries. The data, collected in digital form, can be used

not only to study wildlife but also to influence planning decisions that affect the habitat of these animals.

ARCHAEOLOGY

The protection of archaeological sites is a common mandate of National Park and Forestry agencies. Archaeologists have made use of GIS techniques both to analyze known sites and to predict the location of undiscovered sites. Archaeological measurements, such as site size, location, age, number of artifacts, number of dwellings, together with environmental measurements (such as elevation, slope, aspect, local terrain relief and distance from water) have been used to predict the location of archaeological sites. That these factors are good predictors is not surprising since humans select settlement sites based on the proximity of resources like water and food, a comfortable microclimate, and safety (Custer et. al. 1983).

To develop a predictive model, these measurements are collected for known archaeological sites in a study area. Environmental data are collected for the entire study area as well. The model is then calibrated using a set of locations known to be archaeological sites and a set of locations known not to be sites. Then the model is applied to the entire study area using the environmental data alone. The GIS is used to generate most of the environmental measurements for the area from digital data of the land surface, termed digital terrain data. Reports are produced in map form showing the density and distribution of known and predicted archaeological sites and archaeological and environmental measurements are provided in tabular form.

GEOLOGY

The analysis of the geology of a region, whether for mineral exploration, petroleum exploration, or reconnaissance level mapping, is fundamentally a data integration procedure. The geologist seeks to identify useful geologic patterns in the landscape by relating diverse geologic data sets.

Field sketches are used to record direct observations; the concentration of elements dissolved in local streams provides clues to the composition of rock materials within the drainage basin; aeromagnetic and gravity surveys are used to map subtle changes in the earth's magnetic and gravitational fields that may indicate the presence of significant ore deposits. To be of use, all these data must be analyzed with reference to their geographic location. By providing the capability to display and analyze diverse data sets together, a GIS enables the geologist to work with the data more quickly,

more accurately, and in ways that would not be practical using manual methods.

Many important mineral deposits are not exposed at the earth's surface. To locate them, it is necessary to infer their presence. **Tungsten** is one economically important mineral that is commonly found associated with geologic structures called **plutons**, which are formed deep beneath the earth's surface when a large mass of molten rock is forced into a deeply-buried existing rock unit. As this molten rock or **magma** slowly cools, minerals are formed within and adjacent to it. Depending on the composition of the surrounding rock, the magma, and the physical environment, economically valuable minerals may be formed. Under the right conditions, the fluids migrating away from the magma will form **scheelite**, a mineral rich in tungsten. The mineral is deposited in a layer, called a **skarn**, above the magma.

Over time, the magma cools forming a large massive rock unit, the pluton. If the overlying rock is eroded away, the pluton may be exposed at the earth's surface. Unfortunately, the process of erosion that exposes the pluton also destroys much of the tungsten skarn, the valuable mineral. For this reason, economically worthwhile tungsten ore deposits are more likely to be associated with buried plutons than with exposed plutons. The problem of course is how to identify a buried pluton.

In a recent study, GIS techniques were used to identify sites likely to be shallowly-buried plutons with tungsten skarns. The study area was located in the Nahanni Region of the Yukon and Northwest Territories in Canada. The Mactung and Cantung deposits, two of the largest tungsten skarn deposits in the world, are located in this region. By studying the known ore deposits, geologists were able to define a set of characteristics that might be used to indicate the presence of other deposits in the region. (More detailed discussions of this work can be found in Aronoff et. al. 1986 and in Goodfellow and Aronoff 1988.)

The data sets used in the analysis are shown in Figure 1.7. They included a 1:125,000 scale geological map, stream sediment geochemical data for 11 elements from some 1000 sample points, a catchment basin map showing the drainage area associated with each stream sample, a digital Landsat satellite image of the study area, and the locations of known mineral occurances in the region. The non-digital data sets were converted into digital form and input into the GIS, in this case a system based on image analysis technology. Separate digital data sets were produced for each of the 11 elements in the geochemistry data set, the mineral occurances, and the geology map. The element concentration in a stream sediment or water sample reflects the composition of rocks underlying the entire area drained by that stream (i.e. the entire catchment basin), not just the rock units at the sample site. To show this relationship geographically, in each of the 11 geochemical overlays, the entire catchment basin was assigned the concentration value of the sample point.

The more important factors that were considered to be positive indicators of tungsten skarns were:

1. The presence of high stream sediment tungsten concentrations

2. The presence of curvilinear features visible on the satellite imagery that might indicate ring fractures associated with buried plutons

3. A low Ph value in the stream sample

4. A high concentration of arsenic in the stream sediment sample

5. A high concentration of copper in the stream sediment sample

A coding scheme was developed to indicate the coincidence of favourable indicators as index values from 0 to 127. Plate 4 shows the Landsat satellite image of the study area. Exposed plutonic rock units are shown in red. Five of the higher ranges of index values are displayed. From highest (most favourable) to lowest, the sequence is pink, orange, yellow, green, blue. Three new exploration targets were identified based on this analysis.

Critical to the process was the ability to overlay and mathematically combine selected data layers and the ability to display the results overlayed on the satellite image, the topographic map, or the geologic map, features that require computer-based GIS capabilities. These features enabled the geographic relationships among the different geochemical indicators, the surface geology, and the structural geology visible

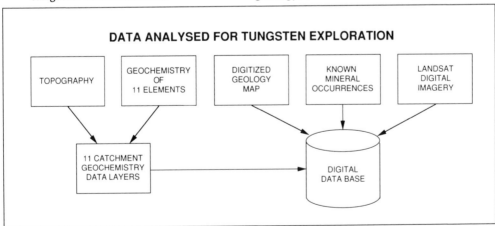

Figure 1.7 Data Sets Used to Identify Potential Tungsten Skarn Exploration Targets.

on the satellite imagery to be analyzed in an integrated manner.

This study is but one of many examples of GIS applications for mineral exploration now appearing in the literature. The recent study by Bonham-Carter et. al. (1988) presents a somewhat different approach based on a Bayesian statistical model.

MUNICIPAL APPLICATIONS

Most of the information needed to operate a municipality is georeferenced, i.e. it is referenced to a specific geographic location. Information about zoning, properties, roads, schools, and parks all pertain to geographic locations. Though computer use is common, the adoption of GISes by municipalities has been slow. In part this has been a result of the high start-up costs of creating the GIS data base. Perhaps more fundamental are the costs of changing the administrative organization of the municipality so that the GIS can be effectively implemented. (Organizational issues are discussed in Chapter 8.)

Municipal GIS applications provide systematic collection, updating, processing, and distribution of land-related data. The capability to handle land survey data is also a common requirement of these systems. Municipal GISes are used for legal, administrative, and economic decision-making, as well as for various planning activities. The terms **Land Information System** (LIS) and **Land Records Information System** (LRIS) are frequently used to refer to GISes that have been specialized for these applications.

The first introduction of computers to municipalities was for accounting applications. Bookkeeping, payrolls, and basic accounting quickly became standard computer tasks and so the computer facilities were organized to serve the finance department. As computer-aided drafting developed in the 1970s, engineering departments began to recognize the value of computer systems for graphics applications such as drafting, map production and updating, and surveying, as well as for cost control and project management. However, the existing computer facilities were commonly oriented to financial functions and the computer facilities staff were either disinterested or were unqualified to support these engineering requirements. As a result, engineering departments tended to develop their own computer facilities to meet their needs.

In a similar way, planning departments developed their own computer facilities to support such activities as statistical analysis, mapping of land use, mapping of neighbourhood demographics, projection of school and recreational needs, and planning for commercial and industrial development. The computer provided a powerful means to develop and compare alternative plans. With the subsequent introduction of microcomputers throughout the administrative organization, municipalities were faced with multiple computer systems that could not share information. This in turn resulted in the storage of multiple copies of information. Since all copies could not be updated at the same time, the same information could be more out-of-date on one system than on another, creating accuracy problems (Liley 1987).

Municipalities have begun to recognize the potential benefits of a more integrated approach to their computerized data in general, and to the organization of georeferenced information in particular. During the 1980s many municipalities did make substantial investments in GIS. The GISes developed for the cities of Minneapolis, Minnesota; Los Angeles, California; Houston, Texas; Calgary, Alberta; Burnaby, British Columbia; and proposed for San Diego, California (to name but a few) are some of the more widely discussed North American examples. They are being used to support such municipal functions as property management, property appraisal, permit

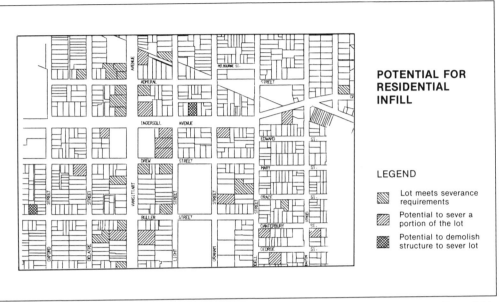

POTENTIAL FOR
RESIDENTIAL
INFILL

LEGEND

▨ Lot meets severance requirements

▨ Potential to sever a portion of the lot

▨ Potential to demolish structure to sever lot

Figure 1.8 Residential Infill Analysis for the City of Woodstock, Ontario. The land parcels potentially available for residential infill development were identified using data on the land parcel, zoning regulations, and the size, age, and condition of the main structure. (Courtesy of the Department of Planning and Development. Oxford County, Ontario.)

and license issuing, subdivision planning, transportation analysis and planning, emergency vehicle routing and dispatching, engineering design, inventory of such facilities as water/sewer systems and electrical cabling, and land use planning.

Figure 1.8 illustrates the results of a residential infill analysis for the City of Woodstock, Ontario. The GIS was used to calculate the area of the main structure in each parcel and compare it to the parcel land area. This determined the amount of land potentially available for residential infilling. The age, size, and condition information for each structure was analyzed to assess whether it was a candidate for demolition. The zoning bylaws that applied to each parcel were also assessed in this manner. A map was then generated showing the three infill classes required for planning and the street network and parcel boundaries.

Figure 1.9 is the conceptual design for a municipal GIS data base. It was developed

by the Environmental Systems Research Institute for the city and county of San Diego, California — a large metropolitan area. The system was designed to meet the needs of some 37 municipal departments for information on land development, public facilities, and the environment. Other

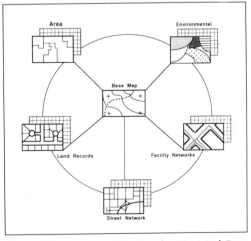

Figure 1.9 Conceptual Design of a Municipal Data Base. (Courtesy of ESRI 1986.)

municipal GISes provide similar functions to those discussed here, though they may differ in the means of implementation and the size of data base. Though the system was designed for a large organization, the principles can be applied to smaller municipalities as well.

In the San Diego study, the data base is divided into six categories of data: basemap, land records, street network, areas, facility networks, and environmental data (see Figure 1.9). For each category, data are stored in the form of maps and tables of data. The two forms of data are linked by a geographic identifier that is a unique ID number. Each map in the data base represents one type of data (such as the road locations) for the entire geographic coverage area of the data base. The associated tabular data contains such information as the name of

each street, the length of each section of street, the address range, etc. Table 1.1 summarizes the maps proposed for each of the six data base categories.

The map layers and tabular data are presented to the user as a continuous geographic data base. That is, any map layer can be used as if it were a single continuous map sheet extending over the entire coverage area. Within the system, the maps are divided into smaller map segments for more efficient computer storage and handling.

The objective of the San Diego GIS design was to provide a fully integrated data base to support administrative and decision-making functions at all levels in the organization. It was to

— provide automated processing of projects and building permits;

Table 1.1   Data Sets Typically Held in a Municipal GIS. (Adapted from ESRI 1986).

Data Category	Example Map Layers
Area Data	Demographic Areas Tax Rate Areas School Districts Emergency Service Areas
Basemap Data	Control Points Topographic Contours Building Footprints
Environmental Data	Soils Map Floodplain Map Noise Level Map Streams and Waterbodies
Land Records Data	Lot Boundaries Land Parcel Boundaries Easements and Right-of-Ways
Network Facilities Data	Sewer System Water System Electrical Cabling Telecommunications
Street Network Data	Road Centerlines Road Intersections Street Lights Street Trees

Table 1.2 Sample Municipal GIS Applications. (Adapted from ESRI 1986).

Type of Application	Examples
Area Mapping and Reporting	Analysis and display of maps
Building Permit	Processing of building permit
Development Tracking	Analyze development trend data Display maps of development trends
Election Management	Provide election map updates Display and analyze election and precinct information
Emergency Response	Provide display of emergency vehicle routing Analyze frequency and location of emergency events
Facilities Management	Support the planning and delivery of maintenance to road, sewer, water, and cable facilities Update, display, and analyze facilities data Plan for facilities expansion
Facility Siting	Selection of optimum locations for new facilities such as parks, police stations, firehalls
Land Development	Updating of lot boundaries Analyze and display land records data Process land development maps
Land Use/Environmental Planning	Display and analyze environmental and land use data Process community land use plan updates
Land Use Regulation	Processing zoning changes Analysis and display of land use regulations
Permitting and Licensing	Processing and tracking Display and analysis of data
Property Management	Inventory and management of publicly owned property Display and analyze data
Safety Inspection/Code Enforcement	Scheduling and tracking safety and code violation inspections Logging and processing complaints Display and analysis of inspection and enforcement data
Transportation Planning	Analysis and display of street, area, and land records information to support transportation planning
Vehicle Routing	Analysis and display of optimum routes for operations such as garbage collection, public transportation, school buses

— provide on-line information about each land parcel to facilitate the processing of projects and permits;

— monitor workloads, development activity, and financial information for planning, budgeting, and fiscal control;

— automate mapping efforts carried out manually by several departments; and

— provide automated support for activities such as public notices and advertising.

Table 1.2 lists some of the applications the GIS would provide. The applications range from the day-to-day transactions of preparing work orders to repair roads and sewers to providing comparative analyses of alternative development plans. In conjunction with the introduction of the GIS technology, proposals were made to adjust the flow of information. These changes would reduce duplication of effort (data would be entered only once into a centralized data base instead of having the same information entered in several data bases), make the data more widely available (a user would be authorized to directly access the data sets he or she needed), and enable the information to be integrated in a virtually unlimited way (any data sets within the data base could be used together). A well-designed municipal data base can improve the effectiveness of the organization in maintaining the information base on which the municipality depends and in making the best use of its information investment.

GLOBAL SCALE APPLICATIONS

Geographic processes do not obey national boundaries. Awareness of the fragility and interrelationships of the global environment have been brought into sharp focus by such events as the drought and famine in the Sahel from 1984–1986, the discovery of the "hole" in the ozone layer over the Antarctic, concerns that increases in the global carbon dioxide concentration may be causing global warming, and the dying lakes and forests of Northern Europe and North America as a result of acid rain.

To understand many of the issues facing the world's societies, we require the capability to analyze geographic information for very large areas. National and international organizations are converting existing data into computer-readable form. Increasingly, new data are being collected in digital form. With the development of remote sensing technology, it is now possible to collect geographic information on a global basis. The earth's weather systems, vegetation, and ocean circulation are monitored on a daily basis using satellites that produce digital images and digital measurements. GIS technology is being used to perform integrated analyses of large global datasets that were originally collected using different resolutions, scales, and projections.

To deal with the global environmental issues facing the human population, it is essential that environmental data be monitored and analyzed at the national and international scale. Geographic information systems can provide some of the analytical tools needed to accomplish this task.

Ecological Effects of Increased Atmospheric Carbon Dioxide

Over the past decade, evidence has been presented showing a steady increase in the concentration of carbon dioxide in the earth's atmosphere. The Canadian Wildlife Service (CWS) is studying how an increase in the atmospheric carbon dioxide concentration and resulting global warming could affect the location and extent of ecosytems in Canada. A model was developed using GIS techniques to assess the sensitivity of Canada's ecosystems to climatic change as a result of a doubling of atmospheric carbon dioxide (Rizzo 1988). At present rates, atmospheric carbon dioxide is expected to double by the year 2050 and possibly as early as 2030. Plate 5a illustrates the current distribution of the Ecoclimatic Provinces of Canada. These geographic divisions represent regional climatic zones as expressed by the vegetation and soils found on standard sites. Shifts in the global weather system would cause changes in the location of these Ecoclimatic Provinces and would ultimately cause a change in the distribution of ecosystems in Canada. A significant shift in the location or extent of Canada's ecosystems would have direct implications for the nation's natural resource base. For example, changes in the extent of agricultural areas and forest lands could significantly affect Canada's ability to grow food and produce forest products.

The classification model used in the climate analysis was based on nine temperature and precipitation parameters. The model was calibrated using existing meteorological data for some 2000 sites. Predicted values were then generated using the General Circulation Model developed by the Goddard Institute for Space Studies in Washington D.C. GIS methods were used to extend the results for the grid of data points to show the results in the geographic context needed.

Plate 5b illustrates the changes in Ecoclimatic Province boundaries that could develop if the earth's carbon dioxide concentration were double what it is today. One of the major shifts predicted by this analysis is the extension of Canada's western grain-producing Grassland Province northward to meet the Arctic Province. Currently these Ecoclimatic Provinces are divided from each other by Boreal Forest, a rich source of forest products. The northern extension of the Grassland Province would probably not be suitable for grain production, however, because the shallow soils and metamorphic geology of the region provide poorer soils than those of the current Grassland Province. A new Arid Ecoclimatic Province can

be seen in the south-central portion of the country. If this region becomes warmer and drier, as indicated by this model, then current agricultural practices would no longer be successful (Rizzo 1988).

Analyses such as these provide valuable predictions of climatic trends that can be used for long range planning at the national level. If changes can be anticipated then there is the opportunity to benefit from them and the possibility of developing mitigation measures in advance. For example, instead of trying to support agricultural methods that are unsuited to a hotter drier climate, the land can be put to alternative uses. Resources can be used to bring new areas under cultivation that are becoming better suited to agriculture.

The Corine System — A GIS for the European Community

Organizations responsible for international planning and decision-making have now begun to implement GISes to support their data collection, analysis, and reporting functions. In many cases it is the power of a computer-based GIS that makes even the contemplation of such projects plausible. Without computer-based techniques to register maps of different scales and map projections, the process of registering the data to a common map base, let alone any analysis, would be prohibitively expensive. The work of the European Community (EC) and the United Nations Environment Program provide examples of this type of GIS application.

In 1985 the EC launched the Coordinated Information on the European Environment (CORINE) program. The objective of this program is to provide a comprehensive integrated spatial data base of environmental data relevant to European policy-making. CORINE is designed to be available throughout the EC and to be relatively easy to use. The facilities will provide the capability to

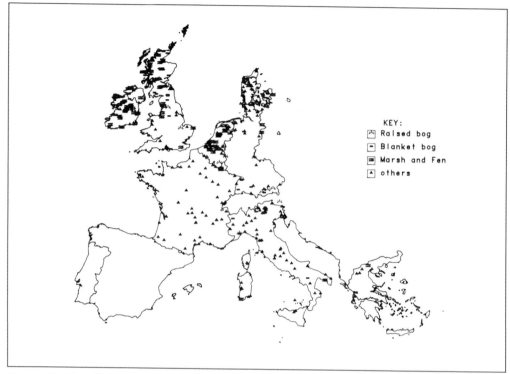

Figure 1.10 Map of Important Peatland Areas in the European Community. (Courtesy of the *International Journal of Geographical Information Systems*.)

inventory resources, analyze diverse geo-referenced environmental data sets, and detect and assess land use change for the geographic area of the entire EC. Wherever possible, existing data sets are used, and collaboration among organizations of different member countries is encouraged. It was recognized that to fulfill these objectives, data at different scales and spatial resolutions would be required ranging from EC-wide coverage down to the 1 km² order of magnitude.

An existing GIS, the ARC/INFO system from the Environmental Systems Research Institute (ESRI), was chosen as a development platform for the CORINE system. Based on the projects completed using this system, the requirements for the CORINE GIS were developed and the invitation for tender to supply the system was issued. The CORINE program initiated an inventory of environmental data currently available for the EC. Some examples are an inventory of ecologically important sites for nature conservation, the collection of water resources data, and an inventory of areas with a high erosion potential. Analyses are being undertaken using these data sets to assess user needs and the operational feasibility of the GIS.

The system has been successfully operated over two computer networks, enabling users in different countries to access the system and exchange data sets. Soils, climatic, topographic, and ecological data sets have been developed, and projects have been undertaken to analyze specific environmental issues related to atmospheric emissions, water pollution, and soil erosion. Figures 1.10 and 1.11 illustrate some of the results. These projects successfully demonstrated the feasibility and usefulness of a

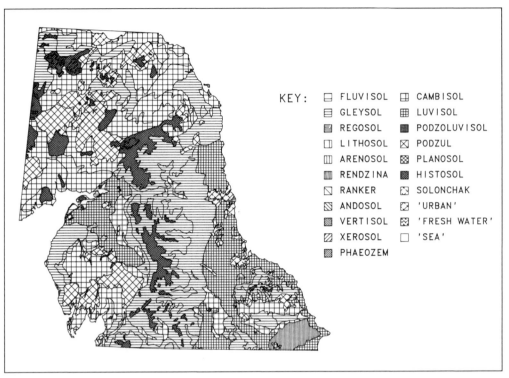

KEY:

⊟	FLUVISOL	⊞	CAMBISOL
⊟	GLEYSOL	⊞	LUVISOL
▦	REGOSOL	▦	PODZOLUVISOL
⊡	LITHOSOL	⊠	PODZUL
Ⅲ	ARENOSOL	⊠	PLANOSOL
▦	RENDZINA	▦	HISTOSOL
◹	RANKER	▨	SOLONCHAK
◺	ANDOSOL	▨	'URBAN'
▨	VERTISOL	⊠	'FRESH WATER'
▨	XEROSOL	□	'SEA'
▨	PHAEOZEM		

Figure 1.11 Sample Soil Map of an Area in Scotland at the 1:1 Million Scale. (Courtesy of the *International Journal of Geographical Information Systems*.)

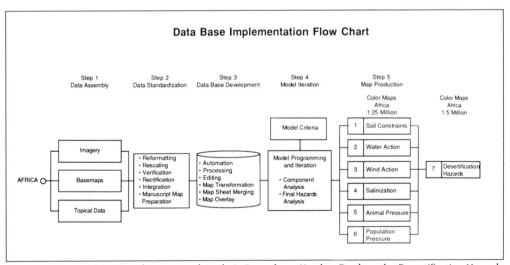

Figure 1.12 Data Base Development and Analysis Procedures Used to Produce the Desertification Hazards Map of Africa. (Courtesy of ESRI. Redlands, California.)

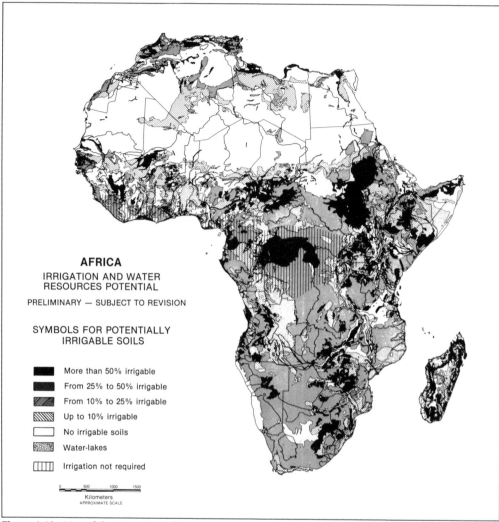

Figure 1.13   Map of the Irrigation and Water Resources Potential of Africa. (Adapted from material provided by ESRI. Redlands, California.)

large GIS for the EC. The program is now continuing with the development of the data base and procurement of a GIS for the full CORINE system (Wiggins et. al. 1987).

GIS Applications at the United Nations Environment Program

The United Nations Environment Program (UNEP) was established to coordinate global environmental assessment and management efforts related to the world's climate, oceans, renewable resources, and pollution. In 1983 the Environmental Systems Research Institute (ESRI) was selected by UNEP to develop a GIS-based system to analyze and map the spread of desert areas on a global scale. The desertification hazard was assessed as a function of soil status, vulnerability of land to desertification processes, and animal and population pressures. These data were obtained from existing maps or were derived from available climate, soils, and vegetation data.

Table 1.3 Degree of Desertification Hazard in Africa By Country. (Adapted from ESRI 1984).

DESERTIFICATION HAZARD RATING

Country	None to Slight % Area	Moderate % Area	Severe % Area	Very Severe % Area
Algeria	4	28	39	29
Angola	86	11	3	0
Benin	79	21	0	0
Botswana	39	61	0	0
Burundi	100	0	0	0
Cameroon	95	5	0	0
Canary Islands	0	28	54	18
Cape Verde	0	0	0	100
Central African Republic	97	3	0	0
Chad	22	29	40	10
Comoros Islands	100	0	0	0
Congo	100	0	0	0
Djibouti	0	90	7	3
Egypt	0	23	36	40
Equatorial Guinea	100	0	0	0
Ethiopia	44	36	15	4
Gabon	100	0	0	0
Gambia	44	56	0	0
Ghana	96	4	0	0
Guinea	98	2	0	0
Guinea Bissau	99	1	0	0
Ivory Coast	100	0	0	0
Kenya	13	64	21	2
Lesotho	27	57	0	16
Liberia	100	0	0	0
Libya	1	28	48	23
Madagascar	91	6	2	0
Madeira	0	0	100	0

DESERTIFICATION HAZARD RATING

Country	None to Slight % Area	Moderate % Area	Severe % Area	Very Severe % Area
Malawi	94	6	0	0
Mali	13	45	6	36
Mauritania	6	17	23	54
Mauritius	100	0	0	0
Morrocco	34	27	36	3
Mozambique	80	20	0	0
Namibia	26	50	24	0
Niger	0	18	53	29
Nigeria	63	31	6	0
Reunion	100	0	0	0
Rwanda	100	0	0	0
Sao Tome and Principe	100	0	0	0
Senegal	27	72	1	0
Sierra Leone	100	0	0	0
Somalia	8	57	34	1
South Africa	11	18	33	38
Sudan	34	34	8	24
Swaziland	70	30	0	0
Tanzania	65	33	1	0
Togo	100	0	0	0
Tunisia	14	26	43	18
Uganda	80	19	1	0
Upper Volta	42	58	0	0
Western Sahara	0	12	70	18
Zaire	100	0	0	0
Zambia	97	3	0	0
Zimbabwe	39	55	6	0

All data sets were rectified to a common map base and then converted into digital form using manual and automated digitizing techniques.

Once the GIS data base had been produced, desertification models developed by the FAO (the United Nations Food and Agriculture Organization) were used to produce the final desertification hazard map and summary statistics. Figure 1.12 illustrates the data base development and analysis process. Figure 1.13 and Table 1.3 are examples of the results: a map of Africa showing the irrigation and water resources potential and a tabulation of the overall desertification hazard by country for the African continent. It is this type of information that is needed to guide policy development and target those areas that could benefit most from remedial action.

In 1985 UNEP implemented the Global Resource Information Data Base (GRID). This project was designed to provide scientists and planners with access to an integrated data base of global environmental information and the GIS technology to undertake integrated analyses. In the course of the two year pilot phase of the program, a GIS capability was developed, data sets entered, and several studies completed illustrating the integrated use of national and global level data for resource management and environmental planning.

Africa was selected as an area of special emphasis during the pilot phase. In 1987, the GRID project developed a national environmental data base for Uganda using existing data from UNEP and other sources. Existing maps were updated and new data sets developed using information interpreted from satellite imagery. Each factor, such as soil types, rainfall, and temperature, was input to the GIS as a separate map. The GIS was then used to model the co-occurrence of conditions significant for land use and planning. The conditions most suitable for different crops were modelled using temperature data, annual rainfall, and soils data. Soil erosion can be a serious threat to a country heavily dependent on agriculture. Erosion potential was first evaluated by analyzing rainfall and soils data. The result was then combined with land use pressures data (population density, and cropping and grazing practices) and re-analyzed to arrive at an erosion hazard estimate. In addition, areas of significant deforestation were mapped using satellite imagery.

These analyses were done relatively quickly (over a 5 month period) at small scales on the order of 1:1 million. They drew heavily on small scale data sets that were, in some cases, continent-wide. Although the level of detail is relatively coarse, this type of data can provide valuable information for national or international scale planning when more detailed information is not available. It can also be used to select areas for more detailed analysis and in this way limited funds can be directed to regions where they will provide the greatest benefit.

The GRID program began a two-year implementation phase in January 1988. Among the projects to be undertaken in this phase are a global assessment of forest resources and a global assessment of land degradation. Both of these studies will depend heavily on the use of remotely sensed data from satellites. During the pilot phase, a 9-parameter global data base was developed and a 17-parameter data base for the continent of Africa was completed (see Table 1.4). During the implementation phase, efforts will be made to increase the number of global data sets and expand the accessibility of the system.

CONCLUSION

Over the past two decades it has become increasingly apparent that resources are becoming more scarce, the effects of human activities more pervasive, and the recognition and prediction of causes and effects more complex. It has also been recognized

Table 1.4 Some of the Data Sets Held by the Global Resource Information Data Base. (Adapted from Mooneyhan 1987).

Parameter	Coverage	Source
Political Boundaries	Global	US State Department
Natural Boundaries	Global	US State Department
Elevation	Global	National Geophysical Data Center, USA
Soils	Global	FAO/UNESCO World Soils Map
Vegetation Index (weekly)	Global	AVHRR satellite data
Precipitation Anomalies (monthly)	Global	Climate Analysis Center NOAA/WMO
Temperature Anomalies (monthly)	Global	Climate Analysis Center NOAA/WMO
Surface Temperature (day/night- monthly)	Global	NASA satellite data
Ozone Distribution	Global	NASA, TOMS satellite
Vegetation Index	Africa	NASA, AVHRR satellite
Vegetation	Africa	FAO
Watersheds	Africa	UNEP/FAO
Annual Rainfall	Africa	UNEP/FAO

that better resource assessment and planning methods yield direct benefits in improved resource management and ultimately in improved quality of life. Georeferenced information has always been critical to the welfare of nations and the welfare of their citizens. It is the technology and the methods to apply them, such as computer-based GIS and remote sensing, that can provide the means to inventory resources and model processes from the local to the global scale. This book presents the fundamental principles on which the applications of this technology are based.

REFERENCES

Aronoff, S., W. Goodfellow, G.F. Bonham-Carter, and D.J. Ellwood. 1986. Integration of Surficial Geochemistry and Landsat Imagery to Discover Skarn Tungsten Deposits Using Image Analysis Techniques. In *Proceedings of the IGARSS '86 Symposium*. European Space Agency, Publications Division. Paris, France.

Beard, M.K., N.R. Chrisman, T.D. Patterson. 1987. Integrating Data for Local Resource Planning: A Case Study of Sand and Gravel Resources. In *GIS for Resource Management: A Compendium*. American Society for Photogrammetry and Remote Sensing. Falls Church, Virginia.

Bobbe, T.J. 1987. An Application of a Geographic Information System to the Timber Sale Planning Process on the Tongass National Forest — Ketchikan Area. In *Proceedings of the GIS '87 Symposium*. American Society for Photogrammetry and Remote Sensing. Falls Church, Virginia. pp.554–562.

Bonham-Carter, G.F., F.P. Agterberg, and D.F. Wright. 1988. Integration of Geological Datasets for Gold Exploration in Nova Scotia. *Photogrammetric Engineering and Remote Sensing* 54(11):1585–1592.

Cowen, David J. 1987. GIS vs. CAD vs. DBMS: What are the Differences? In *Proceedings of the GIS '87 Symposium*. American Society for Photogrammetry and Remote Sensing. Falls Church, Virginia. pp.46–56.

Crain, I. 1987. *The Canada Land Data System: An Overview*. Canada Land Data Systems Division. Environment Canada. Ottawa, Ontario.

Custer, J., T. Eveleigh, and V. Klemas. 1983. A Landsat-Generated Predictive Model for Prehistoric Archaeological Sites in Delaware's Coastal Plain. *Bulletin of the Archaeological Society of Delaware* No.14 pp.19–38.

Dangermond, J. and C. Freedman. 1984. *Findings Regarding a Conceptual Model of a Municipal Data Base and Implications for Software Design*. Environmental Systems Research Institute. Redlands, California.

Dangermond, J. and J. Harrison. 1987. Urban Geographic Information Systems: The San Diego Design Experience. In *Proceedings of the GIS '87 Symposium*. American Society for Photogrammetry and Remote Sensing. Falls Church, Virginia. pp.387–395.

Drinnan, C.H. 1982. Implementation of Land Records Information Systems Using the Informap System. In *Proceedings of the 1982 Annual URISA Conference*. The Urban and Regional Information Systems Association. Washington, D.C..

ESRI. 1986. *San Diego Regional Urban Information System Conceptual Design Study — System Concept and Implementation Program, Volume I*. Environmental Systems Research Institute. Redlands, California.

ESRI. 1984. Map of Desertification Hazards. In *Geographic Information Systems for Resource Management: A Compendium*. American Society for Photogrammetry and Remote Sensing. Falls Church, Virginia.

Gimblett, R., L. Smith, D. Ferguson, and B. Kelley. 1987. A Method for Assessing and Revitalizing Non-Productive or Underutilized Agricultural Land for Alternative Rural Development: A Case Study. In *Proceedings of the GIS '87 Symposium*. American Society for Photogrammetry and Remote Sensing. Falls Church, Virginia. pp.569–576.

Goodfellow, W.D. and S. Aronoff. 1988. Application of Landsat Imagery and Surficial Geochemistry to the Discovery of Tungsten Skarn Deposits Associated with Buried Plutons, Yukon and Northwest Territories, Canada. *Geocarto International* (4):3–16.

Graetz, R.D., R.P. Pech, M.R. Gentle, and J.F. O'Callaghan. 1986. The Application of Landsat Image Data to Rangeland Assessment and Monitoring: the Development and Demonstration of a Land Image-Based Resource Information System (LIBRIS). *Journal of Arid Environments* 10:53–80.

Gros, S.L., T.H.L. Williams, and G. Thompson. 1988. Environmental Impact Modelling of Oil and Gas Wells Using a GIS. In *Proceedings of the 1988 ACSM-ASPRS Annual Convention*. American Society for Photogrammetry and Remote Sensing. Falls Church, Virginia. Vol.5:216–225.

Hanigan, F.L. 1983. Metrocom: Houston's Geographic Information Municipal Management System. In *Proceedings of the 1983 ACSM-ASPRS Annual Convention*. American Society for Photogrammetry and Remote Sensing. Falls Church, Virginia. pp.205–215.

Hart, J.A., D.B. Wherry, and S. Bain. 1985. An Operational GIS for Flathead National Forest. In *Proceedings of Autocarto 7*. American Society for Photogrammetry and Remote Sensing. Falls Church, Virginia. pp.244–253.

Hegyi, F. and P. Sallaway. 1986. *The Integrated Three-Dimensional Forest Land Information System of British Columbia, Canada*. Presented at the 18th IUFRO World Congress held September 1986 in Ljubljana, Yugoslavia.

Hutachareon, M. 1988. Application of Geographic Information Systems Technology to the Analysis of Deforestation and Associated Environmental Hazards in Northern Thailand. In *Proceedings of the GIS '87 Symposium*. American Society for Photogrammetry and Remote Sensing. Falls Church, Virginia. pp.509–518.

Jordan, G. and Leon Vietinghoff. 1987. Fighting Budworm with a GIS. In *Proceedings of the Eighth International Symposium on Automated Cartography*. American Society for Photogrammetry and Remote Sensing. Falls Church, Virginia. pp.492–499.

Liley, R.W. 1987. Integration — The Big Payoff for Municipal Geo-Based Systems. In *Proceedings of the GIS '87 Symposium*. American Society for Photogrammetry and Remote Sensing. Falls Church, Virginia. pp.396–409.

Marozas, B. and J.A. Zack. 1987. Geographic Information Systems Applications to Archaeological Site Location Studies. In *Proceedings of the GIS '87 Symposium*. American Society for Photogrammetry and Remote Sensing. Falls Church, Virginia. pp.628–635.

McGranaghan, M. and Lyndon Wester. 1988. Prototyping an Herbarium Collection Mapping System. In *Proceedings of the 1988 ACSM-ASPRS Annual Convention*. American Society of Photogrammetry and Remote Sensing. Falls Church, Virginia. Volume 5: 232–238.

Mead, R.A. 1989. Personal Communication. US Forest Service, Region 8. Atlanta, Georgia.

Mead, R.A., L.S. Cockerham, and C.M. Robinson. 1988. Mapping Gopher Tortoise Habitat on the Ocala National Forest Using a GIS. In *Proceedings of the GIS/LIS'88 Symposium* held November 1988 in San Antonio, Texas. American Society of Photogrammetry and Remote Sensing. Falls Church, Virginia. Volume 1:395–400.

Merrit, E., L. Heitkemper, and K. Marcus. 1984. CROP-CAST — A Review of an Existing Remote Sensor-Based Agricultural Information System with a View Toward Future Remote Sensor Applications. In *Proceedings of SPIE Volume 481, Recent Advances in Civil Space Remote Sensing*. Institute of Electrical and Electronic Engineers. New York, New York. pp.231–237.

Merchant, J.W. and L.M. Caron. 1986. Geographic Information Systems for Non-Urban Local Level Jurisdictions: A Strategy for Technology Transfer. In *Proceedings of the Geographic Information Systems Workshop*. American Society for Photogrammetry and Remote Sensing. Falls Church, Virginia. pp.119–127.

Montgomery, B.A. ed. 1987. *Gypsy Moth In Michigan, The First Annual Report of the Gypsy Moth Technical Committee*. Michigan Department of Agriculture. Lansing, Michigan.

Mooneyhan, D.W. 1987. An Overview of Applications of Geographic Information Systems Within the United Nations Environment Program. In *Proceedings of the GIS '87 Symposium*. American Society for Photogrammetry and Remote Sensing. Falls Church, Virginia. pp.536–543.

Moyer, D.D., B.J. Niemann Jr., R.F. Gurda, and S.J. Ventura. 1988. Comparing the Costs: Manual Versus Automated Procedures for Handling Land Records. In *Proceedings of the 1988 ACSM-ASPRS Annual Convention*. American Society for Photogrammetry and Remote Sensing. Falls Church, Virginia. Vol.5: 198–206.

Pank, L. F. 1989. Personal Communication. Chief, Branch of Mammals. Alaska Fish and Wildlife Research Centre. U.S. Fish and Wildlife Service. Anchorage, Alaska.

Reisinger, T.W. and C.J. Davis. 1987. Integrating Geographic Information and Decision Support Systems: A Forest Industry Application. In *Proceedings of the GIS '87 Symposium*. American Society for Photogrammetry and Remote Sensing. Falls Church, Virginia. pp.578–584.

Rhind, D. 1987. Recent Developments in Geographical Information Systems in the U.K. *International Journal of Geographical Information Systems* 1(3):229–241.

Rhind, D., P. Armstrong, and S. Openshaw. 1988. The Doomsday Machine: A Nationwide Geographical Information System. *The Geographical Journal* 154(1):56–68.

Rhind, D. and S. Openshaw. 1987. The BBC Doomsday System: A Nation-Wide GIS for $4448. In *Proceedings of the Eighth International Symposium on Automated Cartography*. American Society for Photogrammetry and Remote Sensing. Falls Church, Virginia. pp.595–603.

Rizzo, B. 1988. The Sensitivity of Canada's Ecosystems to Climatic Change. *Newsletter of the Canada Committee on Ecological Land Classification* 17:10–12. Environment Canada. Ottawa, Ontario.

Scepan, J., F. Davis, and L.L. Blum. 1987. A Geographic Information System for Managing California Condor Habitat. In *Proceedings of the GIS '87 Symposium*. American Society for Photogrammetry and Remote Sensing. Falls Church, Virginia. pp.476–486.

Simonett, O., F. Turyatunga, and R. Witt. Environmental Database Development for Assessment of Deforestation, Soil Erosion Hazards, and Crop Suitability: A Joint Uganda-UNEP/GEMS/GRID Case Study. In *Proceedings of the GIS '87 Symposium*. American Society for Photogrammetry and Remote Sensing. Falls Church, Virginia. pp.544–553.

Somers, R. 1987. Development of a Multipurpose Local Government GIS. In *Proceedings of the GIS '87 Symposium*. American Society for Photogrammetry and Remote Sensing. Falls Church, Virginia. pp.410–415.

Somers, R. 1986. Improving Service and Productivity: Applications of Calgary's Geoprocessing and Computer Mapping System. In *Proceedings of the 1986 Annual URISA Conference*. The Urban and Regional Information Systems Association. Washington, D.C.. pp. 85–95.

Stenback, J.M., C.B. Travlos, R.H. Barrett, and R.G. Congalton. 1987. Application of Remotely Sensed Digital Data and a GIS in Evaluating Deer Habitat Suitability on the Tehama Winter Range. In *Proceedings of the GIS '87 Symposium*. American Society for Photogrammetry and Remote Sensing. Falls Church, Virginia. pp.440–445.

Tomlinson, R.F. 1987. Current and Potential Uses of Geographical Information Systems — The North American Experience. *International Journal of Geographical Information Systems* 1(3):203–218.

Tomlinson, R.F. and A.R. Boyle. 1981. The State of Development of Systems for Handling Natural Resources Inventory Data. *Cartographica* 18(4):65–95.

Tomlinson, R.F., Calkins, H.W. and Marble, D.F. 1976. *Computer Handling of Geographical Data*. Natural Resources Research Series XIII. The UNESCO Press. Paris, France.

Tosta, N. and L. Davis. 1986. Utilizing a Geographic Information System for Statewide Resource Assessment: The California Case. In *Proceedings of the 1986 Geographic Information Systems Workshop*. American Society of Photogrammetry and Remote Sensing. Falls Church, Virginia. pp.147–154.

UNEP. 1988. *Meeting of the GRID Scientific and Technical Management Advisory Committee Held January 1988*. GRID Information Series No.15. Global Environment Monitoring System, United Nations Environment Program. Nairobi, Kenya.

UNEP. 1987. *Uganda Case Study: A Sampler Atlas of Environmental Resource Datasets within GRID*. GRID Information Series No.8. Global Environment Monitoring System, United Nations Environment Program. Nairobi, Kenya.

Ventura, S.J., N.R. Chrisman, K. Connors, R.F. Gurda, R.W. Martin. 1988a. Soil Erosion Control Planning in Dane County, Wisconsin. *Journal of Soil and Water Conservation* 43(3):230–233.

Ventura, S.J., B.J. Niemann Jr., and D.D. Moyer. 1988b. A Multipurpose Land Information System Approach to Conservation and Rural Resource Planning. *Journal of Soil and Water Conservation* 43(3):226–229.

Wakeley, R.R. 1987. G.I.S. and Weyerhaeuser — 20 Years Experience. In *Proceedings of the GIS '87 Symposium*. American Society for Photogrammetry and Remote Sensing. Falls Church, Virginia, USA, pp.446–455.

Walsh, S.J. 1985. Geographic Information Systems for Natural Resource Management. *Journal of Soil and Water Conservation* (March–April):202–205.

Wiggins, J.C., R.P. Hartley, M.J. Higgins, and R.J. Whittaker. 1987. Computing Aspects of a Large Geographic Information System for the European Community. *International Journal of Geographical Information Systems* 1(1):77–87.

Young, T.N., J.R. Eby, H.L. Allen, M.J. Hewitt, and K.R. Dixon. 1987. Wildlife Habitat Analysis Using Landsat and Radiotelemetry in a GIS with Application to Spotted Owl Preference for Old Growth. In *Proceedings of the GIS '87 Symposium*. American Society for Photogrammetry and Remote Sensing. Falls Church, Virginia. pp.595–600.

2. WHAT IS A GEOGRAPHIC INFORMATION SYSTEM?

INTRODUCTION

From the earliest civilizations maps have been used to portray information about the earth's surface. Navigators, land surveyors, and the military used maps to show the spatial distribution of important geographic features. Land surveying and map making were an integral part of Roman government. With the decline of the Roman Empire, surveying and map making declined as well.

It was not until the eighteenth century that map making again rose to prominence in Europe as governments realized the value of mapping as a means of recording and planning the use of their lands. National institutes were commissioned to produce map coverage of entire countries. General purpose maps showing the topography of the land and boundaries of national or administrative units were produced. As the study of natural resources developed, thematic maps were used to portray the spatial distribution of such features as geology, geomorphology, soils, and vegetation.

In the twentieth century the pace of science and technology accelerated. This increase created the demand for ever greater volumes of geographic data to be presented in map form more quickly and more accurately. With the development of reconnaissance technologies, such as aerial photography and satellite-based remote sensing, there has been an explosion of geographic data production, wider use, and more sophisticated analyses. Geographic data are now being generated faster than they can be analyzed.

Geographic data have traditionally been presented in the form of a map. Until computers were available, geographic data were represented as points, lines, and areas drawn on a piece of paper or film. They were coded using symbols, textures, and colours that were explained in the map legend or accompanying text. The map and its documentation constituted the geographic data base.

The use of thematic maps of natural resources began as an inventory tool used to record and classify observations. The analysis methods were primarily qualitative. That is, retrieval and analysis of map data depended primarily on visual inspection of the map and an intuitive analysis of the map data. Quantitative map analysis could be done using a scale to measure distances and a dot grid or a **planimeter** to measure areas. (A planimeter is a mechanical or electronic device that calculates the area of a map feature. The outline of a map feature is traced with the planimeter, and mechanical or electronic counters calculate the corresponding area. It is a tedious and slow procedure.)

While it was relatively easy to retrieve small amounts of data or consider the spatial relationships of a few elements, these methods became unwieldy when large volumes of data were involved. It was only in the 1970s with the availability of suitable digital computers that the technology to handle spatial data leapt forward. The computer-based geographic information system was developed to provide the power to analyze large volumes of geographic data.

The physical map can be relatively easy to produce and it stores a considerable amount of spatial information in a compact and accessible form. However, it has a number of important limitations. The data used to make the map usually have to be

generalized (i.e. the data have to be presented with less detail) for the map to be easily read. Areas that are large relative to the map scale have to be represented by a series of maps. Problems arise when maps don't match correctly at the edges and areas of interest extend across map edges. Often these problems are handled by shifting the maps to achieve the best fit for the area of the map being used. For critical applications, re-drafting may be needed.

Updating a map can be an expensive procedure. For changes to be made the film masters of the map sheet must be manually edited and the map reprinted. As a result the physical map is a relatively static document. Retrieving small amounts of information from a physical map is relatively easy; however, the processes of retrieving large amounts of information and of combining the spatial information from several maps are costly and difficult. During the 1960s and 1970s the need to evaluate multiple sets of geographic data was recognized. For example, the rapid and accurate integration of such diverse data as soils, land use, current vegetation, and administrative districts was needed for analyses such as environmental impact assessments. Even local zoning decisions were beginning to require the consideration of multiple geographic factors. The accurate and rapid analysis of diverse geographic data sets was perceived as a requirement for more effective planning.

Data integration could be done using physical maps, as McHarg popularized in his book *Design with Nature*. Map information was combined and integrated by overlaying transparent copies on a light table and visually analyzing the co-occurrence of factors. Then those areas with the desired combination of factors could be delineated by drawing their boundaries on a separate overlay. The transparent maps were usually redrawn for the analysis both to transform the diverse source maps to a common map base and to code each map with different shades of grey to represent different levels of constraint for a particular planning application. But the procedure was time-consuming, and, as the number of factors and number of levels of each factor increased a practical limit was quickly reached.

In North America, work on the first operational computer-based geographic information systems began in the mid-1960s. The Canada Geographic Information System (CGIS), sponsored by the Canadian Federal Government, and the Land Use and Natural Resources Inventory of New York State (LUNR), sponsored by the state of New York, were developed at about the same time. Both systems made extensive use of aerial photography, as well as existing mapping to map resource information. Information layers such as agriculture, forestry, wildlife, soils, and geology, were included. The geographic information on the maps was then encoded into digital form for computer analysis. Although development began in the 1960s, the computer-based portions of these systems became operational only in the early 1970s when computer technology such as random access disks became available (Hardy 1975, Tomlinson et. al. 1976). These early implementations of operational GISes for resource information stimulated the development of technical innovations. They also provided valuable experience on how to manage the creation and operation of large geographic information systems. Papers, such as that by Shelton and Hardy (1974) and the introductory chapters of Tomlinson et. al. (1976), provide valuable advice on managing the implementation and operation of a GIS, advice that is as relevant today as when it was written.

Beginning in the 1960s, the Harvard Graphics Laboratory was one of the most active research groups developing computer-based map analysis programs. The early programs, SYMAP, GRID, and IMGRID, were designed to perform the same overlay procedures as manual methods but with greater speed and flexibility. The rapid development of computer technology over the past two decades has enabled computer-based geographic information systems to develop quickly from those early systems to the full-featured GISes now available. (This brief mention of the development of GIS cannot do justice to the many researchers and organizations who have made important contributions. The July 1988 issue of *The American Cartographer* was devoted to this subject. This and other references cited here provide an introduction to this literature.)

The quantitative improvement in the speed of analysis has provided the means to change the way the analysis of geographic information can be approached. Perhaps the two most important improvements have been the ability to keep georeferenced data current and to integrate multiple data sets efficiently.

The ability to quickly update the geographic data base coupled with the fast and inexpensive production of single maps has meant that a physical map can be used as a snapshot of a continuously changing geographic data base. Since re-analyzing the data is relatively inexpensive and can be rapidly executed, complex planning scenarios can be progressively refined by re-analyzing the plan to assess proposed changes. Decision-makers can propose a number of alternative plans and assess each one by re-running the analysis and comparing the results. This iterative approach would be prohibitively expensive using manual methods.

USING A GIS FOR DECISION-MAKING UNDER UNCERTAINTY

There is a fundamental strategy that underlies all analysis of georeferenced data. An understanding of this strategy will not only lead to better use of the available methods, but is also needed to understand how different levels of data abstraction are related.

We have to make decisions that require knowledge about our complex world. Because we don't have complete knowledge, we are used to making decisions with incomplete information. We select relevant information to remember and record. Using this selection process, we create a conceptual model of our world. The term **model** is used to mean a set of relationships or information about the real world. Our conceptual model of something is our understanding of what it is and how it behaves. When we need to make decisions about the real world, we refer to our model, which is much simpler than the real world itself. It is simpler because we have preselected the information in the model to include the things that are relevant to us. Other details that we don't need tend to be selectively forgotten. Most people have a relatively simple conceptual model of a "car", centered mainly on how to operate it, its appearance, and its performance. Few people have a conceptual model of a car that includes detailed information about all the components of the vehicle, vehicle repair, test procedures, and so on. Our conceptual model of a car is more akin to a brief personalized "user's manual" than to a complete set of specifications and engineering drawings.

The process of using a GIS can be illustrated by a familiar geographic information processing task — planning a vacation. The process is illustrated in Figure 2.1. Notice that our task begins and ends with the real world. We have collected information about the real world. This information is necessarily an abstraction; we couldn't handle and wouldn't want every last detail. We will use this information to make decisions, and finally we will implement those decisions — we will apply our abstract reasoning to the real world, i.e. we will take a real vacation.

The decisions of where to visit and what to do will require that we collect some information. We will want to know about vacation spots that we are likely to enjoy and

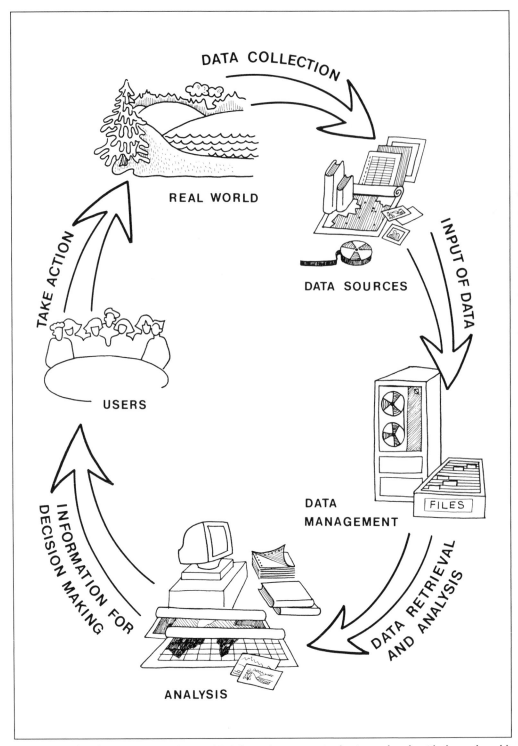

Figure 2.1 The Planning Process. Geographic information processing begins and ends with the real world.

when is the right time to visit. Probably the list of potential vacation spots will come from such sources as our own experience, recommendations from friends, and accounts we have read. Then we'll collect some data about the different destinations. These data about the real world come in the form of maps, books, articles, and even tables of data such as average daytime temperatures.

This collection of information constitutes a data set or data base that will be used in the vacation planning process. If there are several possible destinations and many documents it may become a bit tedious sifting through the pile every time some useful bit of information has to be retrieved, so we may decide to organize the information. The organization, such as filing the literature by destination, makes the process of storing and retrieving items more efficient. The organization of the data base constitutes a simple data base system. (A **data base system** provides for the input, storage, and retrieval of data. The **data base** is the set of data that are stored.)

Having collected and organized our documents, we review the information. This may be done by randomly reading the most attractive brochures or perhaps by systematically thinking about what we enjoy doing on a vacation. Whatever the method of analysis we use, in the end we come to some decision and select a destination. To put the decision into action we implement the plan by going on the trip. If we made the right decision we'll have a good vacation and return satisfied that we made a suitable choice. A "good" information system is one that provides us with the necessary data relevantly organized so that we can make the right decisions about the real world. What is really meant by the "right decision"?

The "right decision" is the one that best achieves the objectives of whoever the system is to serve. To do this it is necessary to know what those objectives are and to be able to correctly predict the results of alternative choices. To make the "right decision" requires that the relevant data be presented in the framework of an appropriate model that is evaluated using true criteria.

The success with which a geographic information system can be used is determined by several factors that can be grouped under four headings: the data set, the data organization, the model, and the criteria. The following sections discuss each of these issues.

GETTING THE RELEVANT DATA

The data used in a GIS represent something about the real world at some point in time. They are always an abstraction of reality because we don't need or want every bit of data, just the ones we think would be useful. The bits we decide to take are the first constraint on the capabilities of the GIS.

You can't use data you don't have.

Then why not take all the data? First, you could never collect all the data, and second, you wouldn't want all the data even if you could get it. Data are costly to collect.

The most cost-effective data collection is to collect only the data you need.

It is costly to collect, store, and sift through large quantities of unnecessary data. Excess data makes it more difficult to use the data you really need. Every expenditure of effort that doesn't contribute to the solution detracts because it represents time, effort, and resources that could have been used elsewhere to improve the analysis. The same argument holds true for the quality of the data.

The optimal data quality is the minimum level of quality that will do the job.

The most important aspects of data quality are accuracy, precision, time, currency, and completeness. Accuracy measures how often, by how much, and how predictably the data will be correct. Precision measures the fineness of the scale used to describe the data. Time indicates at what point or over what period of time the data were collected. Time can often be a critical factor of data quality. Some information may quickly become out-dated. Currency measures how recently the data were collected. In some cases the suitability of the data will depend on the season or the year they were collected. In Canada, for example, summer photography is usually specified for mapping forest cover types. Completeness refers to the portion of the area of interest for which data are available. The term is also used with reference to the classification system that has been used to represent the data. (Data quality is discussed in chapter 5.) There is always a trade-off between higher data quality and higher data cost, and there is a tendency for each unit gain of quality to be more costly than the last. This is illustrated in Figure 2.2.

It costs more and more to gain less and less data quality.

How do these principles apply to our vacation example? Our data base may consist of road maps, vacation brochures, travel guide books, as well as our own personal knowledge of vacation spots. If we have selected an appropriate set of data we should be able to retrieve some information about several

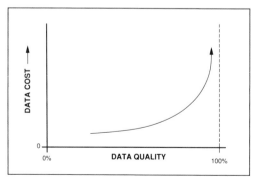

Figure 2.2   The Relationship of Data Quality and Data Cost.

likely vacation spots that meet our criteria. There is of course a trade-off; we could easily swamp ourselves with travel guides, maps, etc., making it difficult to find any particular piece of information, not to mention expensive to acquire. Furthermore, if we are not careful in screening for data quality, we could end up with boxes of material, much of it unreliable.

A more effective approach is to select a set of information that we consider to be reliable and that covers all the regions we might realistically consider visiting on this trip. By judiciously selecting our data in this way, we can make it much easier to retrieve reliable information. We trade off data quality and cost throughout this process. Travel guides can be expensive. We will probably be willing to spend more money on information for a six-week trip to India than for a weekend tour in our area. We would also be more careful in assessing the accuracy of information about the India trip because the consequences of incorrect information could be more serious.

THE DATA ORGANIZATION

The organization of the data is the second major factor for successful use of a GIS. A data base is used to provide this organization. The data base is critical because:

Data is of no value unless the right data can be in the right place at the right time.

Our vacation data might be organized by alphabetical filing or by simply putting the maps in one box and the brochures in another. Depending on the quantity of data and the performance we need from our data base system, these simple forms of organization may suffice. In a computer-based GIS, however, the quantity of data is large enough that the form and performance of the data base are critical to the overall usefulness of the system.

THE DECISION MODEL

A model represents an object or phenomenon that exists in the real world. A good model is the simplest model that correctly and consistently predicts the behaviour of the real world **for the phenomena of interest**. Models are created to predict how certain aspects of the real world will behave. They describe the relationships among data elements in order to predict how events in the real world will occur. The quality of the model is limited by the data that has been selected and the way they are organized. It is also limited by the cost to use it. The more complex the model, the more costly it is to use. The cost may be computer charges, the time of experts. The most cost-effective model is the simplest model that gives results that achieve the minimum required level of accuracy with the minimum cost data.

The most cost-effective model is the simplest model that does the most with the least.

Why this concern with cost? The reason is that there is always a cost. It may not be money; it may be time, it may be incorrect answers. Only by predicting and evaluating these costs can a rational decision be made about performance. There is always a trade-off between cost and performance. Too low a performance level can be too costly in errors caused by inaccuracies, by getting results too late, or by missing better solutions. Too high a performance level can also be too costly by paying for effort that does not improve success.

It is expensive to tolerate performance levels that are too high or too low.

For our vacation, a relatively simple decision model can be used. We want a two-week holiday that is enjoyable to all members of the family and fits within our budget. This may constrain us to areas that can be reached within a two-day drive and limit the types of accommodation and activities. Our model will be used to analyze our data, i.e. to test different candidate destinations proposed after studying the information we collected. The decision process will probably be a fairly informal one based on consensus — with senior family members having power of veto. The relative merits of different destinations will be compared according to some more or less well-defined criteria.

VALID CRITERIA

The fourth major factor in successful GIS applications is the degree to which the criteria used to evaluate the model truly reflect the values of the people to be satisfied. At the end of the information analysis procedure, action is to be taken. The action to be taken must be decided by weighing the alternatives, and by considering the consequences of each alternative as predicted by our models. The decision may be as simple as sending an overdue notice for an unpaid bill, or it may be as complex as deciding to build a dam and drown a valley.

Fundamental to the process are the decision-makers. The decision-makers are those with the mandate, the responsibility for the consequences of the action to be taken. No matter how high the quality of the data, how appropriate the models used, if the wrong criteria are used to evaluate the information produced by a GIS, then the results will probably not be satisfactory. People make decisions guided by criteria.

The criteria used by the people making the decision must be the same ones used by the people who are to be satisfied.

In the context of our vacation example, if one member of the family is designated as the decision maker, then his or her success in choosing a destination will depend on whether the selection criteria used truly represent the values and wishes of the rest of the family, i.e. the people who are to be satisfied.

The success of our vacation depends on the decisions we consciously or not-so consciously made throughout this planning process, just as the decisions made using a GIS depend on the success of each step. We began by deciding what type of information to collect. Though perhaps done in a rather informal way, it is a critical step. Too much data and we'll be researching forever, too little and we might miss a key attraction or make a poor choice of destination. As part of making this trade-off decision, we must accept that we can't review every possible vacation spot, but we will be satisfied with some minimum level of quality. Because it would take too long to find the best possible vacation spot, we'll accept a destination that gives us all a good time and meets our budget of time and money. We defined a simple decision model. These decisions influenced how much data and what kind of data we collected.

The accuracy of our data is important. The level of accuracy we require will depend on the types of decision-making for which the data will be used. Incorrect information could mean choosing a beach vacation at a location and season when it's too cold to swim. The cost of collecting the data and our ability to analyze it are also factors. There are limits to the time and money we can spend collecting and analyzing data for our vacation decision. The effort we are willing to expend depends on how important the decisions are to us.

Our decision model has to be appropriate. It must take into account the factors *we* consider significant for a "good vacation". Naturally these factors are different for each member of the family. If the objective is to make the vacation enjoyable for everyone, then everyone's likes and dislikes need to be taken into account. Teenagers who claim to want a backcountry camping trip when, in fact, opportunities to socialize with peers is their foremost concern, are presenting a non-valid criterion. That is, a camping trip in the backcountry, an unlikely place in which to meet other teenagers, will probably not be satisfactory. The decision model may perform well, and camping in some remote location may be selected — resulting in a very disgruntled teenager. Other factors, such as the time available for travel and budget constraints, must also be correctly assessed. For the decision-making process to be successful, valid criteria must be used.

This discussion has focussed on the decision-making aspect of GISes. While some of these steps are implicit in many applications, fundamentally they are all involved. Throughout this book examples from actual projects are used to illustrate and explain GIS principles. It should be remembered that these examples are excerpts. It is relatively easy to train someone to operate GIS technology. It is far more difficult to educate someone in how to apply the technology effectively in order to satisfy his or her information needs. For GIS applications to be successful, they must be developed in the context of providing a user with the information he or she requires. The success of a GIS application depends on an appropriate selection of the data to be used, the way the data will be organized, the decision model, and the decision criteria.

GEOREFERENCED DATA

Geographic data are commonly characterized as having two fundamental components:

1. the phenomenon being reported such as a physical dimension or class, and

2. the spatial location of the phenomenon.

Examples of a physical dimension might be the height of a forest canopy, the population of a city, or the width of a road. The class could be a rock type, a vegetation type, or the name of a city. The location is usually specified with reference to a common coordinate system such as latitude and longitude.

A third fundamental component to geographic information is time. The time component often is not stated explicitly, but it often is critical. Geographic information describes a phenomenon at a location as it existed at a specific point in time. A land cover map describes the location of different classes of land cover as they existed at the time of data collection. If the area is changing rapidly, this information may quickly be out-dated. The information may then be unsuitable for decision-making that requires the current status of the land. However, the data may be invaluable for analyzing historical trends, such as the conversion of agricultural land to other uses. Similarly, when an area of forest is clear-cut, it loses its tree species composition. Yet the historical information about the tree species composition may be invaluable for preparing the forest regeneration plan.

Geographic data are inherently a form of **spatial data**. Geographic data can be represented on a map or in a geographic information system as either point, line, or area features. **Points** are used to represent the location of geographic phenomena at a point or to represent a map feature that is too small to be shown as an area or line. The location of a city (on a small scale map), a mountain peak, or an airstrip could be represented by a point element. A **line feature** consists of an ordered set of connected points. Lines are used to represent map features that are too narrow to be shown as an area or features that theoretically have no width, such as a political boundary. A shoreline, a contour line, a roadway, or an administrative boundary are examples of line features. An **area feature** is a region enclosed by line features. The geographic extent of a city, a forest stand, or a lake could be represented by an area element. Area elements are often represented in a GIS by polygons. (A **polygon** is a closed plane figure bounded by straight lines. By making the straight-line segments small, curved boundaries can be closely approximated. The polygon shape is produced from curvilinear boundaries when the geographic information is entered in the GIS.) Spatial data that pertain to a location on the earth's surface are often termed **georeferenced data**. In this book, the term spatial data refers to georeferenced data, so the two terms are used interchangeably.

WHAT IS A GIS?

Taken in its broadest sense, a geographic information system is any manual or computer based set of procedures used to store and manipulate geographically referenced data. The definition used in this book is as follows: A GIS is a computer-based system that provides the following four sets of capabilities to handle georeferenced data: 1. input; 2. data management (data storage and retrieval); 3. manipulation and analysis; and 4. output. The restriction to computer-based systems reflects the focus of this book. There are many manual systems that are used routinely to perform these functions and are effective for the tasks they perform and under the conditions in which they operate.

The fundamental components of a GIS and its environment are illustrated in Figure 2.3. A GIS does not operate in a vacuum. To be successful it must reside within a suitable organizational framework. The GIS is operated by staff who report to a management. That management is given the mandate to operate the GIS facility in such a manner as to serve some user community within an industry, business, or government organization. Ultimately, the purpose and justification for the GIS facility is to assist the users in accomplishing the goals of their respective organizations.

Some of the users may be considered internal (i.e. responsible to the same management that controls the GIS), while others are external users reporting to a

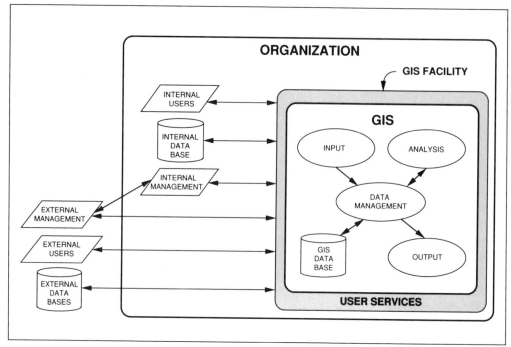

Figure 2.3   The Management Context in Which A GIS Facility Exists.

different management. For example, a national geological survey will use a GIS to support its own research and may also provide analysis services to other departments or the general public. At the same time, the military may develop an even more detailed GIS of the same area, but its use is rigorously restricted and controlled. The nature of a GIS is such that it seems to require this type of institutional setting in order to be successful. The mandate for a GIS facility is almost always in the context of a long-term objective.

CAD (Computer Aided Design and Drafting), DBMS (Data Base Management Systems), and AM-FM (Automated Mapping and Facility Management) systems work with georeferenced data. They can perform many of the same operations as a GIS. What distinguishes a GIS from these other systems is the ability to integrate georeferenced data. This includes operations like **spatial search** and **overlay**. These functions perform

buffer zone generation (e.g. show all areas within 1 km of a stream) and map overlay (e.g. show all areas on well-drained soils, zoned for residential development, and currently forested). (GIS analysis functions are discussed in Chapter 7.)

Often GIS is confused with cartographic systems that store maps in automated form. Again, however, the ability to integrate data is what sets GIS apart from these mapping systems. While the main function of the cartographic system is to generate computer-stored maps, the function of a GIS is, in a sense, to create information by integrating data layers to show the original data in different ways and from different perspectives.

LAND INFORMATION SYSTEMS

A Land Information System (LIS), also termed a Land Related Information System (LRIS), is a special type of GIS. The terms LIS and LRIS have been used rather broadly in the GIS literature to refer to systems that include land ownership information. In this discussion an LIS is considered to be a GIS that is designed to handle detailed land ownership information. This information is commonly recorded on a large scale map (at scales of 1:1,000 to 1:10,000) or more recently, is stored in a computer-based LIS. It is administered and maintained by a government unit that is legally responsible for maintaining the land records in the jurisdiction. (The most complete land records data bases are probably those of the insurance industry and not available to the public.)

The official record of interest in land is termed the **cadastre**. It is the legally recognized registration of the quantity, value, and ownership of land parcels. A cadastral system, the system used to store and maintain the cadastre, consists of three components. These are as follows:

1. Records of the cadastral parcels — the continuous areas of land (i.e. the land units) within which unique and homogeneous interests are recognized.

2. The cadastral record — the graphic (e.g. maps) and text information describing the nature and extent of the land interests.

3. The parcel index — the system for relating parcels and records (NRC 1983).

A GIS used for municipal and county level applications is typically structured as an LIS. The data base will usually include such data layers as the street network, location and attribute records for land parcels and lots, floodplain hazard, a base map, geodetic control (points for which accurate geographic coordinates have been measured), and utility networks (e.g. water and sewage facilities). A municipal LIS may contain on the order of 50 or more separate data layers.

County level LISes tend to include more natural resource and environmental data layers such as wetland conservation areas, ground water recharge areas, soil erosion control plans, land use, and zoning. These broad environmental information categories are more important in describing the actual interests in the land at the county level of government. Other data layers commonly included are land parcels, geodetic control, zoning, land cover, and soils. A number of county, state, or federal jurisdictions may exercise regulatory control over an area. For this reason, the land parcel tends not to be as central a unit of reference in a county level LIS as in a municipal one.

Utility companies often use automated mapping systems to maintain schematic diagrams of their facilities, such as telephone lines, power lines, and gas pipelines. Often the diagrams are used for facility management applications alone and are not accurately referenced to geographic coordinates. These types of facility management/automated mapping systems are not included in the term LIS as used here.

Land information systems integrate property rights information with information on the uses, values, and distribution of natural and cultural resources. They are used for a wide variety of such applications as the maintenance of tax and parcel information, the analysis of changing land use, the scheduling of road maintenance activities, the maintenance of utility inventories, and the management of watersheds. The user groups range from emergency service coordinators and school boards to utilities, public works departments, planners, developers, financial institutions and the general public. Land information systems represent one of the fastest growing GIS application areas, largely because an LIS provides the only practical means to organize and integrate large volumes of land-related information. As more land information is converted to digital form, as legislation defines more stringent information requirements, and as the technology to support an LIS becomes less expensive and more powerful, the automation of land information will become a *defacto* requirement.

THE COMPONENTS OF A GIS

The following is a brief description of the basic components of a GIS. Each component is the subject of subsequent chapters.

Data Input

The data input component converts data from their existing form into one that can be used by the GIS. Georeferenced data are commonly provided as paper maps, tables of attributes, electronic files of maps and associated attribute data, airphotos, and even satellite imagery. The data input procedure can be as straightforward as a file conversion from one electronic format to another, or it can be complex. Data input is typically the major bottleneck in the implementation of a GIS. Construction of large data bases can cost five to ten times that of the GIS hardware and software.

It can take months to years to complete the initial data input. So the expense and time needed to bring the GIS into full operation must be budgeted as part of the overall start-up plan, otherwise pressure to show results can compromise the data input procedure. Cost-cutting compromises at the data input stage are very costly to correct. Those data that may be inaccurate first have to be found, a task that may in itself be more expensive than re-doing the entire data entry. Once the inaccuracies have been corrected, the confidence of the users must then be rebuilt — and the first impressions of users are remarkably resistant to change.

For this reason, data input methods and data quality standards should be carefully considered well before data entry is to begin. They are prone to receiving cursory attention in the midst of a flurry of data entry activity. The various methods of data entry should be evaluated in terms of the processing to be done, the accuracy standards to be met, and the form of output to be produced.

Data Management

The data management component of the GIS includes those functions needed to store and retrieve data from the data base. The methods used to implement these functions affect how efficiently the system performs all operations with the data. There are a variety of methods used to organize the data into computer-readable files. The way the data are structured (data structure) and the way files can be related to each other (the organization of the data base) place constraints on the way in which data can be retrieved and the speed of the retrieval operation. The short- and long-term needs of the users should be identified and used in evaluating performance trade-offs. A person who is expert at GIS data base design and analysis procedures is needed to evaluate these trade-offs.

Data Manipulation and Analysis

The data manipulation and analysis functions determine the information that can be generated by the GIS. A list of required capabilities should be defined as part of the system requirements. What is often not anticipated is that the introduction of a GIS will not only automate certain activities, it will also change the way the organization works. For example, financial and time constraints may force decisions to be made after a study of two or three alternatives. If it becomes less expensive and faster to generate alternatives, it may become feasible to successively refine the plans. The decision method may then change from selecting the best of the few alternatives presented to developing the best alternative by seeking out and evaluating suggested improvements. To anticipate the way in which the data in a GIS will be analyzed requires that the users be involved in specifying the necessary functions and performance levels.

Data Output

The output or reporting functions of GISes vary more in quality, accuracy, and ease of use than in the capabilities available. Reports may be in the form of maps, tables of values, or text in hard-copy (such as paper) or soft-copy (electronic file). The functions needed are determined by the users' needs, and so user involvement is important in specifying the output requirements.

WHY USE A GIS?

A geographic information system is a powerful tool for handling spatial data. In a GIS, data are maintained in a digital format. As such the data are in a form more physically compact than that of paper maps, tabulations, or other conventional types. Large quantities of data can also be maintained and retrieved at greater speeds and lower cost per unit when computer-based systems are used. The ability to manipulate the spatial data and corresponding attribute information and to integrate different types of data in a single analysis and at high speed are unmatched by any manual methods. The ability to perform complex spatial analyses rapidly provides a quantitative as well as a qualitative advantage. Planning scenarios, decision models, change detection and analysis, and other types of plans can be developed by making refinements to successive analyses. This iterative process only becomes practical because each computer run can be done quickly and at a relatively low cost.

It is the spatial analysis capabilities of the computer-based GIS that distinguish it from related graphics-oriented systems like computer-aided design and drafting. The analysis of complex, multiple spatial and non-spatial data sets in an integrated manner forms the major part of a GIS's capabilities. It is a function that cannot be done effectively with manual methods or with computer-aided design and drafting systems. These spatial analysis capabilities of a GIS together enable georeferenced information to be created and used in a completely different context than before.

Not only can diverse data sets be integrated, diverse procedures can also be integrated. For example, data handling procedures such as the data collection, verification, and updating procedure can be integrated instead of compartmentalized into separate operations. At the time that data, such as a change of land registration or land use, are entered in the GIS, the GIS can check the accuracy of the changes, whether zoning and other restrictions would be violated, and update the relevant maps and tabular data. In this way, the users obtain more current information and can manipulate it to meet their specific needs.

THE GIS AND THE ORGANIZATION

As an organization becomes more familiar with a new system, people find new ways of getting a job done. They will develop analysis procedures different from those originally anticipated. While it is not possible to predict what these new methods will be, changes can be expected. The type and variety of functions provided by a specific system will influence the types of innovations that will occur. A system that excels in modelling functions, for example, would encourage the development of new analyses that capitalize on these functions.

There is also a tendency for the principal use of an information system to evolve. New technology tends first to be used by an organization to perform the tasks in the "old" way using the new technology. Later, as familiarity with the technology is gained, new ways of providing the same functions are developed that more fully take advantage of the technology. Finally, new approaches are developed that take full advantage of the potential of the new technology and meet the information needs. The new technology is used to provide new functions. In the case of GISes, the first applications tend to be inventory operations. Later the emphasis shifts to satisfying analysis and, finally, management needs.

For this reason, the management environment of the GIS facility is perhaps the most important single factor in determining its success or failure. It is the organization that in the end determines whether the physical equipment and human resources will function as an effective information system. The provision of effective user services, from training materials to qualified consultants, is critical to the effective utilization of the benefits that a GIS can offer. Budgeting and organizing for user services often receive much less attention than they warrant. The organization must also be able to deal with political, financial, and technical issues on a continuing basis. A successful management environment will enable the organization to be flexible enough to change while continuing to satisfy its mandates. (Issues in the implementation of a GIS are discussed in Chapter 8.)

A GIS is not the solution to all georeferenced information processing requirements. A GIS is expensive to implement. Existing data must be converted to digital form, a task that is usually many times the cost of the hardware and software. A GIS represents a significant overhead cost both to maintain the system and for the considerable degree of expertise required of the personnel who operate it. These costs are more easily justified if the data volumes are large, the data must be frequently accessed, updating of the data is important, and the data will be used repeatedly for a wide range of analyses. If these conditions do not apply, then a GIS may not in fact be a cost-effective solution.

CONCLUSION

This chapter has provided a framework, a context, within which to view the more detailed information presented in the remainder of the book. The approach taken in this book has been to present the fundamental principles and relevant issues needed to understand how GIS technology is used. The technology is developing at such a fast rate that assessments of individual systems and processing technology are out-of-date before they go to press. (There are no reviews of GIS systems included.) However the principles used to apply that technology, the objective in using a GIS, are more fundamental.

The major challenge in acquiring a GIS is the development of analysis approaches that address the problem at hand and make effective use of the technology. Learning to operate a GIS is relatively easy. It is far more difficult to learn how to apply this skill effectively and creatively to satisfy "real world" needs.

GIS technology and a GIS facility can change the objectives that an organization can attain, but it is the organization that must define what those objectives should be. In the case of a public agency, those objectives are ultimately defined by the values of the society. This book focuses on strategies for using GIS technology and identifying those issues that in a given situation will define appropriate objectives. In so doing, an approach to using this technology is developed, whatever the capabilities of a specific system.

REFERENCES

Beard, M.K., N.R. Chrisman, T.D. Patterson. 1987. Integrating Data for Local Resource Planning: A Case Study of Sand and Gravel Resources. In *GIS for Resource Management: A Compendium*. American Society for Photogrammetry and Remote Sensing. Falls Church, Virginia.

Burrough, P.A. 1986. *Principles of Geographical Information Systems for Land Resources Assessment*. Oxford University Press. New York, New York.

Calkins, Hugh M. 1976. Information System Developments in North America. In *Proceedings of the Commission on Geographical Data Sensing and Processing — Moscow 1976*. International Geographical Union. Paris, France. pp.93–113.

Chrisman, N.R. 1987. Fundamental Principles of Geographic Information Systems. In *Proceedings of the Eighth International Symposium on Computer-Assisted Cartography*. American Society of Photogrammetry and Remote Sensing. Falls Church, Virginia. pp.32–41.

Cowen, David J. 1987. GIS vs. CAD vs. DBMS: What are the Differences? In *Proceedings of the GIS '87 Symposium*. American Society for Photogrammetry and Remote Sensing, Falls Church, Virginia. pp.46–56.

Crain, I.K. and C.L. MacDonald. 1983. From Land Inventory to Land Management, The Evolution of an Operational GIS. In *Proceedings of the Sixth International Symposium on Automated Cartography*. University of Ottawa. Ottawa, Ontario. pp.41–50.

Dangermond, J. 1983. *Software Components Commonly Used in Geographic Information Systems*. Environmental Systems Research Institute. Redlands, California.

Hardy, E.E. 1975. The Design, Implementation, and Use of a Statewide Land Use Inventory: The New York Experience. In *Proceedings of the NASA Earth Resources Survey Symposium*. National Aeronautics and Space Administration. Washington, D.C. Vol. 1C: 1573–1577.

Marble, Duane F. 1984. Geographic Information Systems: An Overview. In *Proceedings of the Pecora 9 Symposium*. IEEE Computer Society. Silver Springs, Maryland. pp.2–8.

McHarg, Ian L. 1969. *Design with Nature*. Doubleday & Company. Garden City, New York.

NRC. 1983. *Procedures and Standards for a Multipurpose Cadastre*. Panel on a Multipurpose Cadastre of the National Research Council. National Academy Press. Washington, D.C.

Parent, P. and R. Church. 1987. Evolution of Geographic Information Systems as Decision Making Tools. In *Proceedings of the GIS '87 Symposium*. American Society for Photogrammetry and Remote Sensing. Falls Church, Virginia. pp.63–71.

Parker, H. Dennison. 1987. What is a Geographic Information System? In *Proceedings of the GIS '87 Symposium*. American Society for Photogrammetry and Remote Sensing. Falls Church, Virginia. pp.72–80.

Shelton, R.L. and E. Hardy. 1974. Design Concepts for Land Use and Natural Resource Inventories and Information Systems. In *Proceedings of the Ninth International Symposium on Remote Sensing of Environment*. Environmental Research Institute of Michigan. Ann Arbor, Michigan. Vol. 1:517–535.

Steinitz, C., P. Parker, and L. Jordan. 1976. Hand-Drawn Overlays: Their History and Prospective Uses. *Landscape Architecture* 66 (5): 444–455.

Tomlinson, R.F., Calkins, H.W. and Marble, D.F. 1976. *Computer Handling of Geographical Data*. Natural Resources Research Series XIII. The UNESCO Press. Paris, France.

3. REMOTE SENSING

INTRODUCTION

From time immemorial people have used vantage points high above the landscape to view the terrain below. From these lookouts they could get a "bird's eye view" of the region. They could study the landscape and interpret what they saw. The search might be for good hunting grounds, easy passage through the region, or the best strategy to attack an enemy. The advantage of collecting information about the landscape from a distance was recognized long ago. **Remote sensing** as we know it today is the technique of collecting information from a distance. By convention, "from a distance" is generally considered to be large relative to what a person can reach out and touch, hundreds of feet, hundreds of miles, or more. The data collected from a distance are termed **remotely sensed** data.

Today most natural resource mapping is done using remote sensing. Aerial photography has been used to produce virtually all topographic maps and most forestry, geology, land use, and soils maps. More recently, airborne radar and scanner data as well as satellite imagery are being used for these types of mapping applications. Aerial photography is used to prepare detailed city maps. Municipalities have even used airphotos to identify unrecorded property improvements. The increased tax collection more than pays for the entire analysis!

Remote sensing techniques are used extensively to gather measurements. Measurements of landscape features are used to produce the elevation contours on topographic maps. Aircraft-based remote sensing measurements of gamma radiation and magnetism are used routinely for geological exploration and mapping. Satellite-based systems can now measure phenomena that change continuously over time and cover large, often inaccessible areas.

Weather satellites provide temperature measurements at the earth's surface and at different altitudes above the surface. Patterns of sea surface temperatures show the position of currents and areas of upwelling, important for managing fisheries. Satellite-based systems are used to estimate chlorophyll levels near the sea surface, an important indicator of the availability of the food on which commercial fish stocks depend. Measurements of crop condition can be made at regular intervals through the growing season to identify problem areas and to predict production levels. Remote sensing systems provide the capability to collect uniform measurements in digital form over large areas at very high speed and to analyze phenomena that could not be monitored in any other way.

Despite their wide use, remote sensing techniques are unfamiliar to most users of geographic information. This chapter presents a brief introduction to remote sensing concepts. The emphasis here is on the approach used to produce useful geographic information rather than on the technical details of the sensor systems themselves. However, to understand how remotely sensed data can be used, some basic information is needed about what these data represent. Aerial photography is widely available from provincial, state, and federal agencies, and aerial survey companies can acquire original imagery to customer specifications. Sources of satellite imagery that are publicly available in Canada and the United States are listed in Appendix B.

A BRIEF HISTORY OF REMOTE SENSING

The development of remote sensing as we know it today began with aerial photography. By the early 1860s photographs had successfully been taken from captive balloons. The first military use of aerial photography was in the American Civil War. In June of 1862 photographs taken by the Union Army were used to analyze the defences of Richmond. By the early 1900s photographic technology had improved to the point that smaller cameras and faster lenses and films were available. Photographs were successfully taken using kites and even pigeons as platforms. However, for aerial photography to become practical, it required a navigable platform so that the camera could be positioned where it was needed. This platform was supplied by the piloted airplane.

Wilbur Wright is credited with taking the first photographs from an airplane in 1909. It was soon after that German flying students training at English flying schools began taking airphotos. At the beginning of World War I aerial photography was not in use. During the war, military authorities were at first reluctant to use the new technology, but when semi-official photographic missions produced airphotos of military facilities in German-held territory, they were quickly convinced.

Photo interpreters became recognized as the "eyes of the armed forces" by all countries in the war. The use of aerial photography had a profound effect on military tactics. It was difficult to hide military information from aerial photography, even with the use of decoys and camouflage. Perhaps more important, it was found that photo interpreters became skilled at predicting enemy activity from the information in the airphotos. By studying the quantities of transportation equipment at railheads and ammunition dumps, the construction of new railways and medical facilities, and other indicators, they could correctly anticipate military activities in time for counter measures to be planned. By recognizing several independent factors, interpreters can infer information that is not directly presented. (This use of multiple indicators to deduce information is termed the principle of **convergence of information**. It is central to the effective interpretation of all forms of remotely sensed data.)

By the end of the war, armies had developed the capacity to produce large quantities of aerial photographs rapidly. In 1918, during the Allied Meuse-Argonne offensive, 56,000 prints were produced in four days (Colwell 1983). Military photo interpreters could be depended upon to provide accurate and timely information. But their experience, gained by trial and error, was not formalized.

In the period between the World Wars, the development of military photo interpretation virtually stopped; however, significant advances were made in commercial and scientific applications. Commercial survey companies provided airphoto acquisition, interpretation, and mapping services. The methods of using aerial photographs to make accurate measurements developed into the field of photogrammetry. The techniques for producing topographic maps from stereo airphoto pairs were also developed during this period. Not only did aerial photography reduce the cost and the time to produce topographic maps, it also enabled maps to be made of areas with inaccessible or difficult terrain. Figure 3.1 shows an amphibious bushplane used in Canada for some of the early airphoto missions in the 1920s and 1930s. Photogrammetric methods are now used to produce most topographic and natural resource maps. Papers were published on photo interpretation for archaeology, ecology, forestry, geology, engineering, and other applications and government agencies began using the technology in their mapping operations.

Figure 3.1 Vickers 'Viking' IV flying boat equipped for aerial photography in Manitoba in 1924 (Courtesy of the National Archives of Canada, #C25910).

Germany foresaw the importance of photointerpretation and was the world leader in the field in 1939 at the outbreak of World War II. The German advance was planned using complete airphoto coverage of Allied terrain and installations. During World War II the Allies developed state-of-the-art techniques. For example, on the Pacific front, coastal water depth information was essential for planning assaults by amphibious craft. Photo interpretation methods were developed to estimate water depths to an accuracy of 2 ft for depths up to 30 ft.

After World War II, the photo interpretation techniques developed in war time became the established procedures for civilian applications. Topographic mapping, geologic mapping, and mapping for engineering were routinely done using aerial photography, as they are today. Inventories of natural resources like timber and agriculture used airphotos to derive quantitative estimates of production more cheaply than by ground survey alone. Aerial photography was still the only sensor system in practical use. But experimentation was underway with new types of sensors. Radar technology that had been invented during the war was developed into systems that could continuously scan a swath of terrain to the side of the aircraft. During the 1970s, these instruments became readily available to non-military users.

Infrared film was developed during the war to identify camouflaged military equipment and facilities. This film was sensitive to red, green, and near-infrared wavelengths. It produced a very useful image with non-natural or "false" colours (hence the film is often called false colour infrared film). Green paint strongly reflects green wavelengths;

for this reason it appears green to the naked eye. Green vegetation in addition to reflecting green wavelengths also strongly reflects near-infrared. The human eye is not sensitive to the near-infrared wavelengths and so the vegetation and green paint appear the same colour. However, in the colour infrared photographs, the green vegetation appeared in reddish colours and the green-painted objects appeared blue. Thus a clear distinction could be made between natural vegetation and objects painted green. Green paints have since been developed that reflect both green and infrared wavelengths, as vegetation does.

After the war, military applications also continued to develop. Higher flying aircraft, such as the U-2, that were out of reach of existing weapons enabled the United States to conduct regular reconnaissance over Eastern Bloc countries. The Cuban Missile Crisis of 1962 illustrated the importance of early detection and superior military intelligence. It was the reconnaissance photography from U-2 aircraft that alerted the United States to the new Cuban missile installations and later verified that they had been dismantled. The military requirement for more advanced sensors continues to provide the research impetus to develop new sensor systems and more sophisticated applications.

In addition to its military applications, colour infrared film proved valuable for vegetation analysis. Crop and forest types were more easily identified using this false colour film than on normal colour films, and stressed vegetation could be distinguished from non-stressed (healthy) vegetation. As discussed later in this chapter, near-infrared reflectance is not related to the temperature of the object. Longer wavelengths in the far-infrared region provide this information. (Objects at very high temperatures, such as lava at 1100°C, will emit radiation at near-infrared wavelengths (Lillesand and Kiefer 1987). Most earth objects have temperatures

on the order of 30°C and *emit* virtually no near-infrared radiation. However, earth objects may strongly *reflect* near-infrared radiation that has been emitted from a source such as the sun.) It was not until the 1970s that detectors to record thermal infrared wavelengths were generally available. Thermal infrared scanners create images electronically using solid-state detectors cooled by liquid nitrogen.

The 1960s ushered in a new age for remote sensing. With the development of earth-orbiting satellites, it became feasible to obtain high-altitude images of the earth's surface independent of political boundaries. But perhaps more important was the development of digital electronic imaging systems that could relay image data back to earth. These data could be processed to photographic images using computer-based techniques known as image processing. Image processing could be used to form colour composite photographs from separate digital images that each represented a narrow range of wavelengths of light. Image processing techniques could also be used to sharpen the images and to overlay images taken at different times.

The first non-military satellite designed to collect information about the earth's land resources was the Earth Resources Technology Satellite (ERTS-1) launched in 1972 by the United States. This satellite was later renamed Landsat-1 and was followed by Landsats 2 through 5. The data were received by a network of ground receiving stations around the world. The United States made these data publicly available with virtually no restrictions on its use.

The early Landsat data had a ground resolution of about 80 m and imaged almost the entire earth every 18 days. Later, Landsat satellites produced data with 30 m resolution. A number of other satellite-based sensors have provided data to the international community. Among them are data from the AVHRR sensor with a resolution of about

1 km, the CZCS sensor with a resolution of about 1 km, and the SPOT sensor with resolutions of 20 m and 10 m. These sensor systems are discussed later in this chapter. In addition to France and the United States, a number of other countries, including Canada, India, and Japan, have announced plans for earth resources satellites that would provide data to the international community.

The rapid development of remote sensing technology has provided the capability to generate data at a far greater rate than they can be analyzed and used. In fact, the development of sensor, data receiving, and data storage capabilities has proceeded faster than the development of practical applications. Often, those who could benefit most from the technology don't have the information or the context in which to assess it. For this reason, remote sensing has been given a more prominent treatment in this book than other data sources commonly used in a GIS.

SIX POPULAR MISCONCEPTIONS ABOUT REMOTE SENSING

Remote sensing is unfamiliar to most users of geographic information even though much of the data they use are generated with this technology. When people lack information, the knowledge gap is often filled by hearing about the experiences of others informally "through the grapevine". The information tends to become somewhat distorted along the way to the point where it becomes myth.

A number of myths have developed over the past 15 years or so, particularly about digital remote sensing analysis. To some extent this was a result of the excitement surrounding the first earth resources satellites, the Landsat series. Before the first Landsat was even launched in July 1972, experts were predicting that resource inventory and assessment would be revolutionized by the new data source — digital imagery.

They expected that by using the imagery from satellite-based sensors, accuracies comparable to those achieved with airphotos and field surveys would be achieved at lower cost.

The reality has been somewhat less than those optimistic predictions — but the reality has still provided information-gathering capabilities previously unattainable. Nevertheless, there are many who have tried to use remotely sensed data and were disappointed with the results. There are many information requirements to which remote sensing techniques cannot contribute and many cases where these techniques have been inappropriately or incorrectly applied. However, cars regularly break down and people are frequently injured using them, yet we still consider them a valuable means of transportation. In the same way, remote sensing is now being used as a valuable cost-effective means of acquiring information about earth resources.

Yet many myths remain about the technology. The myths do have a basis in fact, as do the assertions about the dangers of cars. It is the over-generalizations used to dismiss whole categories of applications that are untrue. Anecdotal information is still often used to influence or justify decisions about using remote sensing methods. In an effort to dispel some of these myths, six of the more popular ones are discussed here.

Myth 1. "Satellite-based remote sensing does not have sufficient resolution."

There is often confusion about what spatial resolution is and what needs to be resolved. The term has an imprecise meaning and there are a number of definitions. Forshaw et. al. (1983) provides a review of the definition and significance of spatial resolution in remote sensing. The *Manual of Remote Sensing* (Colwell 1983, p.20) also provides a concise discussion of the complexities of this term and of measuring spatial resolution. The definition that will be used here is as follows: "the spatial resolution of a sensor system is its ability to render a sharply defined image" (adapted from Colwell 1983). A common unit of measure is the number of line-pairs per millimeter that can be distinguished on the image of a standard test pattern. The spatial resolution of a sensor system is a guide to the smallest object that can be "seen". What is usually meant by "seen" is **detected**.

An object is detected if there is an indication that something different from the surroundings is present, e.g. there appears to be an area in the middle of the corn field that is not "corn field". We often would like more than just this detection level of information. We would like to recognize the category of object being detected.

The **recognition level** of resolution categorizes the objects being detected, e.g. is the object in the corn field a different type of vegetation, a pond, a building, or perhaps some type of farm machinery. A more detailed level of resolution is termed the **identification level** or sometimes the **analysis level**. At this level more specific information about the object is discernable. For example, at the recognition level, it may be discerned that an object is not a building or a pond but a different type of vegetation. At the identification level, information on the specific vegetation type might be derived.

To some extent these three levels of resolution are overlapping categories, yet they are useful in evaluating the spatial resolution needed for a specific application. As a rough guide, it takes about a 3X improvement in spatial resolution to move from the detection level to the recognition level, and about a 10X improvement to move from the recognition level to the identification level (Colwell 1983, p.23).

At each resolution level, factors other than the size of the object come into play. Satellite-based digital sensors produce images comprised of cells termed **picture elements** or **pixels**. Each pixel represents a ground area of a specific size. For example, in the Thematic Mapper sensor on-board the Landsat 4 and 5 satellites, each pixel represents a ground area of about 30 m by 30 m. The pixel size gives only a rough idea of the size of objects that might be visible in the image. Contrast is another factor that directly affects resolution. The more contrast between the object and its surrounding, the smaller the object that can be detected. Atmospheric effects, sun angle, and other environmental factors will also affect whether an object will be visible in the image. For these reasons, objects smaller than a pixel will sometimes be detectable and some objects larger than a pixel may not be detectable. It will depend on how well the object contrasts with its surroundings, the atmospheric and illumination effects, and the wavelengths of light used by the sensor.

In my own work in northern Canada, road construction for resource development often occurs so quickly that no up-to-date maps exist of the road network. In these northern areas, the light-coloured sand and gravel of the dirt roads contrast strongly with the surrounding forests and bogs. As a result, roads 20 m wide were clearly visible on colour Landsat MSS imagery that has a pixel resolution of 80 m.

More important than the question of how small an object can be detected is the question of what needs to be detected to generate the information required. Often a surprisingly low resolution will provide the necessary information. In order to gather information about ground features, it is not always necessary to be able to distinguish each individual item on the image. We are all familiar with this principal of remote sensing because we use it daily when we view our surroundings. Every time we look out over a landscape and distinguish agricultural fields from the surrounding forest we do so without being able to resolve individual trees or crop plants.

We might even be able to assess which crops are planted — sometimes before the crop has even sprouted — just by knowing the growing season and agricultural practices of the area. What we are using to generate this crop information are **secondary indicators**, i.e. indicators other than direct observations of the objects to be assessed. For example, rice fields must be flooded at a certain period in the growing season. An assessment of flooded fields at that time can be used to predict the area of cropland being planted to rice even though no rice plants are detectable.

Perhaps the most striking example of the use of secondary indicators was a prediction in the early 1970s of a failure of the Russian wheat crop. Though the crop is harvested in the fall, this prediction was made during the preceding winter! Much of the wheat crop produced in northern countries like Canada and Russia is winter wheat. This type of wheat is planted in the fall, germinates and begins to grow underground, and then must survive the winter and continue growing the following spring. For the plant to survive the winter it must not be exposed to extreme cold. Normally a snow cover during the coldest period of the winter provides sufficient insulation. That particular January the ground was not covered by snow and the temperatures were extremely low. The lack of snow cover was easily detected by weather satellites with spatial resolutions measured in kilometers not meters! Ground temperatures were also detected from the satellite. By knowing that these extreme conditions persisted continuously for too many days, analysts correctly predicted the crop failure — without ever detecting a single wheat plant! Only by using satellite imagery was it possible to continuously monitor such a large agricultural area and to determine that severe weather conditions had occurred over a substantial portion of the winter wheat area.

Private organizations as well as public agencies regularly use satellite imagery (e.g. AVHRR data, discussed later in this chapter) with coarse spatial resolutions on the order of 1 kilometer or more to assess the condition of important food and cash crops, such as coffee, cocoa, wheat, maize, and rice. In fact, data with a finer spatial resolution is often less suitable for crop monitoring. The data sets are much larger and therefore more costly to analyze. Also, the higher resolution data capture fine details such as shadow effects and the transition zones between vegetation types that make automated analyses of crop condition and crop areas *less* accurate.

Finer resolution imagery, such as Landsat and SPOT data (discussed later in this chapter), are used selectively for crop assessment and as the major data source for mapping land use changes and certain map updating programs. Canada uses Landsat data in an on-going program to update cultural features (e.g. roads and urban areas) on the national 1:50,000 topographic map series.

Remotely sensed data from satellites have been found to have sufficient resolution to be used operationally for many earth resource inventory and mapping applications. Visual interpretation of satellite imagery remains a valuable information extraction technique that often cannot be matched by automated methods. However, a major advantage of remotely sensed data in digital format is that computer techniques can be used, where appropriate, to automate the information extraction process and to directly input the derived information into a geographic information system.

Myth 2. "Remotely sensed data, particularly satellite data, are not sufficiently accurate for practical applications."

The nature of "accuracy" must first be considered in addressing this myth. (Accuracy is discussed in more detail in Chapter 5. However, the central issues are presented here.) **Accuracy** is the degree of likelihood that the information provided is correct. This definition focuses on two components of accuracy. The first and more familiar aspect of accuracy is that it predicts the proportion of information that is expected to be correct or the magnitude of error to be expected. The second, and often ignored, aspect of accuracy is that it **involves a probability**.

Few things are known with 100% certainty and maps are never 100% accurate. When a map or other data set is asserted to be 80% accurate it means that when the data set is used, it can be expected that on average 80% of the information will be correct. It will not be 80% correct every time; sometimes the accuracy will be higher and sometimes it will be lower. The measure of this probability of having a higher or lower accuracy than expected is termed the **level of confidence**. The level of confidence is usually quoted as a percent. The 90% level of confidence indicates that the condition is expected to be true 90% of the time. So when a map is rated 80% accurate with a 90% level of confidence it means that if a large number of accuracy tests were done on the map, then 80% or more of the test points would be correct in 9 out of every 10 tests.

The level of accuracy obtainable from any data source, remotely sensed or not, depends on the information to be provided and the level of detail required. For example, a road map with an accuracy of 1 km may be suitable to estimate the driving time between cities. However, engineering drawings of a city street are required to have accuracies on the order of centimetres.

In producing a thematic map, such as a forest map, the same data set may provide a higher level of accuracy if the information is presented at a coarser level of detail. It may be difficult to accurately interpret forest types at the species level of detail from a satellite image. So the accuracy of species level information would probably be quite poor. However, if a coarser level of detail was chosen, a higher level of accuracy could be obtained. Three classes might be defined: deciduous, coniferous, and mixed forest types. These broader classes are easily distinguished on a colour satellite image and could be more accurately interpreted.

Similarly, the example of predicting a winter wheat crop failure (discussed previously) did not require a high level of spatial accuracy to provide for an accurate prediction of wheat production. The information was very valuable even though the level of detail of the prediction was quite coarse, i.e. the wheat production was expected to be "very low" although the actual size of the harvest could not be accurately determined. In fact, whether the ground was 100% clear of snow or only 90% clear, whether the temperature was 1°C warmer or cooler, did not make a difference in the accuracy of the prediction. In specifying a level of accuracy what is important is the level of accuracy that will affect *the correctness of the decisions to be made* when that information is used.

In part, the reason that the coarse level of detail was valuable was because the information was available so early, before the growing season had even begun! Time is an important consideration in the value of information. Different kinds of decisions are possible depending on the time-frame. These different kinds of decisions will be based on information with different levels of accuracy. Frequently the timeliness of information will offset its lower accuracy, so less accurate information obtained earlier can be very valuable.

Accuracy can usually be improved by expending more resources. More money can be spent on field investigations, more time can be spent on analysis, more quality control can be exercised in assembling the data. The greatest benefit for the cost is gained by using the data that are the *least* expensive to acquire and analyze and will produce information with an acceptable level of accuracy. What is acceptable? An **acceptable level of accuracy** is that level where the costs of making the wrong decisions are equal to the costs of acquiring more accurate information. You might not take the time to check your change when it is 25¢ in pennies, but you probably would count $80 worth of $10 bills a second time.

Companies that are large buyers of agricultural commodities use crop predictions, based in part on satellite data, to decide when to buy commodity contracts. Whether a crop yield will be substantially higher or lower than usual is valuable information if known early enough, even though the level of detail may be quite coarse. Companies, such as Earth Satellite Corporation, provide this type of crop reporting service commercially (see for example Merrit et. al. 1984).

Agriculture Canada uses satellite-based remote sensing to monitor wheat production in Canada. During the 1988 prairie drought, this system was used to provide weekly assessments of crop condition and to estimate farm losses. These data were later used to map the severity of the drought so that subsidies to farmers in different regions could be scaled to drought severity.

The California Department of Water Resources has used satellite-based crop area estimates for the prediction of irrigation water demand. The Department uses these estimates to set the water levels to be maintained in reservoirs and the rates at which the water is transported through the irrigation system (Thomas 1988 and Wall et. al. 1984).

The accuracy required of remotely sensed data, as with other sources of geographic information, depends on the application. The critical factor is the level of accuracy needed to make correct decisions. The spatial accuracy of satellite-based remote sensing systems has been shown to be accurate enough to satisfy a wide range of operational information needs.

Myth 3. "Satellite data are too expensive."

"Expensive" is a relative term. It depends on what is being purchased and on the cost of alternative products. The digital data for an AVHRR scene costs $200 to $300 US and provides coverage of an area about 2500 km by about 1500 km. (The exact size depends on how the image is framed by the supplier.) The digital data for a Landsat TM scene provides coverage of an area 185 km by 185 km. It costs about $800 US for a colour print at a scale of 1:250,000 and $3,600 US for the 7 bands of digital data, about 250 Megabytes of data. It would require 1700 standard size airphotos (230 mm by 230 mm or 9 in by 9 in format) at a scale of 1:20,000 to cover the same area with no overlap. About 270 airphotos would be needed at the 1:50,000 scale, and about 45 airphotos at the 1:120,000 scale (the standard scale for NASA high altitude photography). The cost of government-supplied airphotos is heavily subsidized in Canada and the United States. Even so, a single satellite print is considerably less expensive than airphotos for the same area. Acquiring original aerial photography is an expensive undertaking. Airphoto mission costs run in the tens of thousands of dollars.

Where satellite data can provide the required information, they can offer a number of advantages over airphotos. The satellite scene provides coverage of a large area, greatly reducing the number of images to be handled. This reduces both interpretation time and the time needed to assemble the separate interpretations from each photo. A satellite scene is imaged at essentially the same time, providing a consistency difficult or impossible to achieve for large areas with aerial photography. Photo coverage of a large area must usually be done over several days, or at least over several hours. As a result, the images will have different illumination conditions. Where existing photography is used, the photos for a large area often were taken at different times of year or in different years. These inconsistencies tend to make interpretation more difficult and less accurate.

Large scale aerial photographs of mountainous areas present some special problems as a result of relief displacement. **Relief displacement** is the relative shift of points in the image caused by elevation. It has the effect of shifting the relative positions of objects by different amounts, depending on their elevation. Relief displacement is present in all aerial photographs of areas with differences in terrain elevation. In fact, photogrammetric instruments are used to measure this property in airphotos to produce the elevation data for topographic maps. However, when the relief displacement is too severe, it is not possible to perform the necessary corrections. Relief displacement is greatly reduced in imagery acquired with a longer focal-length lens or at higher altitudes. When mapping mountainous terrain, the reduced relief displacement is needed to obtain suitable images.

Satellite imagery provides virtually complete global coverage and is publicly available. There are many countries that do not allow airphotos to be issued to the public or do not have complete airphoto coverage of their territories. In these cases, satellite imagery may be the only source of data available. Oil exploration in less developed countries is commonly done using satellite imagery as base maps because they are more accurate than the current maps, or because current maps are classified and not publicly available or maps simply do not exist. Basic field support, route location, and geophysical data acquisition programs are all planned using these images. The imagery is also used for reconnaissance level geologic mapping. Thus the satellite imagery commonly serves as the only map and a relatively inexpensive one at that.

Imagery in digital form can be enhanced, classified, and used to generate a variety of measurements using specialized computer hardware and software. The field of digital image processing has expanded rapidly over the past decade. The technology has been applied to such diverse applications as enhancing the images from interplanetary satellites to robot vision. Digital image processing is used to enhance satellite images so they can be more accurately interpreted, to produce measurements such as surface temperature and crop condition, to generate estimates of areal extent such as the areas planted to specific crop types, and to produce thematic classification such as land cover maps.

Computer processing enables the data for large areas to be analyzed at high speed. For example, Agriculture Canada can assess the condition of prairie grain crops on a weekly basis. The accuracy of automated analyses depends on the specific application. Often, a well-enhanced image, analyzed by a skilled interpreter, will provide the most accurate information. The method of choice will depend on the size of the area to be analyzed, the type of information to be extracted, the skill of the photo interpreter, and the costs of the different analysis methods.

Where the required information can be produced from satellite imagery, it is a far less expensive data source than flying original aerial photography. Where suitable photography exists, the relative cost of using satellite data will depend on the analysis methods that will be used. Often the wider, more complete, and more recent coverage of the satellite data makes it the less expensive choice.

Myth 4. "Remote sensing other than aerial photography is only experimental."

This assertion is made about satellite imagery, airborne scanners, imaging radar and sometimes even high altitude aerial photography. Large and medium scale aerial photography have been used successfully for so long that the techniques for using them are now accepted as standard. Yet, it was only during and immediately following

World War II that airphoto interpretation became a systematic procedure. A corollary to this general scepticism is that "remote sensing never found a gold mine", or oil field, or other sought-after feature.

In fact, remote sensing has been instrumental in the discovery of many mineral deposits. An early example is the RADAM project in Brazil done in the 1960s. Radar imagery of the Amazon Basin provided the first imagery ever available for much of this region. Complete airphoto coverage had never been obtained because of the almost constant cloud cover for much of the region. The radar image interpretation in conjunction with field surveys led to the discovery of a world class copper deposit. Oil companies like Chevron and Exxon are major users of satellite imagery for exploration. Airborne scanner imagery is regularly used for mineral exploration in the southwestern United States, Australia, and other arid regions.

But has remote sensing alone been responsible for discoveries? Rarely is this the case. Yet rarely is *any* technique used alone for resource exploration, inventory, or monitoring. A combination of techniques is used. Geophysical surveys (geophysics could also be considered a form of remote sensing), field geology, and drilling are used together to locate and develop mineral and petroleum deposits. Weather information, soils data, past crop production records, and field observations are used in crop monitoring.

Resource exploration, inventory, and monitoring are usually more successful and cost-effective when multiple data sources are used. Different data sources provide different types of information, each advancing the investigation by adding new knowledge and serving as an independent check on the interpretation. This "convergence of evidence" approach to using multiple data sets leads to more accurate and confident predictions.

A multi-level approach is commonly used in remote sensing, as in other types of data collection. In the multi-level approach, less detailed and less expensive data are first collected for a wide area. These data are then used to guide the selection of the few, more promising areas for which more expensive and detailed surveys and tests will be done. In this way the maximum effort is focussed on the most favourable areas. It is particularly effective when a few occurrences are being sought over a large area because the entire area is systematically screened. This analysis approach is routinely used in mineral and petroleum exploration. It is also valuable in locating and mapping areas of land cover change. Satellite data and high altitude aerial photography are now used routinely to provide the less-detailed wider view (often termed the synoptic or "bird's eye" view) in a multi-level analysis.

The multi-level approach is also used for estimation. Because it is too costly to measure all of the resource, for many applications a sample is measured to estimate the total amount. Remotely sensed data, such as digital satellite imagery and aerial photography, are used to provide a complete coverage of the area to be sampled. The imagery is interpreted and classified into regions with similar characteristics. Then each region is sampled. Small areas in each region are selected for more detailed and more expensive data collection. Field data may be collected for each sample site, or the sample sites may be further sub-sampled, i.e. a sample of each smaller area might be chosen. (Multi-level sampling is discussed later in this chapter.) For many applications, such as in foresty and agricultural crop inventories, multi-level sampling with remote sensing and field data collection is the only economically feasible method to assess the resource.

Remotely sensed imagery is used to distinguish features and conditions that are difficult or impossible to see by using wavelengths not visible to the human eye. More accurate vegetation mapping and assessment can be done using near-infrared imagery. The far-infrared wavelengths are used to distinguish features by differences in their temperature and emissivity (a measure of how readily the infrared radiation can escape from the surface). Cloud-covered regions can be imaged by radar sensors that use microwave wavelengths that penetrate cloud. Stereo radar images are then used to produce topographic maps.

The objective of all geographic data collection activities, is to use the mix of techniques that provides the required information. Non-photographic remote sensing systems are used routinely to monitor and inventory a wide variety of natural resources.

Myth 5. "Remotely sensed data are too complicated to use."

The effective use of remotely sensed data does require knowledge and skill, as do most other endeavors. Contrary to the myth, the most important knowledge and skills needed are not those in remote sensing but rather are those in the discipline to which the techniques are being applied. It has been found time and again that it is much easier to train a discipline specialist, such as a geologist, a forester, or a planner, in how to use remotely sensed data than to train a specialist in sensors or digital analysis to be expert in geology, forestry, or planning. Remote sensing techniques are always applied to the collection of information about something — "remote sensing" is not itself a resource field.

The amount of training needed for discipline specialists to use remote sensing for useful work need not be long or complicated. The remote sensing center in Nairobi, Kenya, provides basic remote sensing training to agriculture, forestry, geology, planning, and other resource specialists from the 20 country region of Africa that it serves. The courses generally run for 6 weeks. When the trainees return home, their training continues as they use their remote sensing knowledge for practical resource mapping, inventory, and analysis projects in their own work.

At the Centre, the trainees use visual interpretation methods to analyze satellite images. This can be done without computers or other expensive equipment. Just a few simple materials are needed, such as clear mylar to place over the images and the pens to write on it. The colour and black-and-white images for their projects are produced photographically at the Center. The Nairobi Center has imagery for the entire region it serves. In this way, the satellite imagery required by these new users can be provided quickly and inexpensively (Falconer 1986).

In Canada, hundreds of maps from the 1:250,000 national topographic series have been updated using visual interpretation of colour satellite images. The images are projected onto the map being updated and the changes are traced by hand. The method is simple, straightforward, and sufficiently accurate to delineate new roads, towns, and other cultural features. It is also faster and less expensive than computer-based techniques (Gauthier 1987).

Many remote sensing techniques do require highly trained personnel with extensive backgrounds in the physics of remote sensing, computer analysis techniques, and one or more resource disciplines. The higher costs for these specialists and their sophisticated equipment is cost-effective when alternative methods would be too slow, too costly, or when alternatives simply do not exist. However, relatively simple techniques, though less publicized, are widely and successfully used.

Myth 6. "Remotely sensed data are not readily available."

Most industrialized nations have on-going airphoto acquisition programs to support national and regional mapping. This imagery is generally available to the public at the cost of reproduction, a fraction of the full cost to produce it. In Canada, the United States, and most western European countries, airphotos with a range of scales and film types are available for the entire country. But for the majority of the world's nations, complete airphoto coverage either does not exist or is not publicly available.

It was the introduction of earth resources satellites, such as Landsat, and more recently SPOT, that made remotely sensed data readily available to all. After 16 years of satellite image acquisition, enough publicly available Landsat, SPOT, and other satellite images exist to show virtually any area, no matter how inaccessible, for a few hundred dollars or less. The imagery can be ordered directly from the companies or government agencies producing the data. Several sources are listed in Appendix B. Airborne radar and digital multi-spectral imagery are being used more widely and in some cases are made available to the public. For example, Indonesia has used extensive radar image surveys acquired commercially to produce their national topographic map series.

A CONTEXT FOR USING REMOTELY SENSED DATA

Whether it be a person's view of the valley below or a satellite view of the earth, remotely sensed data must be analyzed to extract useful information. When we stand at a lookout we interpret the landscape. We may distinguish different types of trees or different crops. We may draw further conclusions from what we see, such as whether a field is being farmed or has been abandoned. The steps we use informally to interpret our surroundings are the same ones used to systematically analyze remotely sensed data.

The principal steps used to analyze all remotely sensed data are as follows:

1. Definition of information needs
2. Collection of data using remote sensing and other techniques
3. Data analysis
4. Verification of analysis results
5. Reporting of results to those who will use the information
6. Taking action based on the information

These six fundamental steps are used to analyze every form of remotely sensed data and in every application. Not surprisingly, this is basically the same procedure discussed in Chapter 2 for using geographic information in general.

Long ago primitive hunters assessed their territory using this procedure, albeit informally. The information needed was the location of the best hunting areas. The data collection may have been the hunter's view from a lookout. The analysis would have been his classification of the landscape that he saw into areas for hunting different types of game. He would have depended on his experience to judge the accuracy of his assessment. His report would probably be a verbal description to the other members of the hunting party, and the resulting decision would be a plan for the hunt. From the experience he gained during the hunt he would learn the accuracy of his analysis and improve his interpretation techniques for the next time. The technology may have changed and the execution may be more systematic but the approach used is basically the same in modern remote sensing as it was for the hunter.

Before presenting an overview of how remote sensing works, this section presents the basic strategy needed to use remotely sensed data effectively. Often the overall strategy is lost in wrestling with technical details. Technical details are important. But they are not always the prime consideration. Many fundamental decisions are independent of technical details. They are management decisions, such as "What decisions will the information influence ?" or "When is the information needed ?". Perhaps the single factor most responsible for unsuccessful remote sensing applications is the failure to correctly identify the questions to be answered. The six steps to using remotely sensed data provide a framework within which to apply this technology appropriately and successfully.

STEP 1. Defining the Information Needed

The objective of using remotely sensed data is to generate information. Before any data acquisition or analysis can begin, the information needs have to be defined. Only then can the techniques be identified that might best satisfy the requirement. This assessment should take into account such factors as the accuracy needed, how quickly it is needed, in what time period should the information have been collected (e.g. within the past year or in a particular season), the cost to produce it, and the form (e.g. electronic format, paper map, tabulated statistics) in which it is needed.

STEP 2. Data Collection

Remotely sensed data are rarely used as the sole data source. Field observations and measurements as well as existing information such as maps and reports are used together in the analysis. The data requirements must be defined, the available data assessed, and then the new data to be collected must be specified.

Remotely sensed data are collected using a variety of devices that detect energy reflected or emitted from objects. The most familiar of these devices, or sensors, is the photographic camera system. It records reflected light energy in the form of black-and-white or colour images. The detector in this system, i.e. the mechanism for sensing the light energy, is light-sensitive photographic film. Visible light can also be recorded electronically as digital images using sensor systems termed *multi-spectral scanners*. Scanners are also used to detect non-visible energy such as emitted thermal infrared to detect heat, and ultra-violet to detect materials such as oil on the sea surface. Radar systems illuminate the scene using microwave energy and electronically generate images from the reflected signals. Airborne radar systems can produce stereo imagery from which topographic maps can be produced. They have the additional advantage of "seeing through" cloud cover.

The remote sensing data specifications should be planned and integrated with the other data collection activities. Field collection procedures can be modified so that expensive field sample data can be properly integrated into the remote sensing analysis. Similarly, by assessing current field procedures and existing data sets, the most cost-effective remote sensing methods can be chosen.

STEP 3. Analysis

There are three principal types of analyses applied to remotely sensed data. They are measurement, classification, and estimation. These types of analyses may be used individually or collectively in a given application.

Measurement analyses use the values measured by the sensor to calculate environmental conditions like surface temperature, soil moisture, quantity of plant material, or the condition of crops. Measurement results are usually produced as a large number of individual values, one for every sample point.

Classification analyses define regions that have the same characteristics. These results are commonly provided in the form of a map-like image where regions with the same characteristics are shown coded as the same colour or pattern. The image may be produced as a paper map, a digital image, or as a set of boundary outlines for each region. The classification may be used to generalize a measurement analysis for presentation. For example, ranges of temperature could be shown as different coloured classes.

Estimation analyses are commonly applied to classification results. The objective of this type of analysis is to estimate the quantity of a material, such as the quantity of timber or the quantity of wheat for each management area. This type of analysis is not a mapping application, and so precise delineation of boundaries is not usually needed. The type of classification used in this type of analysis serves to divide the area into regions, termed **strata**, that have statistically similar characteristics. However, every sample in a region will generally have a different value. The advantage of using remotely sensed data to stratify the region in this way is that estimates with the same or better accuracy can be obtained with fewer field samples and, therefore, at a lower cost.

STEP 4. Verification of the Analysis Results

To use information effectively we need to know something about its accuracy. If you received stock market information from a friend who regularly loses money in the market, you would probably give it a low accuracy rating. Information need not be 100% correct for it to be useful, so long as the expected level of accuracy is known and is taken into account when the information is used. A flood may be estimated to have left 10,000 people homeless. Whether the actual number is 8,000 or 12,000 probably will not affect the decisions to be made about emergency relief. However, knowing precisely which roads into the area are open will be critical to planning relief operations.

For these reasons, the results of remote sensing analyses, as with any other geographic information, should be accompanied by a report on the quality of the data. The verification step involves testing the results produced in the previous analysis step, in order to verify that the results are of sufficient quality to be accepted for use. (Data quality is discussed in Chapter 5).

People are generally familiar with measurement accuracy and estimation accuracy. Opinion polls are usually reported with an indication of their accuracy. For example, an opinion poll might report that 45% plus or minus 2% would vote "yes" and the confidence level of the survey may be quoted as 95%. However, the accuracy of the classes shown in maps (termed **classification accuracy**) is not as familiar a concept.

Map accuracy depends on many factors. Some are quite obvious. If you know that there has been recent road construction in an area, you would not expect a map made 10 years ago to show today's roads accurately. Other factors are more complex to assess, such as the accuracy trade-offs between the automated classification and visual interpretation of satellite images.

Maps are compiled from various sources of information. Most mapping today uses aerial photography or satellite imagery, either as the principle source of information (e.g. topographic mapping) or to provide the base map and other supporting information. Thematic maps are maps in which the region is subdivided into classes. A soils map or a map showing the countries of the world are thematic maps. Thematic maps represent a classification of source information (such as field samples, airphotos, and satellite images) into map classes. Associated with this classification is an accuracy. Maps are frequently treated as if the information is 100% accurate when the accuracy is actually unknown! Yet when classifications derived from remotely sensed data are presented as 85% accurate, there is often resistance to using the information.

The accuracy of maps may be assessed using a formal accuracy test. National topographic maps are normally assessed in this way. Sample points are selected, measured on the map, and compared with independent measurements like field survey data. The amount of error is then calculated for each sample point, and, if the overall error is too great, the map is rejected.

There is a cost to accuracy assessment that must be weighed against the cost of using incorrect information. The specification of an accuracy level should reflect the minimum accuracy needed for the information to be used as intended. There is no point in demanding position accuracies of vegetation boundaries to the nearest meter if, in the field, they can only be measured to the nearest 10 m. If the information needed is total acreage of each major crop type, it is probably not cost effective to demand high accuracy in the position of field boundaries on a map, but it is important to demand high accuracy in the total acreage counts. The money spent to achieve excessively high accuracy can be as wasteful as specifying too low an accuracy.

Accuracy assessment need not be an expensive procedure, but it should always be considered. The assessment procedure might be informal if the information is used to make decisions that aren't critical. We use stylized roadmaps to navigate the highway system and consider them to be accurate enough for our purposes. However, the maps used to engineer a section of highway are formally tested to ensure that they have the required level of positional accuracy.

A frequently used argument for not assessing map accuracy is that "there is no point in assessing the accuracy of the information because it is the best information available". However, if the accuracy of the information is too low, it may be better not to use it at all. Sometimes the cost of using poor information is too high and the decision that requires that information should either be postponed or based on other factors.

STEP 5. Reporting the Results

Once the quality of the information has been assessed and found to be acceptable, the information can then be assembled into a suitable reporting format. The format may be a paper map, an annotated image (such as a weather map), a computer data file, or a written report with diagrams, maps, and tables. The format selected should convey the required information and take into account the way the data will be used. The map scales and projections, units of measurement, and electronic file formats should be selected for compatibility with the format of other information with which it might be used.

STEP 6. Taking Action

The objective of producing information is for decision-making. Even a decision not to take action is a decision. If information is being produced and not used, it is generally because there is no user, the information never gets to the intended user, or the information is not in a suitable format. Frequently, organizations institute information production programs that are allowed to continue long after the reasons they were started have been forgotten. As with any other service, when there is no clearly defined "customer" to be served and when there is no systematic assessment of the quality of the information produced, the information production tends to become self-perpetuating and the quality of the information tends to decline. Nothing promotes service and quality like having the producers of information answerable to their customers.

THE ART AND SCIENCE OF REMOTE SENSING

Remote sensing is the art and science of obtaining information from a distance, i.e.

obtaining information about objects or phenomena without being in physical contact with them. The science of remote sensing provides the instruments and theory to understand how objects and phenomena can be detected. The art of remote sensing is in the development and use of analysis techniques to generate useful information.

When electromagnetic energy, such as light from the sun, illuminates objects on the earth's surface, it interacts with them. Some of the energy is absorbed (e.g. when an object is heated, it absorbs energy), some is transmitted through the object, some is reflected back from the object, and some is emitted (as when we feel the heat from a hot object), see Figure 3.2. It is the energy reflected and emitted back from an object that is available for detection. By recording and analyzing the way the energy received from an object has been changed, information about that object can be derived. Remote sensing is the art and science of deducing information about objects by analyzing the energy received from them and the energy that illuminates them. The steps used to do this are illustrated in Figure 3.3.

Items (a) to (d) in Figure 3.3 show the processes that generate the energy to be detected by remote sensing systems. Energy from a source like the sun is propogated through the atmosphere, interacts with objects on the earth's surface, and is then re-transmitted through the atmosphere. This energy can then be detected by a sensor system (e), such as a photographic camera, or electronic detector. The energy detected by the sensor is converted into data products, such as airphotos or digital data files stored on a computer tape (f).

These data products must then be analyzed (g). A variety of analysis procedures may be used depending on the types of data available and the information to be produced. Typically, several different types of data are used together. The data from one or more remote sensing systems are usually

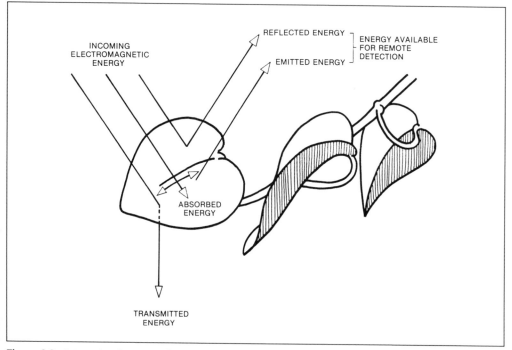

Figure 3.2 Interaction Between Electromagnetic Energy and a Leaf.

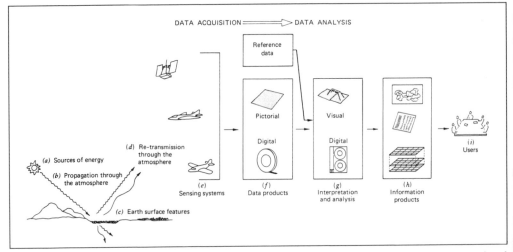

Figure 3.3   The Remote Sensing Process. (From *Remote Sensing and Image Interpretation* by Lillesand and Kiefer 1987, published by John Wiley and Sons.)

analyzed with other data, such as measurements taken at selected ground locations or existing information about field conditions like geology maps, soils maps, and statistical summaries. It is the job of the analyst to design a systematic procedure to extract the desired information from the available data. The information may be a soils map, estimates of crop condition or harvestable timber, or a map of roof lines for municipal mapping.

The information is then assembled into a form suitable for the intended users. Commonly used formats for these information products are the paper map, tabular summary, or a computer file such as for input to a GIS (h). Ultimately, for the information to be of value it must be used by individuals to make decisions (i). The decision may be where to construct a road or how much of a forest to harvest, and the remotely sensed data may have been one of several data sources used to make a map or estimate a quantity. Nevertheless, the justification for the entire procedure is that at some point someone will use the information.

ENERGY SOURCES AND RADIATION PRINCIPLES

Visible light is a form of electromagnetic energy. Radio waves, heat, ultraviolet, and X-rays are other familiar forms. All forms of electromagnetic energy have certain fundamental properties. Electromagnetic energy travels at the "speed of light", 3×10^8 m/sec. It is commonly treated as a wave with both electric and magnetic components, as shown in Figure 3.4. The wavelength is the distance from one wave peak to the next. The frequency is the number of peaks that pass a fixed point per unit of time.

In remote sensing, the type of energy being detected is characterized by its position in the electromagnetic spectrum. As shown in Figure 3.5, the spectrum is commonly divided into units of wavelength measured in microns (also termed micrometers), abbreviated as μm. Each micron is equal to one millionth of a meter (1×10^{-6} m). Historically, the names assigned to regions of the spectrum are more a result of the methods used to detect them than of characteristics of their wavelengths. Ultraviolet, infrared, visible, and microwave regions are some familiar regions of the

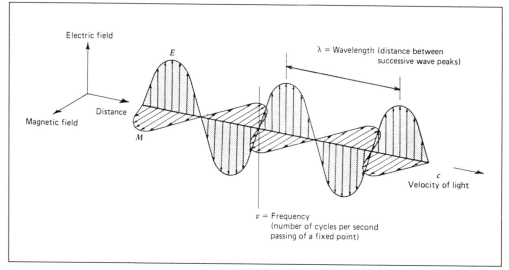

Figure 3.4 Components of an Electromagnetic Wave. (From *Remote Sensing and Image Interpretation* by Lillesand and Kiefer 1987, published by John Wiley and Sons.)

spectrum. There is no precise dividing line between separate regions.

The visible portion of the spectrum, the region to which the human eye is sensitive, has an extremely narrow range of wavelengths. It extends from about 0.4 μm to 0.7 μm (the blue to red regions). The ultraviolet region has the shorter wavelengths next to the visible-blue region. The near-infrared region has the longer wavelengths next to the visible-red portion of the spectrum. The visible, infrared, and microwave portions of the spectrum are the ones most commonly used for remote sensing of earth resources.

The sun is the most familiar source of electromagnetic radiation. However, all matter at temperatures above absolute zero continuously emits electromagnetic radiation. (Absolute zero is 0°K or −273°C). The quantity of energy emitted by an object depends on its physical characteristics, such as its shape, composition, and surface texture. The range of wavelengths at which this energy is radiated depends on the temperature of the object. The higher the temperature, the greater the total energy emitted and the shorter the wavelenghts of emission. Figure 3.6 shows the quantity of energy

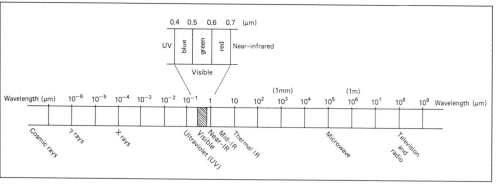

Figure 3.5 The Electromagnetic Spectrum. (From *Remote Sensing and Image Interpretation* by Lillesand and Kiefer 1987, published by John Wiley and Sons.)

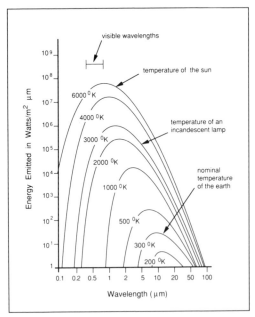

Figure 3.6 Emission of Energy from a Blackbody. The graphs show the spectral distribution of the energy emitted by a standard surface (termed a *blackbody*) at different temperatures. The area under each curve represents the total amount of energy emitted.

emitted and the wavelengths at which the energy is emitted for objects at the temperature of the sun, an incandescent lamp, and a nominal temperature of earth objects (300°K or 27°C). The area under each curve represents the relative amount of energy emitted.

From the Figure it can be seen that the sun's highest energy levels are emitted at a wavelength of about 0.5 μm, in the visible region of the spectrum — a fortunate arrangement for creatures such as humans that depend on light for vision. Photographic film and electronic detectors are used for remote sensing in this region of the spectrum. The energy detected is the solar energy *reflected* from the objects.

Naturally occurring earth features like soil, water, and vegetation radiate much less energy than the sun (hence the smaller area under the curve), and the peak emission is at lower wavelengths. For earth features at a temperature of 27°C, the peak energy

emission is at a wavelength of about 9.7 μm. It is for this reason that wavelengths in this region can be used to measure the temperature of earth objects and this region of the spectrum is termed the thermal infrared. Wavelengths in the 8 μm to 14 μm portion of the spectrum are commonly used for thermal infrared remote sensing.

Wavelengths in the thermal infrared region cannot be detected by the human eye nor by photographic film, although we sense objects emitting high levels of thermal infrared radiation as being warm. Instead, electronic detectors cooled by liquid nitrogen are used to measure the energy emitted at these wavelengths.

Some remote sensing systems use their own source of illumination. A camera with flash is a familiar example. A radar system illuminates the landscape with microwave radiation and detects the energy reflected back from the ground. Since the human eye is not sensitive to microwave radiation, the radar appears to be able to "see in the dark". In addition, microwave radiation is not blocked by clouds, and so, radar imagery can be acquired over cloud-covered regions.

Electromagnetic radiation interacts with the matter it contacts in three ways. It can be absorbed, transmitted, and reflected. Absorbed energy can in turn be re-emitted as illustrated in Figure 3.2. We are most familiar with reflected light. Vegetation appears green because most of the visible light reflected back from the leaves is in the green portion of the spectrum. The leaves are being illuminated by the sun, a source that emits approximately equal quantities of energy in the red, green, and blue portions of the spectrum. Much of the red and blue light is absorbed by the leaf so that, proportionally, more green is reflected and the leaf is perceived by the human eye to be green. Some of the energy is also transmitted. For this reason some light can be seen through the leaf.

The atmosphere also interacts with electromagnetic radiation. For remote sensing, the atmospheric effects of most concern are *absorption* and *scattering*. Ideally the atmosphere would be a perfect transmitter of the energy to be detected. However, the transmission characteristics of the atmosphere are far from perfect since they vary with the wavelength. Some wavelengths are transmitted almost perfectly while others are virtually blocked. Figure 3.7 part B shows the transmittance of the atmosphere at the range of wavelengths used in remote sensing. In some cases the blocking characteristic of the atmosphere is quite beneficial. Without the atmosphere to block most of the ultraviolet radiation, human skin would quickly be sunburned and develop skin cancer when exposed to sunlight.

Atmospheric scattering is the diffusion of radiation by particles in the atmosphere. We see the sky as blue instead of black because sunlight is scattered by the earth's atmosphere and the shorter blue wavelengths are scattered more than the longer green and red wavelengths. At sunrise and sunset the

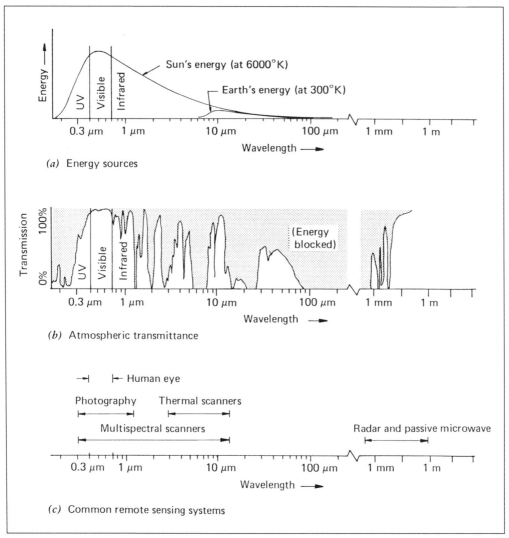

Figure 3.7 Atmospheric Effects and Remote Sensing Systems. (From *Remote Sensing and Image Interpretation* by Lillesand and Kiefer 1987, published by John Wiley and Sons.)

sun's rays travel a longer path through the atmosphere. Almost all the blue portion of the spectrum is scattered or absorbed so that we see only the longer orange and red wavelengths that have penetrated the atmosphere more successfully.

Atmospheric absorption occurs when energy is lost to constituents of the atmosphere. Among the most efficient absorbers are water vapour, carbon dioxide, and ozone. These gases tend to absorb radiation in only specific wavelengths. Those wavelengths that are transmitted well by the atmosphere are termed **atmospheric windows**. It is these regions of the spectrum that are used for remote sensing as shown in Figure 3.7 parts B and C. Note that the visible portion of the spectrum coincides with the wavelength of peak energy emission of the sun (Figure 3.7 part A) and an atmospheric window (Figure 3.7 part B). These are the wavelengths the human eye depends upon for sight.

The peak wavelength of emission from the earth is shown by the small curve in Figure 3.7 part A. Thermal scanners use the atmospheric windows at 3 μm to 5 μm and 8 μm to 14 μm to detect energy in this portion of the spectrum (Figure 3.7 part B). Multi-spectral sensors can simultaneously collect data in several narrow wavelength ranges. The wavelengths selected are usually in the visible and infrared regions, and sometimes ultraviolet wavelengths are used as well. Active microwave sensor systems (known as radar) and passive microwave systems use the 1 mm to 1 m portion of the spectrum.

By using sensors that detect ranges of the spectrum that the human eye cannot detect, images can be created in which features that normally appear identical to the human eye can be distinguished. Two objects that appear to be the same colour in daylight (i.e. reflect the same amount of visible light) may reflect different amounts of another range of wavelengths, such as near-infrared.

In this way sensor systems can, in effect, be used to extend the range of wavelengths that the human eye can see.

SENSOR SYSTEMS COMMONLY USED IN REMOTE SENSING

Photographic Cameras

The photographic camera was the first and is the most widely used remote sensing system. Wavelengths in the visible (0.4 μm to 0.7 μm) and near-infrared (0.7 μm to 0.9 μm) portions of the spectrum can be detected using photographic films. Black-and-white films are sensitive to a single range of wavelengths. Colour films contain three layers, each of which is sensitive to a different range of wavelengths. In a normal colour film, these layers are made sensitive to blue, green, and red wavelength ranges. In false colour infrared films, the three sensitized layers are made sensitive to the green, red, and near-infrared wavelengths.

Films are generally sensitive to a wide range of wavelengths. For example, black-and-white panchromatic film is sensitive to the 0.35 μm to 0.70 μm range. By using filters that pass a narrow range of wavelengths, the film will produce an image showing the reflectance of light in the band of the filter. Multispectral photography can be done by photographing the same scene using multiple cameras simultaneously, each with a different combination of film and filter. For most applications, four bands are used, corresponding to the blue, green, red, and near-infrared. Multispectral photography is not widely used as a result of the difficulty of analyzing the multiple images together. Three separate images can be projected onto a screen using red, green, and blue light sources to produce a colour image. However, the mechanics of doing so tend to be rather time consuming. Instead, multispectral images are generated digitally using scanners. The digital images can be easily manipulated and registered using image processing computer systems.

The scale of an aerial photograph directly affects the size of features that can be seen. The larger the scale, the smaller the ground features that can be seen on the photographs. The scale of a photograph, as for maps, is a measure of the number of units measured on the ground that correspond to each unit as measured on the map. A scale of 1:10,000 means that 10,000 units on the ground correspond to each unit on the map. So a distance of 10,000 cm on the ground (equivalent to 100 m) would be shown on the map as a distance of 1 cm. A scale of 1:50,000 is a *smaller* scale than 1:10,000, because the fraction is a smaller number. At a scale of 1:50,000, 1 cm on the map would represent 500 m on the ground. The terminology used to categorize airphoto scales is not precisely defined. Small scale airphotos are considered to be about 1:50,000 or smaller, medium scale is between 1:12,000 and 1:50,000 and large scale airphotos are 1:12,000 or larger (Lillesand and Kiefer 1987).

In general, small scale airphotos are used for reconnaissance level mapping, large area resource assessments, and general resource planning. Geological mapping, land use planning, agricultural monitoring, topographic mapping, and forest monitoring are commonly done using small scale imagery. Medium scale airphotos are used for identification, classification, and mapping of such features as forest types, agricultural crop types, vegetation communities, soil types, surface materials mapping, and geology. Large scale airphotos are used for intensive monitoring and detailed measurements, such as engineering surveys for road construction, surveys of damage caused by natural disasters, surveys of diseased vegetation, and hazardous waste spills.

Photographic film can resolve very fine details. Using high precision instruments, measurements as fine as .01 mm can be reliably made from aerial photographs. On a 1:50,000 scale airphoto .01 mm corresponds to a ground distance of 0.5 m. On a 1:6,000 airphoto .01 mm corresponds to a ground distance of 0.06 m or 6 cm.

Most aerial photography is taken using sophisticated photogrammetric camera systems with automated film advance and exposure controls and using long continuous rolls of film. Photographs are taken at a precisely controlled rate along the path directly beneath the aircraft so that there is an overlap area, usually 60%, between successive images. By viewing successive pairs of photographs through a stereoscope a 3-dimensional image is perceived. Figure 3.8 is an airphoto stereopair at a scale of 1:6,000. The images have been positioned for viewing with a pocket stereoscope.

Human eyes can use the slightly different perspectives of the two images to perceive differences in distance in the scene, this is termed the **stereo-effect**. So, when looking vertically down at the landscape, one can perceive differences in the heights of features giving a 3-dimensional view. The strength of this stereo-effect depends on the magnitude of the difference in the two views. The closer the objects, the more different are their relative positions within the image and the closer they appear.

When the landscape is viewed from far away, such as from an aircraft, the terrain appears much more level than when viewed from the ground. This is because both eyes see the far away ground objects as being in about the same relative position. In an airphoto stereopair, the positions at which the two photographs were taken were hundreds of metres apart, much farther apart than the few centimetres between a person's eyes. As a result, the difference in relative positions of objects is greater in the photographs than would be seen by human eyes, and so, the relative heights of features appear to be exaggerated. This vertical exaggeration in aerial photographs improves the accuracy with which the elevation of features can be

Figure 3.8 Airphoto Stereopair of Parliament Hill in Ottawa. This pair of photographs can be viewed in stereo using a pocket stereoscope. The complex of buildings in the upper portion of the scene are the Canadian Houses of Parliament overlooking the Ottawa River. In the lower portion of the photographs the locks of the Rideau Canal can be seen. When viewed in stereo, the relative heights of features appear to be exaggerated, as explained in the text. The ground distance across each image is 360 m. (Reproduced with the permission of the National Airphoto Library. Energy, Mines, and Resources Canada. Ottawa, Ontario. Photo ID: A25961–44 and A25961–45. ©1982)

measured and improves image interpretation. The stereopair in Figure 3.8 has a vertical exaggeration of approximately 20,000 times when viewed with a pocket stereoscope.

When the two aerial photographs in Figure 3.8 are compared, tall features like the towers on the Centre Block (the large building at the top of the photo) slant in different directions. For an individual tower, the top is shifted differently relative to its base. It is this difference in the amount and direction that a feature is shifted on the two photos that determines how high it will appear when viewed in stereo. Photogrammetric instruments, such as the stereoplotter shown in Figure 3.9, are used to measure this shift precisely to provide accurate elevation measurements. Today most topographic maps are produced from aerial photographs using photogrammetric instruments such as this.

As a rule-of-thumb, the aerial camera systems and photogrammetric instruments used to produce topographic maps provide elevation accuracies on the order of a 10,000th of the flying height. The airphotos in Figure 3.8 were flown at an altitude of 900 m above ground, so elevation measurements would be expected to have an accuracy of approximately .09 m RMS (root mean square error). Aerial photography at a scale of 1:50,000 is commonly flown at an altitude of 7600 m. Elevation measurements from this photography would be expected to have an accuracy of about .76 m RMS. (These calculations are for airphotos that have a standard 60% overlap and were photographed using a 152 mm lens.)

When viewing a stereopair the brain assumes that an object that looks the same and is in about the same position in both images is stationary, in which case all shifts in relative position are a result of differences in the elevation of the object. (Cars that were moving appear in such different positions on the two photographs that when viewed in stereo they cannot be fused.) In some cases this "visual assumption" is incorrect and the 3-dimensional perception of the viewer is fooled. When viewing Figure 3.8 in stereo, the foam line in the river at the top of the image appears to be at a much higher elevation than the road along the shore. This illusion is created because the foam line in the water had moved downstream in the short time between the exposure of the first and second photos. Because the shape of the foam pattern changed very little, the eye is able to fuse the images of the foam line. But because the relative position has shifted, the eye interprets this shift as height, and so the feature appears to be a huge wave!

The stereo view greatly increases the quantity and quality of information that can be

Figure 3.9 A Stereoplotter. Stereoplotters are used in making topographic maps from aerial photographs. An operator views the stereo image through the eyepieces of the instrument. Using two hand wheels and a foot wheel, a measuring mark appears to float within the 3-dimensional image of the terrain. The measuring mark is moved so as to appear to rest on the surface of the terrain while tracing the objects of interest. The resulting motions can be transferred to the plotting table and directly drawn onto a map or the data can be stored in digital form for later processing. Elevation contours, boundaries of regions, and other features are mapped in this way. (Courtesy of Wild-Leitz.)

extracted from remotely sensed data. For example, the 3-dimensional shape of a land-form is one of the factors used to identify its composition. Gravel, sand, silt, and clay materials can be accurately interpreted from 1:40,000 scale black-and-white aerial photographs viewed in stereo. As well as providing the 3-dimensional perspective, stereo viewing also improves the perceived resolution of the photos. Airphotos tend to appear sharper when viewed in stereo than the same photographs viewed singly.

Electro-Optical Scanners

Electro-optical scanners use an optical system similar to a telescope to view the terrain below. Instead of using film, the image is created by light-sensitive detectors that produce electrical signals proportional to the brightness of the light energy. A single detector can be made to view a strip of terrain by using a rotating mirror to direct its field-of-view across the landscape below. This process, termed **scanning**, is illustrated in Figure 3.10. The usual scanning pattern is in swaths perpendicular to the flight path. The scanning rate is timed to the forward motion of the aircraft or spacecraft so that successive scans view adjacent swaths of terrain.

In practice 20 to 40 detectors or more are used to scan the terrain and simultaneously detect energy in several narrow wavelength bands. The brightness of the detected radiation is output as a voltage level. The voltage

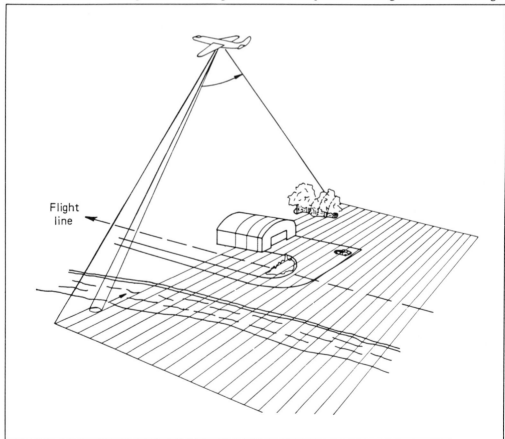

Flight line

Figure 3.10 An Electro-Optical Scanner in Operation. (From *Remote Sensing and Image Interpretation* by Lillesand and Kiefer 1987, published by John Wiley and Sons.)

level changes continuously during a scan as the detector's view sweeps across the terrain. However, to efficiently process the data, the signal level is recorded at regular intervals during each scan. Each value recorded is a sample of the voltage level at a particular instant in time. The area of ground being detected at that instant is termed the **instantaneous field of view** or **IFOV**. The IFOV is the smallest ground area for which a digital measurement is collected. The size of the IFOV determines the theoretical limit of resolution of the sensor. The size of the IFOV is determined by the design of the sensor and the altitude at which the sensor is flown. The smaller the angle of view of the sensor and the lower the altitude, the smaller the resulting IFOV and the higher the spatial resolution. (Other factors, such as contrast, affect spatial resolution, and under certain conditions features smaller than the IFOV can be detected.)

To view the digital images created by scanners, a special type of computer system, termed an **image analysis system**, is used to reconstruct and display the data as a picture-like image. The reconstructed image is composed of a large number of small picture elements, termed **pixels**, each of which is assigned a colour or shade of grey. Usually the image is displayed so that a pixel represents an area about the size of the IFOV.

Airborne electro-optical scanners flown at low altitudes can produce imagery of high spatial resolution. Figure 3.11 is a near-infrared scanner image of an urban area. Each pixel of the image represents a 0.4 m by 0.4 m ground area (i.e. a pixel resolution

Figure 3.11 A High Resolution Image from an Electro-Optical Scanner. This near-infrared image was produced by the MEIS scanner operated by the Canada Centre for Remote Sensing. Green vegetation is light-toned in the image. Each pixel in the digital image represents a ground area of 0.4 m × 0.4 m. The swath width (the narrow dimension of this image) is approximately 1000 pixels wide, corresponding to a ground distance of 400 m. (Courtesy of the Canada Centre for Remote Sensing, Ottawa, Ontario.)

of 0.4 m). Civilian earth resources satellites, such as the Landsat series launched by the United States and the SPOT satellite launched by France, produce imagery with pixel resolutions on the order of tens of meters. Military satellites produce imagery of much higher resolution.

One of the principal disadvantages of using a rotating mirror to scan the terrain is that it introduces complex geometric distortions that must be corrected in the imagery. Also, the detectors can only dwell on each area for a very short time during the scan. Pushbroom scanners overcome these difficulties by avoiding the use of a moving mirror. Instead, a large number of detectors are arranged in a line (termed a **linear array**) across the field of view so that the full width of the swath can be detected at one time (see Figure 3.12). The image data are produced one line at a time by reading the output values from the entire line of detectors. The forward motion of the aircraft or spacecraft moves the linear array over successive

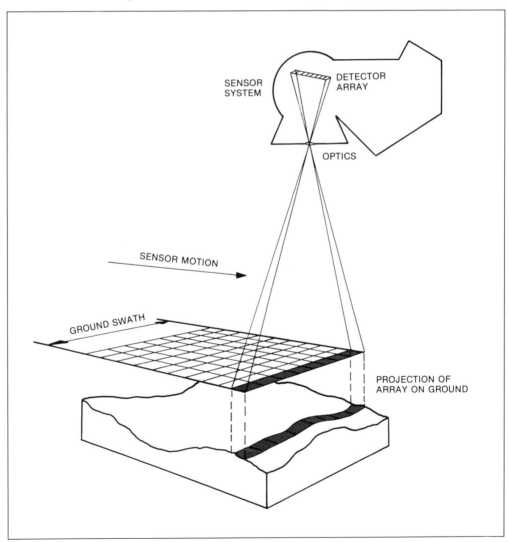

Figure 3.12   The Principle of Pushbroom Scanning.

swaths of terrain. The SPOT satellite sensors use pushbroom scanners while the older Landsat system uses a moving mirror design.

The pushbroom scanner has several advantages over the mirror scanning systems. Pushbroom scanners are generally smaller, lighter, and consume less power. With no moving parts, they have a higher reliability and longer life expectancy. The imagery they produce has a more predictable geometry and better spatial and radiometric (i.e. measurement of the brightness levels) accuracy. The principal disadvantage of pushbroom scanners is that a much larger number of detectors (thousands) have to be calibrated.

Current scanner systems commonly detect 3 to 11 wavelength bands simultaneously. Experimental scanners are now being tested that simultaneously record hundreds of adjacent bands. Scanners of this type are termed **imaging spectrometers** because reflectance values are collected for a large number of very narrow bands (commonly 0.01 μm band widths are used) over a continuous portion of the spectrum for every pixel. The data values pertaining to a single pixel can be used to construct a spectral plot for that pixel. In this way the ground condition at each location can be identified from its spectral plot.

The rationale for collecting this enormous volume of data is that some features, such as geologic minerals or stressed vegetation, can be distinguished by the fine details in the shape of their spectral plots that cannot be seen using the wider bands (which are usually .1 μm to .2 μm wide) of more conventional scanners (Goetz et. al. 1985, Vane and Goetz 1988). Once a procedure is well-researched using an imaging spectrometer, it may only be necessary to collect data for the few narrow bands needed to identify the materials of interest rather than for all the bands.

Scanners are used to sense wavelengths in the ultraviolet, visible, and infrared bands. The strategy used in selecting bands is to choose a set that together will best distinguish the features of interest. The more bands used, the more data there is to store and process, and hence the more expensive it is to handle and process the data. In the case of airborne systems, the large quantities of digital image data can be stored on tape during the flight. For satellite-based systems, the data must be transmitted back to earth. The time and electrical power needed to transmit the data thus becomes a limiting factor. In addition, as higher data transmission rates are used, the equipment needed to transmit and receive the data becomes more complex and expensive.

Airborne scanner systems have been used widely for geological mapping and for heat loss detection. Geological investigations have been most successful in arid areas. The sparse vegetation cover enables the light reflected from the exposed rock and soil material to be directly measured. Analysis of reflectance data for the visible and infrared bands has been used to identify the mineral composition of the exposed materials (see for example Goetz et. al. 1985).

Airborne thermal image surveys are often conducted to analyze heat loss from buildings, to detect and monitor heated fluids being ejected into lakes and rivers, and to support repair and maintenance activities by detecting unusual heat sources like steam escaping from ruptured underground pipes (see Aronoff and Ross 1982). Figure 3.13 is a thermal infrared image of a residential area in a Canadian city. Warm areas on the image appear lighter in tone. Since the image was taken during the winter and at night, areas that appear warm in the image indicate possibly excessive heat loss. This image was used to identify homes that appeared to have warm roofs, indicating a high heat loss. If the homeowners were interested, then a more intensive site investigation was done to verify that there was excessive heat loss and to suggest improvements. See also Plate 15.

Figure 3.13 Thermal Infrared Image of a Residential Area. This thermal image was acquired on a winter night. Warmer objects, such as the city streets, appear lighter in tone. The roofs of buildings are generally dark in tone, indicating a relatively cool temperature and low heat loss. Lighter-toned roofs may indicate excessive heat loss. (Courtesy of the Canada Centre for Remote Sensing. Ottawa, Ontario.)

In some cases thermal infrared imagery has been used to distinguish terrain types. Differences in temperature, moisture content, and surface texture can make natural features appear different on a thermal infrared image. The thermal infrared imagery has been shown to differentiate soils and rock units that appear similar on visible and near-infrared imagery (see for example Kahle and Goetz 1983, and Gillespie and Abbot 1984).

Satellite-based scanners have produced most of the publicly available satellite imagery in use today. The principal earth resources satellites are the French SPOT and United States' Landsat, NOAA, and GOES satellites (discussed below). The following is a brief discussion of the major satellite scanner systems currently in operation.

Landsat

The Earth Resources Technology Satellite (ERTS-1, later renamed Landsat-1), was the first unmanned satellite designed to provide systematic global coverage of earth resources. Launched by the United States

on July 23, 1972, it was primarily designed as an experimental system to test the feasibility of collecting earth resources data from satellites. Originally there were to be six satellites in the ERTS series. In 1975, the ERTS program and satellites were renamed Landsat.

In addition to its scientific aspects, the Landsat program was also an expression of an ideal. The data were made publicly available world-wide. This non-discriminatory access to data was termed the *Open Skies Policy*. The data collected from space, in particular the Landsat data, were to be available to all and not limited by national interests. The resulting world-wide experimentation with Landsat data produced overwhelmingly favourable results. In fact, that the satellites were designed to be experimental not operational systems led to demands for levels of service better than NASA, and later NOAA, had planned to provide.

The first three Landsat satellites carried a multispectral scanner (MSS) and a return beam vidicon (RBV) camera. (An RBV is similar to a television camera but is calibrated to high standards of geometric accuracy.) The RBV systems quickly became the secondary sensors; in part, this was because of technical malfunctions. More importantly, the MSS system for the first time provided digital images suitable for computer analysis for most of the earth's surface.

The MSS on Landsats 1, 2, and 3 imaged a 185 km wide swath in four bands designated bands 4, 5, 6, and 7. (The first three bands were assigned to the RBVs.) The MSS covered two bands in the visible and two in the near-infrared, as shown in Table 3.1 and Figure 3.14. (The eighth band flown on Landsat 3 failed shortly after launch and is not indicated in the figure.) Each picture element or pixel in an MSS image represents a ground area of approximately 79 m by 56 m. Each scene shows a 185 km by 185 km area and consists of approximately

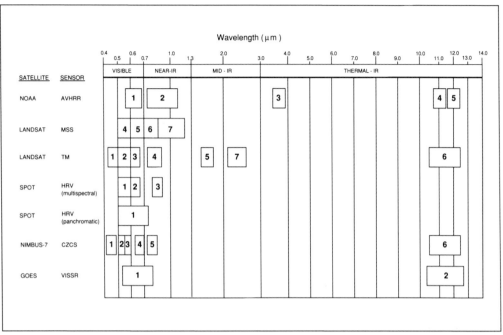

Figure 3.14 Wavelengths Detected By Operational Satellites. The wavelength bands detected by sensors on several operational satellites are shown. The numbered boxes represent the individual channels, and the width of the box indicates the range of wavelengths detected. (Adapted from Lillesand and Kiefer 1987.)

Table 3.1 Wavelength Bands Used in Landsats 1 to 5. (Adapted From Lillesand and Kiefer 1987.)

Sensor	Missions	Band	Wavelengths in μm	Resolution 1-3	4,5
MSS	1 to 5	4	0.5–0.6 (green)	79	82
		5	0.6–0.7 (red)	79	82
		6	0.7–0.8 (near-IR)	79	82
		7	0.8–1.1 (near-IR)	79	82
	3*	8	10.4–12.6 (thermal IR)	240	
TM	4,5	1	0.45–0.52 (blue)	30	
		2	0.52–0.60 (green)	30	
		3	0.63–0.69 (red)	30	
		4	0.76–0.90 (near-IR)	30	
		5	1.55–1.75 (mid-IR)	30	
		6	10.4–12.5 (thermal IR)	120	
		7	2.08–2.35 (mid-IR)	30	

* Failed Shortly After Launch

2340 scan lines with 3240 pixels per line, a total of 7.5 million pixels per scene. With four values per pixel (one for each band), a scene contains over 30 million pixel values. This volume of data was collected in about 25 seconds, a data rate that tested the technology at that time. Satellites now collect data at a rate several times faster.

The satellite platform used for Landsats 1, 2, and 3 is shown in Figure 3.15. These satellites were about 3 m high, 1.5 m in diameter, and weighed about 950 kg. They were launched into circular orbits with altitudes of 900 km. Circling the earth every 103 minutes, the satellites completed 14 orbits per day. The spacing between successive orbits was set to be 2760 km at the equator with a north-south path that came within 9° of latitude of the north and south poles. With this orbit the satellites kept pace with the sun's westward progress as the earth rotates. The result was that the satellites always crossed the equator at the same local sun time (i.e. the time determined by the sun's position on the horizon). An orbit that always passes over the earth at the same local sun time is termed a **sun-synchronous** orbit. Sun-synchronous orbits are usually used for earth resources satellites so that the sun illumination conditions are consistent. Sun elevation, relative position, and intensity still vary with the seasons, but every scene has the illumination of the same time of day. The orbit provided a

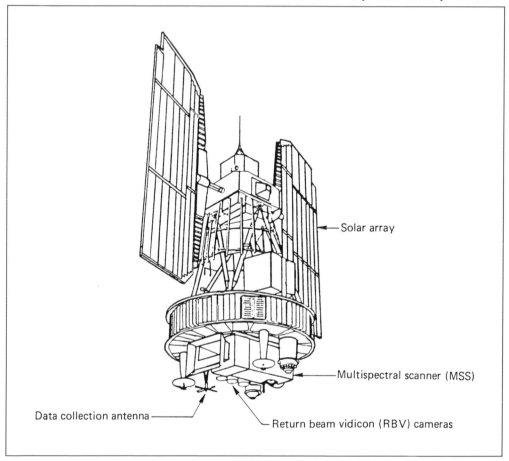

Figure 3.15   The Satellite Platform Used for Landsats 1, 2, and 3. (From *Remote Sensing and Image Interpretation* by Lillesand and Kiefer 1987, published by John Wiley and Sons.)

return period of 18 days (i.e. every 18 days the orbit path would repeat itself). Landsats 1 through 3 operated from 1972 to 1983.

The Landsat 4 and 5 satellites, were launched into lower sun-synchronous earth orbits of about 700 km and have a return period of 16 days (Figures 3.16 and 3.17). System engineers deactivated Landsat 4 shortly after launch as a result of electrical problems. It remains in orbit and available for limited use if Landsat 5 fails. Landsat 5 was launched in 1984 and was in operation as of January 1989. The 2000 kg satellite is somewhat larger than the earlier Landsats

and carries two multispectral scanner systems, the MSS and TM sensors. The MSS sensor was included to provide continuity with the previous Landsat data. It images a swath 185 km wide, as before, and the data is processed so that it is compatible with the earlier MSS data. The other sensor is the Thematic Mapper (TM), a more advanced multispectral scanner than the MSS system. As shown in Table 3.1 and Figure 3.14, this sensor provides 7 bands ranging from the visible blue to the thermal infrared. (Band 7 is out of sequence because it was added late in the design stage.) The TM sensor also

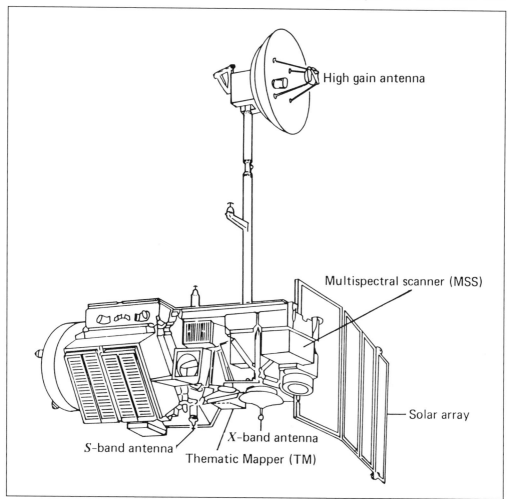

Figure 3.16 The Satellite Platform Used for Landsats 4 and 5. (From *Remote Sensing and Image Interpretation* by Lillesand and Kiefer 1987, published by John Wiley and Sons.)

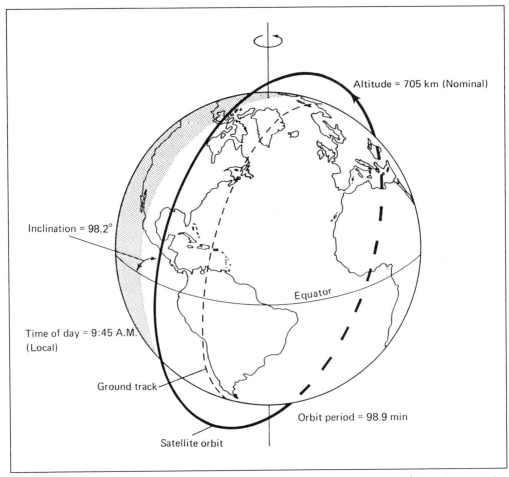

Figure 3.17 The Sun-Synchronous Orbit of Landsats 4 and 5. (From *Remote Sensing and Image Interpretation* by Lillesand and Kiefer 1987, published by John Wiley and Sons.)

provides a higher spatial resolution; each pixel represents a 30 m by 30 m ground area (except in the case of band 6 which uses a larger 120 m by 120 m pixel).

Plates 6a, 6b, and 6c are TM images of an area south-west of Toronto, Ontario. Three different colour composite images have been produced by using different combinations of bands and assigning one band to each of the red, green, and blue channels of the display. (Display hardware is discussed in Chapter 4.) Plate 6a is composed of the red, green, and blue visible bands (TM bands 1, 2, and 3 respectively). It is a "natural colour" representation in that

features have about the same colour as they would appear to the human eye — from an altitude of 700 km. The images have low contrast because the atmosphere scatters and absorbs a significant proportion of the reflected light, especially in the blue band.

Plate 6b is a "false colour" image. It is composed of the green and red visible bands, and a near infrared band (TM bands 2, 3, and 4). In this image the colours of the features are "false" i.e. different from the way they would normally appear to the human eye. They are similar to those produced using false colour infrared photography.

This false colour image has more contrast than the normal colour representation shown in Plate 6a. It appears clearer because there is much less scattering and absorption of the near infrared wavelengths used in this false colour image than of the blue wavelengths used in the natural colour image. Healthy vegetation is represented in shades of red using this band combination. This type of image is often preferred over the natural colour image because it can provide better discrimination among vegetation types. However, as shown here, the street network in urban areas may be more difficult to interpret.

Plate 6c is a different false colour image produced from the visible red band, a near-infrared, and a mid-infrared band (TM bands 3, 4, and 5). The images have good contrast because the atmospheric effects at these wavelengths are less than for the bands used in the previous two images. Healthy vegetation appears bright green and fallow fields and cleared areas are purple. There is relatively little detail in the urban areas.

The choice of band combinations and assignment of these bands to the three colour output channels greatly changes the types of information that can be derived from the image data. A discussion of band selection is beyond the scope of this introduction. More detailed discussions can be found in the remote sensing literature and in texts such as Campbell (1987) and Lillesand and Kiefer (1987).

The TM sensor provides a dramatic improvement in spatial resolution and improved spectral resolution with seven narrower bands instead of the four wider bands used in the MSS. This additional resolution comes at the cost of a greatly increased quantity of data. A Landsat TM scene contains about 36 million pixels compared with the 7.5 million in an MSS scene. The four bands of data in an MSS scene contain about 30 million data values whereas seven bands of TM data contain about 250 million values! The cost of the data reflects the increased quality and quantity — $660 US for four bands of MSS digital data compared with $3600 US for 7 bands of TM data (EOSAT prices as of January 1989).

SPOT

The SPOT (Système Pour l'Observation de la Terre) program was begun by France in 1978. It was designed from the start to be a long-term, operational, commercial system, and subsequent satellites in the series are already being manufactured. SPOT has developed into an international program with data available from outlets in more than thirty countries. SPOT-1 (Figure 3.18) was launched in early 1986. It carries two identical HRV pushbroom scanners (HRV for High Resolution Visible). Each can operate in one of two modes. In the panchromatic mode, a single visible band (0.51 μm to 0.73 μm) is detected and an image with 10 m by 10 m pixels is produced. In the multispectral mode, three images with 20 m by 20 m pixels are produced, one for each of three bands: 0.50 μm to 0.59 μm (green), 0.61 μm to 0.68 μm (red), and 0.79 μm to 0.89 μm (near-infrared). Plate 7 is a SPOT image produced from the three band (20 m pixels) image data of the Ottawa, Ontario area.

SPOT has a sun-synchronous orbit at an altitude of 832 km and a return period of 26 days. When both of the SPOT sensors are pointed vertically, they together image a 117 km wide swath. Separately they each image a 60 km swath. However, because the sensors are pointable, there is the potential to view a location more frequently, from adjacent satellite paths. This capability to view the same area from two widely separated locations also enables full scene stereo images to be produced (see Figure 3.19). To produce topographic maps, SPOT stereo images can be analyzed with photogrammetric instruments in a manner similar to that used for stereo aerial photographs. The elevation

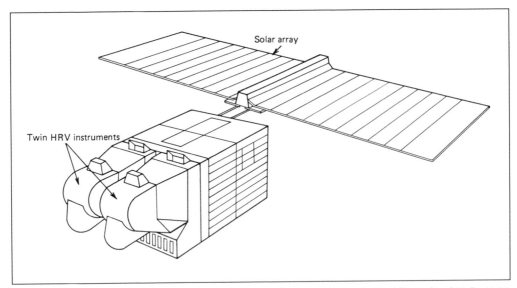

Figure 3.18 The SPOT Satellite. (From *Remote Sensing and Image Interpretation* by Lillesand and Kiefer 1987, published by John Wiley and Sons.)

information produced from the SPOT data is not as accurate as that produced from aerial photographs; however, the imagery is much less expensive than flying new

Pass on first coverage day

Pass on second coverage day

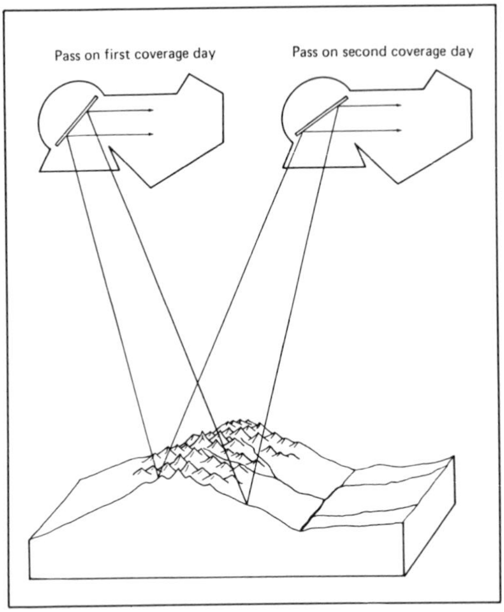

Figure 3.19 Stereoscopic Imaging with the SPOT Satellite. (From *Remote Sensing and Image Interpretation* by Lillesand and Kiefer 1987, published by John Wiley and Sons.)

photography. More recently, automated methods have been developed to calculate the elevation data directly from the digital imagery by using advanced computer-based image processing.

Plate 8 is a portion of a 1:50,000 scale topographic map of an area in Malaysia produced from SPOT stereo images by Digim Incorporated of Montreal, Quebec. The map has a contour interval of 20 m. The elevation contours in the map were generated entirely by computer analysis of the imagery, and the remainder of the map was produced by interpreting the SPOT images visually and using conventional cartographic methods. The elevation data were generated directly from panchromatic image data (10 m pixels) using specialized stereo correlation software. Plate 9 is an image produced from a SPOT multispectral image (20 m pixels) of the same area acquired from a satellite position 4.2 degrees to the west. This image is an **ortho-image**, meaning that it has been corrected for all geometric distortions caused by earth rotation and curvature, satellite motion, attitude, and viewing perspective as well as relief displacement.

The high degree of correction of the ortho-image enables it to be directly superimposed on the map to accurately delineate the boundaries of features like the vegetated areas.

In Canada, 1:50,000 scale topographic maps are produced from aerial photographs with scales of 1:60,000 to 1:80,000 using conventional photogrammetric methods. The NATO standards for class A-1 topographic maps are 0.5 times the contour interval RMS (root mean square error) for elevation contours and 0.25 times the contour interval RMS for spot heights, both at the 90% confidence level. Using 1:60,000 scale photography of terrain similar to that of the Malaysian example and conventional photogrammetric methods, the elevation of spot heights would be mapped with an accuracy on the order of 1 m to 2 m RMS and the contours with an accuracy on the order of 2 m to 4 m at the 90% confidence level. It is estimated that the topographic map shown here, produced using elevation data calculated from the SPOT imagery, has a contour accuracy of about 10 m at the 90% confidence level.

Conventional photogrammetric mapping methods are considerably slower than using computer-based stereo correlation. Highly skilled staff are needed to continuously operate the equipment. With the automated methods, the elevation data are generated more rapidly, and the human operator is needed only to review and edit the results. Digim Inc. has estimated the cost to produce topographic maps using automated methods to be on the order of $25 to $35 CDN per square kilometer compared with about $50 CDN using conventional methods.

The use of digital imagery and stereo correlation for topographic mapping is still being refined. It is a computationally intensive procedure and requires good quality digital stereo images (difficult to obtain for cloud prone areas) and terrain that is not too rugged and not too uniform (so that the correlation program does not "get lost"). Though they cannot yet provide the same level of accuracy as conventional methods, these automated methods are an attractive solution to the desperate need for mapping of the many poorly mapped regions in the world.

Uses of Landsat and SPOT Data

Landsat and more recently SPOT imagery are now routinely used for a wide range of applications. In geology, satellite imagery is used for reconnaissance level mapping in mineral and petroleum exploration as well as national mapping programs. In agriculture, this imagery has been used to estimate the area planted to important food and cash crops and the condition of the crops through the growing season. Landsat imagery is used routinely to estimate crop areas and crop condition. Examples of these applications are given later in this chapter.

In forestry, SPOT and Landsat data have been used to estimate forest losses caused by fires, clearcutting, and disease (see for example Sader 1987 and Vogelmann and Rock 1988). They have also been used for quick, comparative forest land valuation. In some cases, satellite data have been used with digital terrain data to provide forest inventory data (see for example Strahler et. al. 1983).

SPOT and Landsat data are widely used for land use planning. Current land cover, the detection of changes in land use over time, and route location planning (see Aronoff and Ross 1984) are common applications. Landsat data have been used to monitor rangeland condition, wildlife habitat, identify water pollution (see Plate 14), and to aid in the assessment of damage caused by natural disasters. Landsat and SPOT photographs of the explosion and fire at the Chernobyl nuclear reactor site in Russia put imagery from these satellites into newspapers around the world. With the advent of higher resolution imagery, satellite imagery, such as 10 m resolution SPOT data, is being used as a substitute for high altitude aerial photography.

The list of successes is a long one and the development of applications of these data is continuing. There have been failures too. In fact, in the early days of the Landsat program, the overselling and subsequent disappointments of many Landsat applications tended to overshadow the successes. It has been shown that to achieve accurate, cost-effective results, it is critical that remotely sensed data be used in an appropriate context (as discussed earlier) and that the data must be supported by appropriate reference information. An understanding of *when* to use a particular type of image data is as important as the knowledge of *how* to use those data.

Ocean Monitoring Satellites — The Coastal Zone Colour Scanner

The oceans are an important natural resource that are difficult to map and monitor over large areas or for extended periods of time. Traditional oceanographic techniques depend on the collection of widely spaced samples usually collected at different times. To provide complete coverage of large areas at regular intervals, satellite-based remote sensing systems are the only practical data collection method. Two examples of satellite-based sensors designed specifically for ocean monitoring are the Coastal Zone Color Scanner (CZCS) carried on the Nimbus-7 satellite and the sensors carried on the Seasat satellite. (Seasat is discussed later, in the section on microwave sensors). Landsat and SPOT satellite data are also used extensively to monitor sediment and chlorophyll concentrations (see Plate 14), phytoplankton, and pollution in marine and fresh-water environments and to map water depths (see for example Alfoldi and Munday 1978, Hallada 1984, Jupp et. al. 1985, and Lindell et. al. 1985). (**Phytoplankton** are the single-celled plants living in surface waters that form the base of the marine food chain. They are critical to the biological productivity of the ocean, including the production of commercial fish and shellfish.)

The CZCS was launched by the United States in 1978 and operated until June 1986. It was designed as a proof-of-concept mission to measure ocean color and temperature in coastal zones. The imagery has a 1600 km swath width with a pixel resolution of 825 m. Single images, representing 2 minutes of data collection, provide coverage of a 1600 km across-track by 800 km along-track area. The six channels include four narrow bands in the visible portion of the spectrum, a near-infrared band, and a thermal infrared band (see Table 3.2 and Figure 3.14).

Data from the CZCS have been used to map suspended sediment and phytoplankton in coastal regions. See for example Clark and Maynard (1986) and Wrigley and Klooster (1983). CZCS studies have provided valuable physical oceanographic information about such processes as the formation of eddies and ring-shaped currents, changes in ocean current locations, sea surface temperatures and the position of upwelling areas that are critical to commercial fisheries. CZCS data have also been used for environmental monitoring of coastal regions, such as estimating the sediment content in coastal waters (Tassan and Sturm, 1986) and the detection of acid waste pollution (Elrod, 1988).

Table 3.2   Wavelength Bands Used in the Coastal Zone Color Scanner (CZCS).

Channel	Wavelength in μm
1	0.43–0.45 (blue)
2	0.51–0.53 (green)
3	0.54–0.56 (green)
4	0.66–0.68 (red)
5	0.70–0.80 (near-IR)
6	10.5–12.5 (thermal IR)

Meteorological Satellites — NOAA and GOES

Meteorological satellites provide more frequent coverage but coarser resolution imagery than the Landsat and SPOT satellites. Although designed primarily to collect weather data, they are used routinely for earth resources monitoring over large areas, such as crop and vegetation condition assessment and the monitoring of desertification. The coarser resolution reduces by a substantial amount the volume of data to be processed and the data are much less expensive per unit area than Landsat and SPOT imagery. Many countries have launched meteorological satellites. Perhaps the data most widely available are those of the NOAA and GOES satellites operated by the United States.

The NOAA Satellites. A series of sun-synchronous polar orbiting satellites was launched by the U.S. National Oceanic and Atmospheric Administration (NOAA). They were designed as operational civilian satellites to provide visual and infrared observation and measurement of the earth's surface, atmosphere, and cloud cover. Starting with NOAA-1 in January 1970, the first five satellites flew at altitudes of about 1,450 km. NOAA-6, launched in October 1978, and subsequent satellites in the series (NOAA 7 through 10) were placed into lower orbits of about 850 km and provided more frequent coverage. Beginning with NOAA-6, these satellites have carried a sensor called the Advanced Very High Resolution Radiometer (AVHRR).

AVHRR data have a spatial resolution of 1.1 km (i.e. each pixel represents an area of 1.1 km by 1.1 km). The imagery is received as a continuous image covering a 2400 km wide area (the swath width). It is generally distributed as individual images covering an area 2500 km by about 1500 km. Plate 10 is a colour composite AVHRR image of Western Canada produced from the visible and near-infrared bands. These image data were used for crop condition assessment, an application discussed later in this chapter.

Each NOAA satellite provides daily global AVHRR digital image coverage in the visible and near-infrared spectrum, and twice daily coverage in the thermal infrared. The data are made available in two formats, a Local Area Coverage (LAC) at the full 1.1 km spatial resolution and a Global Area Coverage (GAC) at a 4 km spatial resolution. The sensors produce data in four or five wavelength bands depending on the satellite, as shown in Table 3.3 and Figure 3.14. The even-numbered satellites cross the equator north to south in daylight (7:30 AM local sun time), the odd-numbered satellites at night (2:30 AM local sun time).

The data, in both digital and photographic images, are used operationally in a variety of time-critical and large area applications. AVHRR data are used to monitor snow coverage, assess snow depths and melting conditions, monitor floods, detect and map forest fires, monitor crop conditions, monitor dust and sandstorms, identify geologic events like volcanic eruptions, and

Table 3.3 Wavelength Bands Used in the AVHRR Sensor.

Channel	NOAA 6, 8, 10 in μm	NOAA 7, 9 in μm
1	0.58–0.68 (red)	0.58–0.68 (red)
2	0.72–1.10 (near-IR)	0.72–1.10 (near-IR)
3	3.55–3.93 (mid-IR)	3.55–3.93 (mid-IR)
4	10.5–11.5 (thermal IR)	10.5–11.5 (thermal IR)
5	(channel 4 repeated)	11.5–12.5 (thermal IR)

map regional drainage networks, physiography, and geology (Colwell 1983, Lillesand and Kiefer 1987).

The daily coverage has proven particularly valuable for regional crop condition monitoring and rangeland assessment (see for example Paltridge and Barber 1988). The data are used as an input to crop models for the prediction of the crop area and the level of production for important cash and food crops. Less time critical regional analyses include drought and desertification monitoring and water current mapping. The water temperature patterns shown on the thermal infrared imagery and the patterns of water turbidity shown in the visual red channel have been used to detect algal blooms and analyze water currents and water mixing in lakes and coastal areas (Holben 1986, Stumpf and Tyler 1988).

GOES Satellites. The Geostationary Operational Environmental Satellites (GOES) provide continuous monitoring of temperature, humidity, and cloud cover for weather forecasting. The satellites orbit the earth at an altitude of 36,000 km in the same direction as the earth's rotation. They maintain a stationary position relative to the earth and for this reason are termed **geo-stationary**. The two GOES satellites that provide coverage of North America are operated by the United States. GOES-East is positioned at longitude 75° West and GOES-West at longitude 135° West. Two others are operated by Europe and Japan.

GOES digital images are produced in two bands, a visible band (0.55 μm to 0.75 μm) and a thermal infrared band (10.2 μm to 12.5 μm). The visible imagery can be provided with resolutions of 1, 2, 4, and 8 km. The thermal infrared imagery has a resolution of 8 to 14 km. Images in the visible band are generated during daylight hours, the thermal infrared images are produced day and night. GOES full earth disk images show the earth as a circular disc, covering approximately 100° of longitude between latitudes 70° N and 70° S (see Figure 3.20). They are produced at the rate of 2 per hour, one in each band. As an alternative, smaller areas can be imaged more frequently (as short as 3 minute intervals) to monitor the development and movement of severe storms. In addition to the imaging sensors, the GOES satellite has an instrument that detects infrared radiation in 12 separate wavelength bands. This instrument does not produce an image. Rather, it collects data for one selected area at a time. The data are computer-processed to generate vertical temperature and moisture profiles through the atmosphere that are used in the analysis and prediction of weather conditions.

The principal application of GOES satellite data is for weather forecasting and the detection and tracking of severe storm events. They are used by the US National Weather service to provide a wide range of weather warnings to serve public, aviation, and marine interests. GOES images are distributed electronically in near-real time (i.e. almost as soon as they are received) for use in local weather forecasting and are commonly seen on television weather broadcasts.

Quantitative analyses of GOES data provide a range of information. Maps of sea surface temperatures show the location of major ocean currents. GOES data are used to estimate precipitation for areas where direct measurement is not possible (see for example Meisner and Arkin 1984). Snow cover mapping from GOES data are used to forecast snowmelt run-off, to predict flood hazards, and for other water resource planning applications. During the winter, cloud-free GOES imagery and other satellite data over the Great Lakes are used to chart navigable passageways through ice-fast areas (NOAA 1985).

Microwave

Microwave imaging systems use antennas as detectors. They operate in bands that have

Figure 3.20 GOES Image of North and South America. This image was produced from the visible wavelength band (0.55 μm to 0.75 μm) with a pixel resolution of 8 km. (Courtesy of the Satellite Data Services Division, National Oceanic and Atmospheric Administration (NOAA). Washington, D.C.)

been designated by letters (see Table 3.4). Passive microwave systems are commonly used on weather satellites (such as the Nimbus series flown by the USA) to detect microwave emissions from the earth. These data are used to produce temperature profiles of the lower atmosphere (the troposphere), atmospheric liquid and water vapour content, and surface and near-surface wind conditions. These passive microwave systems commonly have pixel resolutions of 25 km or larger.

In active microwave remote sensing, a pulse of microwave energy is sent from the

Table 3.4 Wavelength Bands Used in Microwave Remote Sensing.

Band Designation	Wavelength in cm
Ka	0.75 – 1.10
K	1.10 – 1.67
Ku	1.67 – 2.40
X	2.40 – 3.75
C	3.75 – 7.50
S	7.50 – 15.00
L	15.00 – 30.00
P	30.00 – 100.00

sensor towards the target. The energy reflected back to the sensor is measured and used to produce images. This technology, commonly known as RADAR (RAdio Detection And Ranging), was developed during World War II to detect enemy aircraft. Because the system provides its own microwave source of illumination and the microwave wavelengths used are not blocked by clouds, radar sensors can provide an all-weather day-and-night imaging capability.

In radar imaging systems, the strength of the reflected energy depends on the surface roughness and the orientation of the terrain and the electrical properties of the surface material. For example, metal is a much better reflector than soil, and wet soil reflects better than dry soil. Features that reflect more of the microwave energy back to the sensor will appear brighter in the image.

The operating principle of a Side Looking Airborne Radar system (SLAR) is shown in Figure 3.21. This system produces imagery of the area to either side of the flight path, hence the term "Side Looking". An antenna is mounted lengthwise along the belly of the aircraft. Short pulses of energy at a specific wavelength are transmitted from the antenna in a fan shape, perpendicular to the flight line. When the pulse strikes a target, a signal is returned and detected by the receiver using the same antenna. The signal reflected from objects further away take longer to return. The difference in this return period and the strength of the reflection are used to produce the image. The image data is recorded directly onto film or as digital data onto magnetic tape. As the aircraft moves forward, successive swaths of terrain to one or both sides of the aircraft can be recorded, producing a continuous strip of imagery. The area directly beneath the aircraft cannot be imaged using this type of system.

In a SLAR system, one of the factors limiting the resolution is the physical length of the antenna. For this reason, these systems are termed **real aperture** radars. The larger the antenna, the better the spatial resolution. A 15 m antenna is about the longest practical antenna length. A SLAR flown at about a 5 km altitude produces imagery with a spatial resolution on the order of 15 to 20 m. To acquire high quality imagery from higher altitudes or from satellites, a different technology was developed to overcome the limitation of antenna size.

Synthetic Aperture Radar (SAR) uses a short antenna about 1 m to 2 m in length that is electronically made to perform as if it were hundreds of times longer. In this way, a SAR system can provide higher resolution imagery at far greater altitudes than real aperture SLAR systems. SAR systems operated from satellites have achieved spatial resolutions of 25 meters. Aircraft-based SAR systems achieve resolutions of a few meters from altitudes on the order of 6 km, and stereo SAR imagery can be produced by imaging the same area from different viewing positions. Photogrammetric methods can then be used to produce elevation measurements from the stereo imagery. Figure 3.22 shows a rectified radar image generated from a SAR stereopair with elevation contours produced in this way. (See also Leberl et. al. 1987, and Mercer and Kirby 1987.)

SAR systems provide higher spatial resolution than SLARs at the cost of more complex digital signal processing. The equipment and computer time make SAR systems considerably more expensive to build and operate. However they provide a unique all-weather, relatively high resolution imaging capability.

Seasat-1 was launched in June of 1978 by the United States. Though it failed only 99 days after launch, it convincingly demonstrated the value of a satellite-based radar system. Seasat produced radar imagery with a 100 km swath width and a 25 m spatial resolution. It also carried a radar altimeter that measured the sea surface height profile and wave heights with a relative accuracy

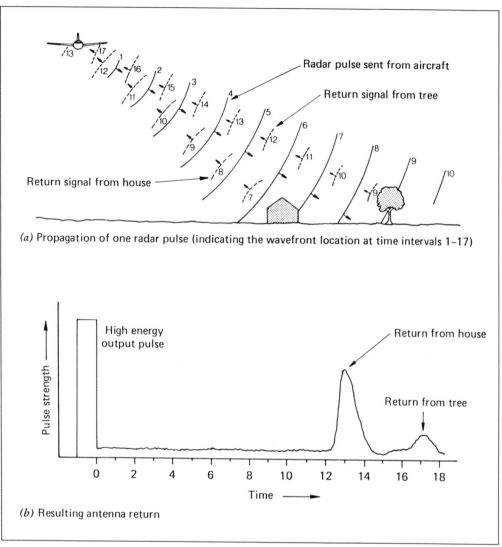

(a) Propagation of one radar pulse (indicating the wavefront location at time intervals 1–17)

Radar pulse sent from aircraft

Return signal from tree

Return signal from house

High energy output pulse

Return from house

Return from tree

Pulse strength

0   2   4   6   8   10   12   14   16   18

Time ⟶

(b) Resulting antenna return

Figure 3.21 Operating Principle of a Side-Looking Airborne Radar. (From *Remote Sensing and Image Interpretation* by Lillesand and Kiefer 1987, published by John Wiley and Sons.)

of 10 cm. It was found that the ocean surface topography was a function of the ocean bottom topography and so the Seasat data were used to create a map of the ocean floor! Other Seasat sensors measured sea surface temperature, rain rate, and the water vapour content of the atmosphere. The data have been used to map ocean currents, wave heights and direction, and arctic ice conditions. Over land areas, the data proved particularly valuable for mapping geology and water resources. Figure 3.23 is a Seasat image of the Rocky Mountain region of British Columbia, Canada. The image is striking for the texture and details shown in the steeply dipping rock strata of this rugged terrain.

Imaging radar systems are routinely used to map areas with persistent cloud cover, such as many tropical and sub-tropical

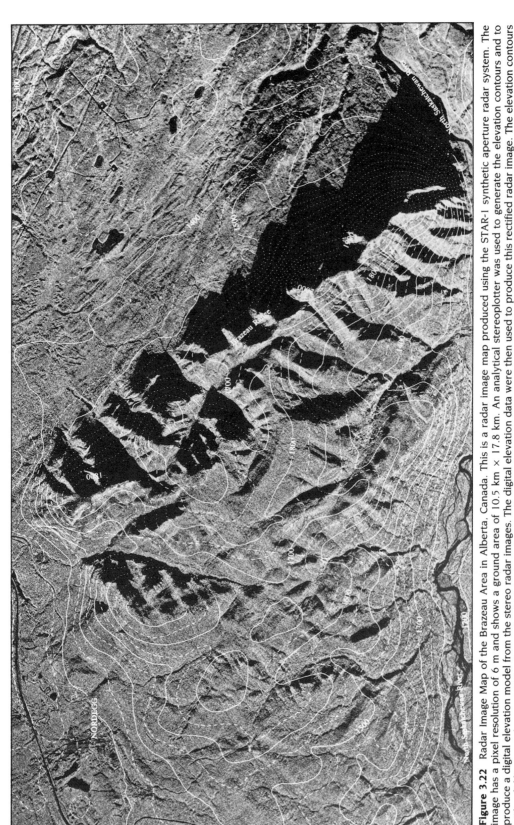

Figure 3.22 Radar Image Map of the Brazeau Area in Alberta, Canada. This is a radar image map produced using the STAR-1 synthetic aperture radar system. The image has a pixel resolution of 6 m and shows a ground area of 10.5 km × 17.8 km. An analytical stereoplotter was used to generate the elevation contours and to produce a digital elevation model from the stereo radar images. The digital elevation data were then used to produce this rectified radar image. The elevation contours with solid lines are drawn at 100 m intervals which is about 3 times the RMS elevation accuracy achieved. The intermediate contour lines are dashed to indicate a reduced level of confidence. The dotted contours in areas of shadow have been interpolated. The planimetric accuracy of this image map is ± 25 m RMS and the elevation accuracy

dates for the same area. This is carried out on a dedicated image analysis system at the Manitoba Remote Sensing Centre in Winnipeg.

For this crop condition monitoring application, the seven daily images produced over a week are combined to produce a weekly composite image. The fourteen digital images (a red and near-infrared image for each day) are combined by using computer-based image processing. For each pixel location, the seven NDVI values (one for each date) are calculated and the highest value is used to form the composite image. Areas of clouds and poor atmospheric conditions have low NDVI values. So, unless an area had poor conditions during all of the seven satellite passes, it is shown cloud-free on the composite image. Also, since the pixel with the highest NDVI value is used, the maximum reading for vegetation development is measured. (Holben 1986 reviews the use of composite AVHRR imagery.)

Plate 10 is a composite AVHRR image created in this way. The colour rendition is produced by representing the NDVI values in shades of red and the red wavelength values as shades of blue and green. The more developed the vegetation, i.e. the greener the vegetation, the redder it appears in this image. The predominantly blue colour of the districts in the lower right portion of the image (southern Manitoba) indicates poor crop conditions as compared with normal years. There was a severe drought during the 1988 growing season and the wide-spread effects are illustrated well in this image.

Although the crop condition measurements are derived from the AVHRR image data, the information is usually reported in tabular form. For each crop and each reporting district, the average NDVI (i.e. average NDVI for all the pixels within the district boundary) is reported along with the value that would be expected for that date in a normal year. Different crop types have different development patterns. So, in order to compare the crop condition for different years, areas with the same distribution of crop types must be used.

Figure 3.24 illustrates the change in the NDVI value for an area over a normal growing season and for a drought year. The shape of the curves are similar, but the curve for 1988 is substantially lower than the others. The relative timing of the crop development can be compared using the date at which the curve begins to rise. (The dates are shown in Julian days, the number of days from January 1.) In 1988 crop development was slower compared with a normal year and relatively little growth was achieved before declining. By the middle of the growing season it was known that there would be serious shortfalls in wheat production, and by comparing NDVI crop development curves for each crop reporting district, estimates of the size of the shortfall could be made.

The information provided by this monitoring program was used to identify areas where remedial efforts might be worthwhile, to identify areas where no further effort should be expended, to predict the size of the shortfall for a number of grain crops, and to set compensation levels so that higher drought relief payments were made to farmers in areas that were more severely affected.

Classification Analyses

Conditions as varied as soil texture, crop types, forest species composition, geologic strata, and human activity are all routinely assessed by visually interpreting airborne and satellite images. This procedure of identifying and mapping areas with similar characteristics is the most widely used form of classification.

The example presented here illustrates some of the computer-based classification procedures that have been developed over the past 15 years. The application of these

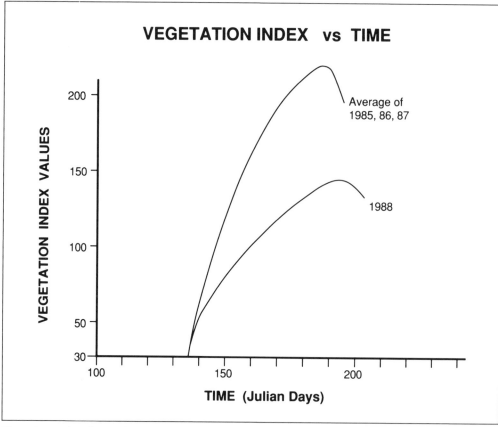

Figure 3.24 Plot of Normalized Difference Vegetation Index (NDVI) Values for Agricultural Areas in Manitoba. The graph shows the change in NDVI values through the growing season for one agricultural area in Manitoba. Time is measured in Julian days, the number of days from January 1. During the drought of 1988, crop development was poor compared with the average NDVI values for the three previous years. (Courtesy of the Canada Centre for Remote Sensing. Ottawa, Ontario.)

automated classification methods complements those of visual interpretation. The use of patterns, tones, and shapes to deduce information tends to be more efficiently done by a human interpreter than by computer analysis. However, computer techniques are used to derive quantitative measurements from the reflectance data, to simultaneously analyze the data from more than three bands, to classify large volumes of data at high speed, and to quantify the areal extent of each class.

Hydroelectric projects commonly involve the construction of dams and the flooding of significant areas of land. In North America,

environmental assessments and studies are usually required at various stages in the development and presentation of the project plan. If the project is accepted, it is often necessary to conduct further studies during and after construction to monitor environmental effects. Remote sensing technology has been used to collect and analyze environmental information to monitor water quality, assess soil erosion potential, quantify environmental effects such as the types of land that will be flooded, and monitor post-development conditions.

Ontario Hydro, the power utility for the province, had proposed construction of a hydroelectric generating facility on the Little Jackfish River in northern Ontario. A tentative dam site had been selected and information was needed to prepare the environmental assessment. One information requirement in the planning process was a quantitative estimate of the area of each type of land cover that would be flooded. Landsat MSS satellite data were used to generate a classification of the river basin into land cover classes and a tabulation of the amount of each class that was expected to be flooded.

The Landsat data were classified by the Ontario Centre for Remote Sensing using image processing methods. A supervised classification procedure was used in which sample areas, termed **training areas**, were delineated for each land cover class to be mapped. These samples, areas for which the land cover was known (e.g. from field observations), were drawn interactively on the image displayed on the colour monitor of the image analysis system. A fraction of a percent of the total area was entered as training areas. The classification program then used a statistical procedure to classify each pixel in the image. For each pixel, the reflectance values in the four MSS wavelength bands were compared to those of the training areas. Each pixel was assigned the class of the training area to which its reflectance values were most similar. The data for any number of wavelength bands and for images of different dates could be simultaneously analyzed using this procedure. The classification results were then registered to the topographic map base being used for the study, and output as a colour map. A portion of the map is shown in Plate 11.

From the location of the dam and the proposed water level for the reservoir, the elevation of the shoreline was determined and plotted on a topographic map of the area. The shoreline boundary was then digitized and entered in the image analysis system. The shoreline boundary of the proposed reservoir is shown as a yellow line in Plate 11. The image analysis system was then used to do an overlay analysis to calculate the area of each land cover class that would be flooded. The computer essentially did a tally of the number of pixels of each class that was within the proposed reservoir and multiplied the number of pixels by the ground area represented by each pixel. Table 3.5 is a tabulation of the results.

Table 3.5 Land Cover Classes to be Flooded by Proposed Reservoir. (Courtesy of the Ontario Centre for Remote Sensing.)

Land Cover Class	Area (in ha)
Coniferous	471
Low Density Coniferous	11
Mixed Coniferous	488
Deciduous	195
Mixed Deciduous	783
Wetland — Treed Bog	241
Wetland — Open Bog	14
Total Area to be Flooded	2,203

OCRS has found the accuracy of this type of classification for areas in northern Ontario to be 80% to 95%. A relatively high accuracy can be achieved by defining relatively broad classes. For the purposes of the study, the broad classes provided sufficient information for the intermediate scale river basin study for which it was used (Patterson and Sears 1986).

The accuracy level achieved depends on such factors as the resolution of the image data, the accuracy of the field verification data, the classification algorithm used, and also the way the classes were defined. The definition of the classes is an extremely important consideration. In general, the classification accuracy tends to be lower when a finer division of classes is specified. For example, the land cover classes used in this study were fairly broad; e.g. deciduous forest, coniferous forest, mixed coniferous, and water. However the same image data will give poorer classification results if the classes are defined more finely, such as pine forest, spruce forest, wheat crop, barley crop, and so on. To achieve an acceptable level of accuracy, the fineness of the class divisions must be matched to the distinctiveness of their reflectance values in the remotely sensed data.

The level of classification accuracy needed will depend on the application, and specifically on the *decisions* that will be made using the classification information. In this example, if the information needed was whether a significant amount of agricultural land would be flooded, then the level of accuracy that was provided would probably have been sufficient. However, if a precise inventory of land cover losses was needed, then the level of accuracy for this classification might not suffice.

Another trade-off is cost. By expending greater effort, and therefore increasing the cost, classification accuracies can be improved. As discussed in Chapter 2, the most cost-effective analysis method is not necessarily the most accurate one; it is the least expensive method that provides sufficiently accurate information with which to make a correct decision. For many applications, as in this study, the automated classification of satellite imagery provides the most cost-effective solution. Large areas can be classified quickly and relatively inexpensively. Also, the classification results are in a digital image form that can be analyzed together with other georeferenced data using image processing or GIS equipment.

Estimation Analyses

Remote sensing-based estimation analyses are treated here as a separate category. This type of analysis is used for such assessments as the area planted to a particular crop type, the expected crop production for a region,

or the forest resources inventory of a region. The information is usually provided in the form of estimates of quantities for each administrative or management unit.

The objective of an estimation analysis is to produce estimates of total amounts for each of the specified regions. Although the estimation procedure generally uses a classification of the remotely sensed data, the classification is incorporated into a statistical sample rather than being used as a map. For this reason, the position accuracy of individual field boundaries or the overall accuracy of the classification of the image data are less of a problem.

It is important to note that field data collection is an integral part of most surveys of this kind. The use of remotely sensed data does not eliminate the need for field observations. A remote sensing-based estimation procedure generally uses other data sources like field observations, farm surveys, and existing survey information together with the image data. Field data collection is one of the most expensive operations in a survey. Compared with using conventional surveys, using remotely sensed data can reduce the quantity (and therefore the cost) of field data that must be collected to achieve a given level of accuracy and may also decrease the amount of time needed to produce the estimate. Of course there are additional data and processing costs associated with the image data. Remote sensing methods are therefore appropriate if they provide an overall reduction in cost to achieve estimates of the same or better accuracy, if the information can be provided more quickly, or if there is no other reliable method available. In some cases, crop area estimates are produced without direct field observations if these data are unobtainable.

The example presented here is an operational application of satellite-based remote sensing to estimate potato crop area in New Brunswick and Prince Edward Island, Canada. In these provinces, potato is the single most important crop under cultivation. Large fluctuations in the annual potato production are a major problem for economic forecasting and for planning product marketing. A remote sensing-based estimation procedure was developed to improve the crop production estimates that are used to plan the handling and marketing of the crop (Ryerson et al. 1983). For the potato crop area information to be useful, the estimates were required to be within 6% of the true value and to be available by the end of August. (The accuracy required was actually a coefficient of variation less than 6%. The confidence level associated with one coefficient of variation is about 70%, as discussed below.) The remote sensing-based sampling and field verification procedure that was developed satisfied these time and accuracy requirements at a lower cost than the conventional farm questionnaire and field survey method used previously.

Figure 3.25 illustrates the sampling principle used to generate crop area estimates from satellite image data and field data. In practise, the estimation procedure is considerably more complex than the example presented here. Additional steps are used to take into account such factors as regional differences in productivity and missing data for areas obscured by clouds. By incorporating these factors into the calculation, the accuracy of the results are improved.

In this application, the objective was to estimate the total potato crop area using satellite imagery and field data. The satellite imagery was computer classified using an image processing system. This was done by defining training areas in the image that were known from field observations to be potato cropland. The classification program then generated a land cover class image for the entire area. Plate 12 is a portion of a Landsat satellite image showing the potato land cover class (red).

Next, a suitable *sampling frame* was selected for the study. A **sampling frame**

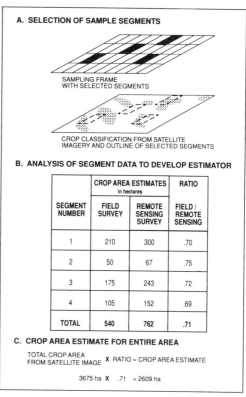

Figure 3.25 Crop Area Estimation Using Remotely Sensed Data.

defines the set of all sampling units (entities on which observations are made) in the study population. In this example the sampling frame consisted of a grid of 1 km by 3 km areas, termed **segments**, that covered the entire study area. A sample of these segments was selected for field verification according to a strict statistical sample design. The size of the sample used and the segments chosen depended on such factors as the accuracy level required, the expected variability in the data, and the analysis procedure that would be used.

In Figure 3.25 part A, four segments have been selected for field verification (in an operational survey a much larger number of segments would be selected). Qualified personnel would be sent to each selected segment to identify the crop types and field boundaries from the ground. They would determine the boundaries, area, and type of crop for each field in the segment and record their observations on the aerial photograph for the segment.

The area of potato crop in each segment would also be measured from the classified satellite image using the image processing system. The area estimates obtained for the four segments using the two measurement methods are shown in Figure 3.25 part B. Notice that the crop area as calculated from the imagery is a significant overestimate of the crop area as determined by field observation. A map produced from the classified image would contain significant classification errors. However, the overestimate is quite consistent. For the four segments, the ratio of the crop area determined by field investigation to the crop area as determined by classification of the satellite image is 0.71 (i.e. the average crop area by field investigation divided by the average crop area by satellite classification). The ratio for individual segments ranged from 0.69 to 0.75.

Because the ratio between the field and satellite image measurements is relatively consistent, an estimate of the total crop area based on this ratio should be fairly accurate. An estimation calculation based on this type of ratio is termed a **ratio estimator**. (Chhikara et. al. 1986 discusses the use of this and other commonly used estimators.) By using an estimator such as this one, the number of field observations needed and thus the cost to obtain an estimate with a given level of accuracy can be reduced.

The calculation of the ratio estimate of the total crop area is shown in Figure 3.25 part C. The total area classified as potato crop in the satellite image was obtained using a program in the image processing system to count all the pixels in the image that are classed as potato cropland. In this example the potato cropland pixels represented an area of 3675 hectares. The estimate of the area planted to potatoes was then calculated by multiplying the image area estimate by 0.71.

This gave a crop area estimate of 2609 hectares.

The assumption made in using this method is that the ratio would be consistent among all the sample segments, including the ones that were not selected for field measurement. If that assumption were true then a field enumeration of the entire area would have given a result equal to 0.71 times the area classified as potato cropland on the satellite image. (Of course a field enumeration would involve measuring every segment on the ground, a prohibitively costly undertaking.)

The accuracy of the image-derived crop area estimate depends on the *consistency* of the image classification rather than on the *classification accuracy* (i.e. the proportion of the area correctly classified). The level of consistency is evaluated from the individual sample segment measurements. Inconsistencies in the classification contribute to the non-sampling error in the estimate.

In crop reporting applications, the (estimated) error of an estimate is generally expressed as a coefficient of variation (CV). (The coefficient of variation is the standard error of an estimator divided by the expected value of the estimator.) This measurement, given as a percentage of the estimate, is used to determine the range of values within which the true value is expected to occur for a given confidence interval. The greater the inconsistency in the classification the higher the expected level of error and the larger the estimated CV.

The range of values determined by one CV has a confidence level of about 68%; that is, if the crop estimate had a CV of 5%, it would mean that the actual crop area should be within ±5% of the estimate 68% of the time. In other words, if this procedure were repeated many times then, on average, 68 times out of 100 the true crop area would be within ±5% of the estimate.

For the example in Figure 3.25, the interval would be from (2609 − 5% of 2609) hectares to (2609 + 5% of 2609) hectares, that is

Table 3.6 Comparison of Potato Crop Area Estimates Obtained Using Landsat Data and by Conventional Estimation Methods. (Source: Agriculture Division, Statistics Canada).

NEW BRUNSWICK ESTIMATES

	1980	1981	1982	1983	1984	1985	1986	1987	1988
Published Area Estimate (acres)[1]	52,000	54,000	54,000	53,000	54,000	55,500	51,000	50,500	47,000
Census Area (acres)[2]		53,793					48,466		
Landsat Regression Estimate (acres)	51,119	53,827	55,233	51,773	cloud	58,734	51,948	cloud	46,612
CV of Landsat (%)[3]	5.1	5.5	5.2	7.4		5.8	6.6		2.2
% Difference (Landsat – Published Estimate)	-1.7	-0.3	2.3	-2.3		5.8	1.9		-0.8
% Difference (Landsat – Census)		0.1					7.2		

PRINCE EDWARD ISLAND ESTIMATES

	1983	1984	1985	1986	1987	1988
Published Area Estimate (acres)[1]	70,000	72,000	71,000	70,000	67,000	68,000
Census Area (acres)[2]				64,219		
Landsat Regression Estimate (acres)	70,871	69,161	cloud	65,390	64,692	69,693
CV of Landsat Estimate (%)[3]	11.5	10.9		11.6	9.6	6.6
% Difference (Landsat – Published Estimate)	1.2	-3.9		-6.6	-3.5	2.5
% Difference (Landsat – Census)				1.8		

1. Official Statistics Canada published estimate of area planted to potato.
2. Census of Agriculture Data from the Agriculture Division, Statistics Canada.
3. The coefficient of variation (CV) is calculated for only that portion of the study region on the satellite image.

from 2479 hectares to 2739 hectares. On average, the true value would be expected to fall within this range 68 times out of 100. A more detailed discussion of confidence levels and intervals can be found in Chhikara et al. (1986) and can also be found in texts on sample survey design.

Table 3.6 shows a comparison of the potato crop area estimates obtained using a regression-based remote sensing survey approach and Statistics Canada published estimates. The remote sensing-based estimate was in most cases within $\pm$ 3% of the official published estimate. Furthermore, the CV for the remote sensing based survey was usually well within acceptable limits, about the same as obtained using conventional survey methods.

One draw-back in using the remote sensing-based method is that there can be a year in which no suitable imagery is obtained. However, the remote sensing-based method has proven itself to be a viable, cost-effective and accurate approach to the collection of potato crop area estimates.

REMOTE SENSING AND GEOGRAPHIC INFORMATION SYSTEMS

This chapter has provided an introduction to remote sensing principles and techniques. Remote sensing provides much of the information that is input to a GIS, from global scale vegetation and climatic data to the roof outlines entered into a municipal GIS. The objective for including this chapter in a GIS book is to make GIS users more aware of the wide range of information that can be produced using this technology. As with other fields, to make effective use of remote sensing technology requires technically skilled personnel as well as the appropriate technology. In Canada and the United States both the data and the expertise to use them are readily available.

Remote sensing and GIS technology developed separately. In part this was a result of the use of different types of equipment and the need for different technical skills. While a user of remote sensing technology may develop expertise in sensor systems and image processing methods, the expert GIS user may become more familiar with principles of map projections, spatial analysis, and the design of spatial data bases. Although the technology may encourage different technical orientations, in both cases the user must understand the nature of the information being collected — the forestry, geology, building structures, roadway design, and so on.

Ultimately, remote sensing and GIS technology are both used to collect, analyze, and report information about the earth's resources and the infrastructure we have developed to use them. The two technologies provide complementary capabilities. Remote sensing analyses are improved by the verification data retrieved from a GIS, and GIS applications can benefit from the information that remote sensing can generate. Often the image data are the most current spatial information available for an area. The use of digital image data offers the additional advantage of a computer compatible format that can be input directly to a GIS.

The integrated use of remote sensing and GIS methods and technology can not only improve the quality of geographic information but also enable information previously unavailable to be economically produced. Over the past few years manufacturers have developed more sophisticated technology for integrating remote sensing systems and geographic information systems. The effective use of these tools, however, depends on users sufficiently knowledgeable to apply them.

REFERENCES

Alfoldi, T.T. and J.C. Munday Jr. 1978. Water Quality Analysis by Digital Chromaticity Mapping of Landsat Data. *Canadian Journal of Remote Sensing* 4(2):108–126.

Aronoff, S. and G.A. Ross. 1984. Use of Remotely Sensed Data in Environmental Planning — A Case Study in Environmental Analysis for Gas Field Development Planning. *Journal of Environmental Management* 19:1–14.

Aronoff, S. and G.A. Ross. 1982. Detection of Environmental Disturbance using Colour Aerial Photography and Thermal Infrared Imagery. *Photogrammetric Engineering and Remote Sensing* 48(4):587–591.

Aronoff, S., G.A. Ross, and W.A. Ross. 1982. Environmental Monitoring of the Athabasca Oil Sands Region Using Landsat Data. *Photogrammetria* 38:77–86.

Campbell, J.B. 1987. *Introduction to Remote Sensing*. The Guilford Press. London, U.K.

Chhikara, R., J. Lundgren, and G. Houston. 1986. Crop Acreage Estimation Using Landsat-Based Estimator as an Auxiliary Variable. *IEEE Transcations on Geoscience and Remote Sensing* Volume GE-24 (1):157–167.

Clark, D.K. and N.G. Maynard. 1986. Coastal Zone Color Scanner Imagery of Phytoplankton Pigment Distribution in Icelandic Waters. In *Proceedings of SPIE Volume 637 Ocean Optics VII* held March 31 to April 2, 1986. The International Society for Optical Engineering. Bellingham, Washington. pp.350–357.

Colwell, R.N. (Editor-in-Chief). 1983. *Manual of Remote Sensing*. American Society of Photogrammetry. Falls Church, Virginia.

Colwell, R.N. (Editor). 1960. *Manual of Photographic Interpretation*. American Society of Photogrammetry. Falls Church, Virginia.

Dahlberg, R.E. 1986. Creation of a Satellite Image Map of Illinois — A Case Study of Technology Transfer and Linkage. *Cartographica* 23(4):14–28.

Elrod, J.A. 1988. CZCS View of an Oceanic Acid Waste Dump. *Remote Sensing of Environment* 25:245–254.

Falconer, A. 1986. Personal communication. Regional Centre for Services in Surveying, Mapping, and Remote Sensing. Nairobi, Kenya.

Forshaw, M., A. Haskell, P. Miller, D. Stanley, and J. Townshend. 1983. Spatial Resolution of Remotely Sensed Imagery: A Review Paper. *International Journal of Remote Sensing* 4(3):497–520.

Gauthier, J. 1987. Topographic Mapping from Satellite Data: A Canadian Point of View. *Geocarto International* 2(3):61–66.

Gillespie, A. and E. Abbot. 1984. Mapping Compositional Differences in Silicate Rocks with Six-Channel Thermal Images. In *Proceedings of the 9th Canadian Symposium on Remote Sensing*. Canadian Aeronautics and Space Institute. Ottawa, Ontario. pp.327–336.

Goetz, A., G. Vane, J. Solomon, and B. Rock. 1985. Imaging Spectrometry for Earth Remote Sensing. *Science* 228 (4704):1147–1153.

Goetz, A., B. Rock, and L. Rowan. 1983. Remote Sensing for Exploration: An Overview. *Economic Geology* 78(4):573–684.

Hallada, W.A. 1984. Mapping Bathymetry with Landsat 4 Thematic Mapper, Preliminary Findings. In *Proceedings of the 9th Canadian Symposium on Remote Sensing*. Canadian Aeronautics and Space Institute. Ottawa, Ontario. pp.277–285.

Harris, R. 1987. *Satellite Remote Sensing: An Introduction*. Routledge & Kegan Paul Ltd. London, UK.

Holben, B.N. 1986. Characteristics of Maximum-Value Composite Images from Temporal AVHRR Data. *International Journal of Remote Sensing* 7(11):1417–1434.

Jupp, D., K. Mayo, D. Kuchler, S. Heggen, S. Kendall, B. Radke, and T. Ayling. 1985. *Landsat Based Interpretation of the Cairns Section of the Great Barrier Reef Marine Park*. Natural Resources Series No. 4. Division of Water and Land Resources. Commonwealth Scientific and Industrial Research Organization of Australia. Canberra, Australia.

Kahle, A. and A. Goetz. 1983. Mineralogic Information from a New Airborne Thermal Infrared Multispectral Scanner. *Science* 222 (4619):24–27.

Leberl, F., G. Domik, and B. Mercer. 1987. Methods and Accuracy of Operational Digital Image Mapping with Aircraft SAR. In *Proceedings of the 1987 ASPRS — ACSM Annual Convention*. American Society of Photogrammetry and Remote Sensing. Falls Church, Virginia. Volume 4: 148–158.

Lillesand, T.M. and R.W. Kiefer. 1987. *Remote Sensing and Image Interpretation*, Second Edition. John Wiley and Sons. New York, New York.

Lindell, L., O. Steinvall, M. Jonsson, and Th. Claesson. 1985. Mapping of Coastal-Water Turbidity Using Landsat Imagery. *International Journal of Remote Sensing* 6(5):629–642.

Lo, C.P. 1986. *Applied Remote Sensing*. Longman Incorporated. New York, New York.

Meisner, B.N. and P.A. Arkin. 1984. The GOES Precipitation Index: Large Scale Tropical Rainfall Estimates Using Infrared Data. In *Proceedings of the 15th Conference on Hurricanes and Tropical Meteorology*. American Meteorological Society. Boston, Massachusetts. pp.203–206.

Mercer, J.B. and M.E. Kirby. 1987. Topographic Mapping Using STAR-1 Radar Data. *Geocarto International* (3):39–42.

Merrit, E., L. Heitkemper, and K. Marcus. 1984. CROPCAST — A Review of an Existing Remote Sensor-Based Agricultural Information System with a View Toward Future Remote Sensor Applications. In *Proceedings of SPIE Volume 481, Recent Advances in Civil Space Remote Sensing*. Institute of Electrical and Electronic Engineers. New York, New York. pp.231–237.

NOAA. 1985. *Envirosat-2000 Report: GOES-Next Overview*. National Oceanic and Atmospheric Administration. U.S. Department of Commerce. Washington, D.C.

Paltridge, G.W. and J. Barber. 1988. Monitoring Grassland Dryness and Fire Potential in Australia with NOAA/AVHRR Data. *Remote Sensing of Environment* 25:381–394.

Patterson, W.M. and Sears, S.K. 1986. Remote Sensing as a Tool for Assessing Environmental Effects of Hydroelectric Development in a Remote River Basin. In *Proceedings of the Symposium on Remote Sensing for Resources Development and Environmental Management*. ITC. Enschede, Netherlands.

Ross, G. and S. Aronoff. 1984. Use of Remotely Sensed Data in Environmental Planning — A Case Study of

Environmental Analysis for Gas Field Facilities Development Planning. *Journal of Environmental Management* 19:1–14.

Ryerson, R., J-L. Tambay, R. Plourde, and J. Harvie. 1983. *The Use of Landsat, Ground Data, and a Regression Estimator for Potato Area Estimation*. CCRS Report 83-2. Canada Centre for Remote Sensing. Ottawa, Ontario.

Sader, S.A. 1987. Digital Image Classification Approach for Estimating Forest Clearing and Regrowth Rates and Trends. In *Proceedings of the IGARSS '87 Symposium*. Institute of Electrical and Electronic Engineers. New York, New York. pp.209–213.

Strahler, A., T. Logan, J. Franklin, and H. Bowlin. 1983. *Automated Classification and Inventory in the Eldorado National Forest*. United States Department of Agriculture, Forest Service. Washington, D.C.

Stumpf, R.P. and M.A. Tyler. 1988. Satellite Detection of Bloom and Pigment Distributions in Estuaries. *Remote Sensing of Environment* 24:385–404.

Tassan, S. and B. Sturm. 1986. An Algorithm for the Retrieval of Sediment Content in Turbid Coastal Waters from CZCS Data. *International Journal of Remote Sensing* 7(5):643–655.

Thomas, R.W. 1988. Personal Communication. Research Fellow. Remote Sensing Research Program. University of California. Berkeley, California.

Vane, G. and A. Goetz. 1988. Terrestrial Imaging Spectrometry. *Remote Sensing of Environment* 24:1–29.

Vogelmann, J.E. and B.N. Rock. 1988. Assessing Forest Damage in High-Elevation Coniferous Forests in Vermont and New Hampshire Using Thematic Mapper Data. *Remote Sensing of Environment* 24:227–246.

Wall, S., R. Thomas, C. Brown, and E. Bauer. 1984. Landsat-Based Inventory System for Agriculture in California. *Remote Sensing of Environment* 14:267–278.

Wrigley, R.C. and S.A. Klooster. 1983. Coastal Zone Color Scanner Data of Rich Coastal Waters. In *Proceedings of the 1983 International Geoscience and Remote Sensing Symposium*. Institute of Electrical and Electronics Engineers. New York, New York. Volume II: 2.1–2.5.

Yates, H., J. Tarpley, S. Schneider, D. McGinnis, and R. Scofield. 1984. The Role of Meteorological Satellites in Agricultural Remote Sensing. *Remote Sensing of Environment* 14:219–233.

4. DATA INPUT AND OUTPUT

INTRODUCTION

For a GIS to be useful it must be capable of receiving and producing information in an effective manner. The data input and output functions are the means by which a GIS communicates with the world outside. In the past, when a GIS was implemented virtually all of the data to be input had to be specially converted into a digital form and structured in the format specific to the system. Over the past five years, standard digital geographic data sets have become more widely available. Automated methods of data conversion, such as scanning, have been improved and digital methods of data collection can be used that generate digital files directly. For example, GIS-compatible data sets can be generated directly from digital satellite imagery. Large, complete data bases at regional and global scales can be produced rapidly and economically in this manner.

Output technology has also seen rapid advances. The price of colour hardcopy devices has dropped substantially. Colour graphics terminals have become increasingly sophisticated with substantial computing power now built into the device. As a result, system performance has been enhanced as more of the graphics display generation is performed by the display hardware instead of the computer running the GIS software.

Each GIS installation will have a somewhat unique set of requirements. No one device or approach is optimum for all situations. The objective in defining GIS input and output requirements is to identify the mix of equipment and methods needed to meet the required level of performance and quality. This chapter provides an overview of the principal methods used to input and output data in a GIS environment and an overview of existing digital geographic data sets. A list of sources for purchasing these data can be found in Appendix B.

DATA INPUT

Data input is the procedure of encoding data into a computer-readable form and writing the data to the GIS data base. Data entry is usually the major bottleneck in implementing a GIS. The initial cost of building the data base is commonly 5 to 10 times the cost of the GIS hardware and software. The creation of an accurate and well-documented data base is critical to the operation of a GIS. Accurate information can only be generated if the data on which it is based were accurate to begin with. Documentation is needed that describes the quality of the data in order to assess their suitability for a specific application. Data quality information includes the date of collection, the positional accuracy, classification accuracy, completeness, and the method used to collect and encode the data. Data quality is discussed in Chapter 5.

The data to be entered in a GIS are of two types — spatial data and associated non-spatial attribute data. The spatial data represent the geographic location of features. Points, lines, and areas are used to represent geographic features like a street, a lake, or a forest stand. The non-spatial attribute data provide descriptive information like the name of a street, the salinity of a lake, or the composition of a forest stand. During data input the spatial and attribute data must be entered and correctly linked (i.e. the attributes must be logically attached to the features they describe). Suitable verification procedures are needed to check that data quality standards are met.

There are five types of data entry systems commonly used in a GIS: keyboard entry, coordinate geometry, manual digitizing, scanning, and the input of existing digital files. Keyboard entry, as its name implies, involves manually entering the data at a computer terminal. Attribute data are commonly input by keyboard, whereas spatial data are rarely entered in this way. In coordinate geometry procedures (COGO), the survey data are commonly entered by keyboard. From these data the coordinates of the spatial features are calculated and a GIS-compatible data file created.

Manual digitizing is the most widely used method for entering spatial data from maps. The map is mounted on a digitizing table and a hand held device, termed a puck or cursor, is used to trace each map feature (Figure 4.1). The position of the cursor is accurately measured by the device to generate the coordinate data in digital form.

Scanning, also termed **scan digitizing**, is a more automated method for entering map data. A raster digital image of the map is produced after which additional computer processing is done to improve the quality of the image and to convert the raster data to vector format. Operator-assisted editing and checking is then done to generate the final GIS-compatible data file.

Spatial and attribute data sets are becoming more widely available in digital form, often in a format that can be directly input to a GIS. In Canada and the United States, numerous government programs are underway to convert existing map and tabular data to digital form and to change

Figure 4.1   Manual Digitizing Using a Digitizing Table. (Courtesy of the Ontario Centre for Remote Sensing. Toronto, Ontario.)

data collection and mapping procedures so that new data are produced in a digital format. These digital data sets are priced at a fraction of the cost of digitizing existing maps. Over the next decade, the increased availability of digital data should reduce the current high cost and lengthy production times needed to develop digital geographic data bases.

KEYBOARD ENTRY AND COORDINATE GEOMETRY PROCEDURES

Most attribute data are entered by keyboard. In many cases these data can be obtained in digital form from an existing data base into which they were keyboard entered. Field observations can be recorded in digital form by keyboard entry of the data in the field using small hand-held computers. These portable computers use a bubble memory or other rugged storage hardware to collect the data. The data files are then periodically downloaded to another computer or copied onto diskette for storage.

Keyboard entry can be used during manual digitizing to enter the attribute information. However, this is usually more efficiently handled as a separate operation in which the attributes are entered with a code to indicate the spatial element (such as the line or polygon feature) that they describe. The attribute file is subsequently linked to the spatial data.

Coordinate geometry (COGO) procedures are used to enter land record information. A very high level of precision is obtained by entering the actual survey measurements. The high level of precision may be needed when the maps must represent the land cadastre exactly as it is expressed in the legal description.

For a city with 100,000 parcels, it would cost on the order of $1.50 per parcel or $150,000 to digitize the parcels manually. COGO procedures are commonly 6 times

and can be 20 times more expensive than manual digitizing. Paying this high a premium for COGO data entry can be difficult to justify. Surveyors and engineers want the higher accuracy of COGO for their computational applications. However, planners and most other users of the digital files are quite willing to accept the lower accuracy provided by manual digitizing. For this reason local governments have tended not to use COGO (Dangermond 1988).

MANUAL DIGITIZING

In manual digitizing the map is affixed to a digitizing table and a pointing device is used to trace the map features (see Figure 4.1). Digitizing tables can be as large as 1 m × 1.5 m or more. A smaller device, termed a digitizing tablet (usually equipped with a mouse instead of the more precise cursor) is commonly used as a device to operate the GIS. The digitizing table electronically encodes the position of the pointing device with a precision of fractions of a millimetre. The most common table digitizer uses a fine grid of wires embedded in the table. The cursor normally has a crosshair for precise positioning and 16 or more control buttons that are used to operate the data entry software and to enter attribute data. As the map elements are traced, the coordinate data generated from the digitizing table are either processed immediately by the GIS or are stored for later processing. The digitizing operation itself requires little computing power and so can be done off-line, ie. not using the full GIS. A smaller and less expensive computer can be used to control the digitizing process and store the data. The data can later be transferred to the GIS for processing.

The efficiency of digitizing depends on the quality of the digitizing software and the skill of the operator. The process of tracing lines is time-consuming and error prone. The software can provide aids that substantially reduce the effort of detecting and correcting errors. Attribute data may be entered during the digitizing process, but usually only an identification number is coded. The attribute information referenced to the same identification numbers are then entered separately. If the attribute data are already in GIS-compatible files, they may be entered directly.

Manual digitizing is a tedious job. Operator fatigue can seriously degrade data quality. Work scheduling should limit the hours per day that an individual spends digitizing and suitable quality assurance procedures should be used to ensure that the digitized data and associated attribute data satisfy the required accuracy standards. A commonly used quality check is to produce a verification plot of the digitized data that is visually compared with the map from which the data were originally digitized.

SCANNING

Scanning (or scan digitizing) provides a faster means of data entry than manual digitizing. In scanning, a digital image of the map is produced by moving an electronic detector across the map surface. Two scanner designs are commonly used. In a **flat-bed scanner** the map is placed on a flat scanning stage over which the detector is moved in both X and Y directions. In a **drum-scanner** (Figure 4.2), the map is mounted on a cylindrical drum. The detector is moved horizontally across the drum as it rotates. The sensor motion across the drum provides the movement in the X direction. The drum rotation provides the movement in the Y direction.

The output from the scanner is a digital image. The fineness of detail captured by the scanner depends on the size of the map

Figure 4.2   Map Digitizing Using A Drum Scanner. (Courtesy of the US Geological Survey, Reston, Virginia.)

area viewed by the detector, termed the **spot size**. Spot sizes on the order of 20 microns (.02 mm) are commonly used. Scanners can record colour information by scanning the same document three times using red, green and blue filters. Usually only a black-and-white image is produced. The raster image is computer processed to improve the image quality and is then edited and checked by an operator. If the data are required in vector format, additional raster-to-vector conversion processing is done. During the editing procedure or after conversion to vector format, each spatial element is tagged and assigned an identification number. The attribute data are linked to the spatial data by means of these identification numbers.

Scanning Versus Manual Digitizing

Scanning has been adopted by many organizations as the principal means of spatial data entry, yet the subject is controversial. One reason for the controversy is that rigorous trials are few and of necessity are specific to the organization and application. Another organization will face a different situation that could completely change the cost-benefit trade-offs.

Data entry using scanning is claimed to be 5 to 10 or more times faster than digitizing (see for example Blakeman 1987, Chrisman 1987). Organizations like Environment Canada (for the Canada Land Data System), the US National Parks Service, and the US Forest Service use scanner-based data entry systems operationally.

Maps normally must be re-drafted before they can be scanned. Re-drafting is often considered to be a major disadvantage of the scanning option. However, though re-drafting may be an additional step in the digitizing process it does not necessarily add to the overall cost. In fact, re-drafting can *reduce* the total cost of *both* scanning and manual digitizing. Studies by the US Forest Service have shown that by including a map preparation step (with re-drafting) before the manual digitizing is done can reduce the overall digital encoding costs by as much as 50%. The reasons for this are:

1. The re-drafting process is not done on a computer system, it is done manually. Therefore computer operating costs and the higher salaries paid to computer operators are not incurred.

2. The digitizing operation proceeds much more quickly and requires less editing if the map has fewer errors and inconsistencies. Polygons that do not close, lines that don't correctly meet at adjoining map boundaries, boundaries that are inconsistent on different map layers (such as the shoreline of a reservoir that fluctuates with the water level) can be reconciled and corrected during the re-drafting stage. Faster completion of the digitization and editing functions reduces the amount and therefore the costs of expensive computer system and computer operator time.

3. Where many changes must be made to a complex map and many inconsistencies must be resolved, manual drafting is considerably faster than digitizing. Drafting and digitizing are very different operations requiring different skill sets. They are not equivalent tasks.

4. It is very time-consuming and hence very costly to make large numbers of changes to a map once it is in digital form. Not only must the changes be entered, the map coverage normally must be re-processed and re-integrated into the GIS data base.

While a scanning system is for the most part automated, and so requires less highly trained personnel for some tasks, more complex equipment must be maintained, more sophisticated software must be written or purchased, and there are more steps in the process. Scanners are more expensive than digitizing tables. A 150 cm by 100 cm (60 inch by 44 inch) digitizing table will cost on the order of $12,000 US. A high quality scanner will cost on the order of $100,000 US or more. Whereas virtually all GISes include some form of manual digitizing capability, separate special-purpose software is needed to operate a scanning system. A scanner also requires sophisticated software to edit and clean the raster data and then to convert the raster data into vector form (for input to a vector-based GIS).

Scanning works best with maps that are very clean, simple, and do not contain extraneous information, such as text or graphic symbols. It is most cost-effective for maps with large numbers of spatial elements (such as 1,000 or more polygons) and maps with large numbers of irregularly shaped features (such as sinuous lines and irregular-shaped polygons). The volume of production is also a factor; the higher equipment cost is more easily justified if the sustained production of large numbers of maps is needed. Tailoring the data entry system to take maximum advantage of a scanner is also more easily justified in a high volume production environment.

Manual digitizing tends to be more cost-effective when there are relatively few maps that are not in a form that can be scanned. Maps that contain a lot of extraneous information, require interpretation or adjustment during the encoding process, or have a small number of features to be encoded are generally not worth scanning (Dangermond 1988).

There is a strong demand for faster, more cost-effective data entry methods than armies of operators tediously digitizing mountains of paper maps. Scanning will never completely replace manual digitizing. However, scanners are being used more widely and the technology has improved significantly.

Direct Use of Raster Scanned Images

Much of the difficulty in using raster scanning to enter map information is in the extraction of the points, lines, and polygons from the raster image. In some cases, the map need only be used as a background on which to overlay other geographic information. The map information can be read by the user directly from the image without being explicitly entered as spatial features or attributes in the data base. Airphotos, satellite images, and other images can be stored and presented in this way.

At first this idea met with resistance by the GIS and cartographic communities. Raster images were considered aesthetically inferior to drafted maps. In a coarse resolution raster image, lines appeared as staircases. However, a high resolution raster image can provide as smooth a line as desired, and eliminate the blocky appearances. The cost, however, was that the image required large amounts, often megabytes, of data storage. More recently, with improved data compression techniques, lower cost storage devices, and new storage technology (such as video disks and optical disks), there has been renewed interest in the use of raster images within a GIS. Also, the more recent satellite systems, such as Landsats 4 and 5 and SPOT, have provided imagery with resolutions of 30 m to as fine as 10 m, a resolution that is comparable to high altitude aerial photography.

Some vendors of image processing and GIS systems have developed systems capable of simultaneously displaying vector and raster data in correct registration. These systems enable the raster satellite image to be used as a background over which a vector map is displayed (see Plate 2). The vector map can then be updated or a new map created by interactively drawing over the image. In a similar manner, a paper map can be captured as an image and used for interactive interpretation and updating.

Using the raster image as a background can be an effective solution when a relatively small amount of data needs to be extracted but a large area must be scanned to find that data. For example, changes in land use, such as the clearing of land or construction of roads, are generally a few isolated occurrences at unpredictable locations. By using the imagery directly as a background, the changes can be found visually and the GIS data base updated as required.

Raster images may be an attractive option when the spatial information need only be viewed. A data base can be provided that serves as a map library, retrieving the image with the desired city, street, or other feature and overlaying the appropriate graphic symbols, such as political boundaries or a road network. The Doomsday system mentioned in Chapter 1 is a micro-computer-based GIS that uses raster images for spatial data storage. The 500 megabyte data base is held on two video disks and consists of 30 million words, 21,000 spatial data sets, 24,000 topographic maps, statistical tabulations, picture libraries, and even film clips. These data cover a wide range of environmental, socioeconomic, and demographic fields. The special video disk player developed for this system is capable of storing and overlaying information held in both analogue and digital form. The system is extremely easy to use, designed for school children to learn in a couple of hours, and is very inexpensive at under $5000 US to schools (about $6000 US to the public) for the hardware, software, and data base.

This system is based on the direct use of raster-scanned images. Maps, photographs, and pictures are stored in raster format. Attribute data and vector boundaries are stored in a compacted form linked to the raster data set. The user can view maps, scroll across the entire country at the national level of resolution, and zoom in on more local data by moving from small to large scale maps of the same area. Area and distances can be measured interactively. Data can be retrieved by area name, coordinate position, or by a pre-defined window. Satellite image maps, air photos, slides, text, and maps can be retrieved as desired. Figure 4.3 illustrates a map displayed on the BBC Doomsday system from its videodisk image data base. The display can be interactively interrogated to obtain a value or name for a point of interest. Spatial data can be analyzed, plotted, and overlayed. There is also a floppy disk so that users can enter their own data or output selected data from the video disk (Rhind and Openshaw 1987).

Although this system has limited analytical capabilities (it was not designed to be a full-featured GIS), it does illustrate the effective use of raster scanned images to inexpensively produce a very large data base. The system also shows that new technology can economically store the large image data sets required.

Figure 4.3 Map of an Urban Area as Displayed on the BBC Doomsday System. The map is stored as a colour raster image on a video-disk. (Courtesy of Birkbeck College. London, UK.)

REMOTELY SENSED DATA

Remote sensing, and in particular aerial photography, has long been used in map production. In fact, most of the spatial data now used in a GIS are produced with remote sensing technology. For example the information presented on standard topographic maps is generated by photogrammetric means using stereoplotters to map elevation contours, natural and cultural (built) features. (**Photogrammetry** is the art and science of obtaining reliable spatial measurements from aerial photography and other remotely sensed images. It is here considered to be an aspect of remote sensing.) Aerial photography is also used extensively to map natural resources such as soils, forest types, land use, and geology.

Since 1972, with the launch of Landsat-1, high resolution digital remotely sensed data have become widely available. These satellite-based systems now produce publicly available data for virtually all of the earth's surface at resolutions as high as 10 metres, equivalent to high altitude aerial photography. Airborne digital systems produce imagery with resolutions measured in centimetres. These systems can sense radiation that photographic films cannot record, such as thermal infrared and microwave wavelengths.

The images produced by these sensor systems represent measurements of the reflected and emitted radiation from the earth's surface. Because the imagery is in digital form, the data for a single wavelength or multiple wavelengths used in combination can be computer processed to directly generate geographic information. Surface temperature, land use, crop condition, flooding, water quality, and forest harvesting are a few examples of geographic data that can be produced using computer methods. Computer enhancement of digital imagery enables other resource information to be visually interpreted and used to update the information in a GIS.

As illustrated in Chapter 3, digital elevation data are now being generated directly from SPOT digital satellite imagery using stereo correlation methods. Topographic maps at the 1:50,000 scale are being produced with computer-generated contours (see Plates 8 and 9). Contour intervals of 20 m to 25 m with accuracies on the order of 10 m RMS at the 90% confidence level can be achieved (see for example Gugan and Dowman 1988, and Simard et. al. 1988). The procedure is computationally intensive, requiring large amounts of expensive computer-processing power. However it can be less expensive than using aerial photography and conventional photogrammetric methods. Several companies are now producing digital elevation data directly from digital SPOT image data on a commercial basis. It is not yet being used operationally by mapping agencies in Canada and the United States.

The fields of digital remote sensing and GIS have tended to remain separate. Yet this division is an artificial one. Ultimately, the information produced using remote sensing methods is geographic information. Much of that information cannot be obtained as quickly or as inexpensively using other methods. The information generated from satellite imagery is sometimes dismissed as having insufficient spatial resolution or as being too expensive. Yet geographic information, computer-generated from digital satellite imagery, has been used cost-effectively to provide information for planning at even the county level. Niemann and Sullivan (1987), for example, report the use of digital satellite data to assess soil erosion conditions at the local level, and numerous examples of natural resource mapping applications can be found in the remote sensing literature. What is needed now is greater collaboration between the GIS and remote sensing communities. The basic technology is available.

EXISTING DIGITAL DATA

In Canada and the United States, low-cost digital geographic information is becoming more readily available. At the federal level, these data sets are being produced by the national mapping agencies and agencies responsible for census and other nation-wide statistical data. At the state and provincial level, data conversion programs vary widely. Automation of land ownership records is a high priority. Natural resource information is being converted to digital form at both the federal and state or provincial levels. At the county and municipal levels the conversion of data to digital form is usually done in order to implement a GIS installation. The land ownership information is usually a central component of these GIS data bases.

Digital data sets are produced to satisfy a wide range of users. The cost of the data, currency, and accuracy vary. The accuracy with which boundaries are drawn, the date of the information, and the method of compilation may be sufficiently different to create errors when different data layers or adjacent map sheets within a data layer are used together. Dulaney (1987) provides examples of these types of problems. Figure 4.4 is a map produced from the US Geological Survey 1:250,000 Land Use/Land Cover digital data set. To generate this map coverage, the data for two adjacent data sets were joined. Notice the abrupt change in the land use categories along a horizontal line across the center of the map. This change coincides with the boundary between two map sheets from which the data were digitized. The differences may be a result of discrepancies in airphoto interpretation or of the three year difference in the source dates of the aerial photography used. This type of anomaly can produce erroneous results, depending on the application.

Ideally, problems such as these are corrected at the map preparation stage, before the data are digitized. However, no data set can be perfect. Problems such as these may occur in any digital data set and must be identified and taken into account. Similarly, different data sets produced for the same area may not register accurately enough to be used together. Operational problems, such as differences in data formats or long delivery times may also be a factor.

As digital data sets become more widely used, data formats should become more standardized. Private companies are also beginning to provide off-the-shelf data base products. Although there may be difficulties, the cost of existing data is usually a fraction of the cost of creating a new data set. For this reason, the availability of inexpensive data sets will make GIS technology economically more attractive and easier to implement.

Numerous digital data sets are currently available or under development. In the United States there has been considerable effort made to coordinate and standardize the production and distribution of digital geographic data. At the federal level, the Federal Interagency Coordinating Committee on Digital Cartography (FICCDC) was formed for this purpose in 1983. Some 14 organizations participate in the Committee, which holds regular meetings and produces a newsletter and a variety of reports. In Canada, some provinces have instituted coordinating committees. At the federal level, Energy, Mines, and Resources Canada recently (1988) organized the Inter-Agency Committee on Geomatics to coordinate such activities as quality and format standards.

Examples of the more widely used data sets available from federal agencies in Canada and the United States are presented here to illustrate the range of existing digital information. These data sets can be broadly grouped into four categories: base cartographic data, natural resource data, digital elevation data, and census-related data.

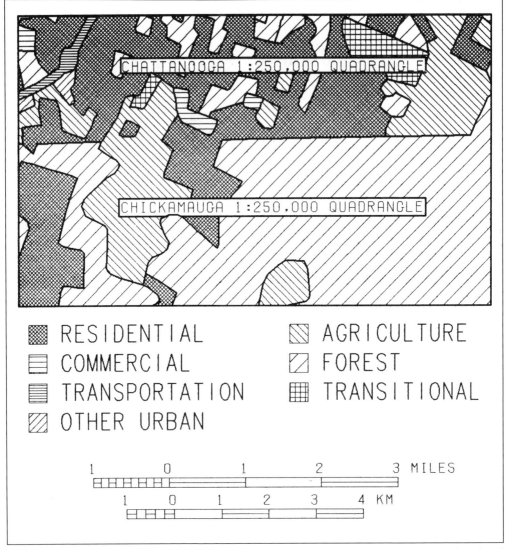

Figure 4.4 Illustration of An Artifact Boundary Between Map Sheet Quadrangles in 1:250,000 Land Use/Land Cover Data. (From Dulaney 1987, courtesy of the American Society for Photogrammetry and Remote Sensing.)

Base Cartographic Data

Base cartographic data include the topographic and planimetric information usually portrayed on a map. (**Topographic data** are those data that portray relief, such as elevation contours and spot heights. **Planimetric data** include roads and streams, as well as cultural data such as administrative and political boundaries, cities, and towns.) Often these data sets are digitized versions of an existing map series with each type of information, such as the elevation contours, assigned to separate data layers. Base cartographic data sets are produced in two basic formats: graphics and topologically-structured.

Graphics Format

The graphics format data set is designed for plotting maps. It is essentially the line and point features digitized in vector format. In this digital form, the map can be easily updated or modified to produce special purpose maps. However, the absence of topology severely limits their use for spatial analysis. (A discussion of graphics and topologically-structured formats is provided in Chapter 6.)

These data sets are well-suited for the Computer Aided Drafting systems used in digital mapping. A commonly used interchange format developed by the digital mapping industry is the Standard Interchange Format (SIF). It was designed primarily for transferring graphics elements like lines, points, curves, and symbols. Drawing information like shading, line weights, and line colour is an integral part of these data sets.

Graphics format data sets can be imported into a GIS. However, there can be considerable difficulty in using the data. The data files often have not been checked for topological consistency. They may contain such inconsistencies as lines that do not meet precisely, that overshoot or undershoot the correct connection point. There may be missing lines or gaps that create polygons that are not closed. For these data to be used by a vector-based GIS, the files generally have to be run through an automated "cleaning" procedure to correct these problems. Resolving these errors may require considerable operator-assisted editing as well (Dangermond 1988).

In Canada, the Canada Centre for Mapping (Energy, Mines, and Resources Canada) is compiling a digital topographic data set from the 1:250,000 National Topographic Map Series. Due for completion in 1990, this data set is being created by scanning the separate map layers that were used to produce the original printed maps. These data are distributed in SIF and other formats. Plans have also been announced to provide more detailed coverage of southern Canada and selected northern areas by scan digitizing map layers from the 1:50,000 National Topographic Map Series. Digital stereo compilation will be used to produce digital topographic data of the highest level of accuracy for areas of special interest.

In the United States, the US Geological Survey provides base cartographic data sets (discussed below) in graphics format, as well as the topologically-structured Digital Line Graph (DLG) format.

Topologically-Structured Format

The topologically structured format is designed to encode geographic information in a form better suited for spatial analysis and other geographic studies. Most GISes are now designed to use this topological information. (Topology is discussed in Chapter 6.)

In Canada, topologically-structured cartographic data sets designed for GIS use are being developed at the provincial level. For example, the Ontario Ministry of Natural Resources is compiling digital mapping at scales of 1:10,000, 1:20,000, and 1:50,000 for the province. These data sets are topologically-structured in DLG format. Eighteen map layers including vegetation, hydrology, transportation, cadastre, contour lines, and elevation are being produced. A standardized attribute coding system is used for each of the data layers.

US Geological Survey Digital Line Graph Data. The US Geological Survey has developed a digital cartographic data set called Digital Line Graph (DLG) data. The data are topologically-structured and supplied in Digital Line Graph Version 3 format. This exchange format, developed and supported by the US Geological Survey, supports non-spatial attribute as well as spatial data.

The DLG data set has been developed from previous mapping at the 1:2 million, and more recently at the 1:100,000, and 1:24,000 scales. The older 1:2 million data set includes transportation, hydrography, and political boundary maps for the entire country. The 1:100,000 scale data sets for transportation and hydrography have been completed for the entire United States. The political boundary and land net (Public Land Survey System) data layers are still being developed. The 1:24,000 series will include the Public Land Survey System, political boundary, transportation, hydrography, and contour data layers. Figure 4.5 is a map produced from the transportation data layer of the 1:24,000 DLG series. Only limited coverage is currently available in the 1:24,000 series.

These data sets represent a comprehensive, standardized, inexpensive, and publicly available source of digital information. The complete coverage (at the 1:100,000 scale) makes it possible to assemble large-area data bases quickly and at low cost. The spatial accuracy of the data is designed to satisfy urban area planning and network analysis requirements.

Natural Resource Data Sets

In the United States, major federal agencies have adopted GIS methods to analyze their

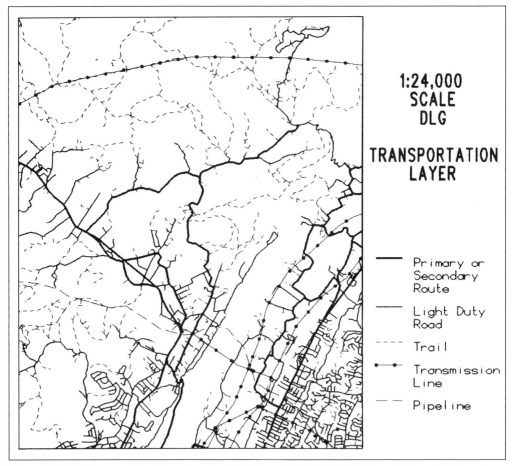

Figure 4.5 The Transportation Layer from the 1:24,00 scale DLG Data Set. (From Dulaney 1987, courtesy of the American Society for Photogrammetry and Remote Sensing.)

natural resource information. The US Geological Survey, Bureau of Land Management, National Parks Service, and National Oceanic and Atmospheric Administration are a few examples. Numerous data sets are available with different restrictions on distribution. One of the more widely available data sets is the Land Use/Land Cover data set.

Land Use, Land Cover, and Associated Data

The US Geological Survey has developed a Land Use/Land Cover data set compiled primarily from 1:58,000 colour infrared aerial photography and mapped at the 1:250,000 scale. The digital data are generated by both manual digitizing and scan digitizing. The land use and land cover classes include urban areas, agricultural land, rangeland, forest, wetlands, barren land, and tundra. Associated maps show 4 categories of information: political boundaries, hydrologic units (watershed boundaries), federal land ownership, and census subdivisions (NCIC 1985). Data are available for about 75% of the United States. A separate Land Cover digital data set is being developed for Alaska. This data set uses a different classification scheme and is being produced by automated classification of digital satellite imagery. The Land Use/Land Cover data sets are available from the National Cartographic Information Center.

In Canada, the Canada Land Data System (incorporating the Canada Geographic Information System) of Environment Canada and the Canada Soils Information System of Agriculture Canada can provide natural resource digital data at the national level. (These organizations are listed in Appendix B.) However, natural resources are the responsibility of the provinces. For this reason, most natural resource digital data sets are now being developed by the provincial governments. Forestry is perhaps

the most widely used application. Most provincial agencies responsible for forest resources have implemented geographic information systems for forest mapping and inventory.

British Columbia has perhaps the most developed provincial forestry GIS operation (discussed in Chapter 1). The Ministry of Forests is responsible for maintaining 6,600 forestry maps at a scale of 1:20,000 and the associated non-spatial attribute information (including species, crown closure, age, height, and stand density). Re-inventory using aerial photography is done on a 10 year cycle. However, depletion updating is done using automated classification of digital satellite data. Vector boundaries can be generated directly from the satellite data using automated edge-detection image analysis methods. Currently, 1,500 maps are updated annually. The system has been designed to reach a production level of 3,300 maps annually, giving a 2 year cycle for depletion updates for the entire map base.

Provincial programs are producing digital data sets for agriculture, geology, soils, vegetation, water, and other resource types (Tomlinson Associates 1984). The public availability of these natural resource data vary. In some cases, such as geologic data from drill logs, the data remain confidential for a specified period of time. Unfortunately, neither the format nor the mapping accuracy have been standardized. As a result, using these data sets together can be a difficult undertaking.

Census-Related Data Sets

In Canada and the United States the agencies responsible for disseminating census data provide a number of digital data sets that can be input to a GIS. Census and other statistical data are provided in the form of attribute data sets coded by geographic location. Enumeration districts, street addresses, postal codes, census tracts and similar area codes are used. Spatial data sets are provided that can be linked to the attribute data sets by means of these area codes. Street networks in metropolitan areas, census tract boundaries, and political boundaries are examples of the spatial data sets commonly available. Typical attribute data sets are population and household characteristics, such as income, employment, age distribution, and ethno-cultural data, and dwelling characteristics like type, tenure, age, size, equipment, and facilities. (The data are provided at an appropriate level of aggregation and confidentiality procedures, such as random rounding and cell suppression, are used so that data are not traceable to the individual respondent.)

The spatial and attribute data sets are used together to produce special purpose maps and to retrieve information for selected geographic areas. They are also used for more specialized analyses, three of which are address matching, district delineation, and route selection. These functions are discussed in Chapter 7 and are briefly noted here.

Address-Matching. Address-matching is the technique of linking data from separate files by means of a common attribute, the street address. Address-matching can be used to add geographic information from the spatial data file to an attribute data file that contains street addresses. For example, welfare case records include the name and address of each recipient but may not include the census tract. The census tract information could be retrieved from the spatial data files by using the address as a key to find the data in the other file. Through an address-matching program, the census tract number could be retrieved from the spatial data file and added to the attribute file. The welfare recipient records could then be analyzed by census tract groupings (Teng 1983).

District Delineation. District delineation is a procedure that defines compact areas based on one or more attributes. For example, district-delineation can be used to divide an area into electoral districts that each have about the same population. Conceptually, this involves starting at one point and enlarging the area until it encompasses the specified number of people. Then a new district is started and the process repeated. The population information would be retrieved from the attribute data file and the information needed to define and enlarge the district boundaries would be retrieved from the spatial data file. District-delineation is used to define police and fire service districts, school districts, and commercial market areas.

Network Analysis. Network analysis is used to optimize transportation routing, such as bus routes and emergency vehicle dispatching. This procedure takes into account the length of each transportation segment and factors that affect the speed of travel or the quantity of material that can be carried. Sophisticated systems can take into account, for example, the effects of rush hour traffic, road closures, and vehicle availability in order to make the best assignment of delivery vehicles and routings.

Data Availability

In the United States, the Bureau of the Census provided street network spatial data in the form of GBF/DIME-Files. These data sets have now been replaced by the TIGER file data set.

In Canada, Statistics Canada provides in digital form the boundaries of several levels of administrative and statistical areas, including municipality, county, and census tract for the entire country. These digital data sets, called CARTLIBs, are available in several formats, some of which are topologically structured. The Area Master File (AMF) data set, also from Statistics Canada, provides digital street network data for over 300 major municipalities across the country. A range of other geographic products are available, as listed in Statistics Canada's publication *1986 Census Products and Services* (1988).

GBF/DIME-Files and TIGER Files

The United States Bureau of the Census developed a geographic coding system to automate the processing of census questionnaires. This system, called the GBF/DIME system, has been in use since 1970. The acronym stands for Geographic Base Files (GBF) created using the Dual Independent Map Encoding (DIME) system. GBF/DIME-Files are topologically structured. They were produced for some 350 major cities and suburbs across the United States and were current to 1980. The spatial data included street networks, street addresses, political boundaries, and major hydrographic features.

One of the benefits of geographic encoding was that the census data could be easily aggregated by geographic region for reporting purposes. Local governments found the GBF/DIME-Files to be a relatively inexpensive data source for their GIS, and GIS vendors provided software to read the GBF/DIME-File spatial and attribute data. Digital street maps could be produced from the data and after editing could be used as digital base maps for municipal applications.

However, GBF/DIME-Files were not designed to be used as a digital map base and have some serious limitations for this application. The data do not accurately show the shape of streets because each street segment is represented as a straight line connecting two adjacent intersections. A curved street segment thus becomes a straight line. The address range is provided for each street segment but the geographic position of each address location is not included. So the location of individual addresses has to be estimated by assuming that they are evenly distributed along the street.

In preparation for the 1990 census, the Bureau of the Census developed the TIGER (Topologically Integrated Geographic Encoding and Referencing) System to replace the GBF/DIME System. The TIGER files overcome many of the limitations of the earlier GBF/DIME-Files. The TIGER Files cover the 50 states, the District of Columbia, Puerto Rico, the Virgin Islands of the United States, and the outlying areas of the Pacific over which the United States has jurisdiction. Figure 4.6 is part of a map produced from TIGER File data. It shows the transportation and hydrology networks and census block numbers.

The TIGER File spatial data originated with either the USGS DLG-3 hydrography and transportation data sets or with the 1980 GBF/DIME-Files; these data sets were merged into a single, seamless data set that is topologically structured. Attribute data in the TIGER File include feature names, political and statistical geographic area codes (such as county, incorporated place, census tract, and block numbers), and potential address ranges and ZIP codes for that portion of the file that originated with the 1980 GBF/DIME-Files. With the release of the TIGER Files, the Census Bureau no longer supports or sells the GBF/DIME-Files.

From the decennial census (1980 and 1990), the Census Bureau distributes other attribute data, such as population and housing unit counts, income, occupation, racial distribution, and housing values. These data are also available in digital form. Not all of the attributes are available for every geographic level; for example, only the total population and housing unit counts are available at the census block level. Since all of these data are geographically referenced using the census geographic area codes contained in the TIGER File, they can be easily integrated into an existing GIS data base by file matching, using the geographic area codes as the match keys. In this way, a relatively inexpensive topologically-structured digital data set is available that can be analyzed using the full range of GIS functions.

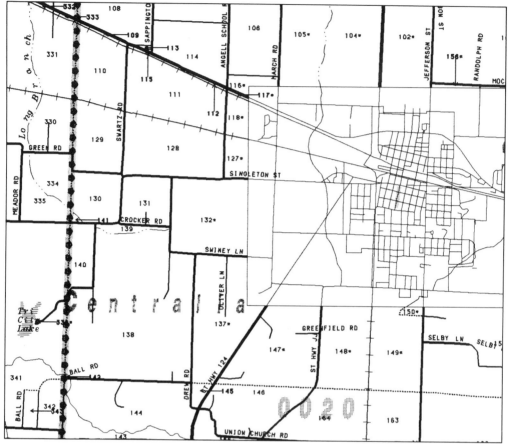

Figure 4.6   TIGER File Data. A section of the Boone County, Missouri, line data plotted at an original scale of approximately 1:45,000. The transportation and stream network and census block numbers are shown. (Courtesy of the United States Bureau of the Census. Washington, D.C.)

CARTLIB Data

Statistics Canada recently made available a digital cartographic data set called CARTLIB. These data are topologically-structured boundary files. Five data sets are available; federal electoral districts, and four levels of census reporting areas from the census divisions to the census tract. Figure 4.7 illustrates the Census Tract boundary file for the Ottawa-Hull Census Metropolitan Area.

CARTLIB data are intended for use in thematic mapping applications and for spatial analyses of census information. Census boundaries as detailed as the Census Tract level are available for all of Canada. The more detailed Enumeration Area boundary files are under development and available only for limited areas. CARTLIB files use the same geographic codes as the national census data. Integrated spatial analyses can be done using the attribute information from the national census data files and the spatial data from the CARTLIB files for the region of interest. Applications have ranged from groundwater analyses using drill hole data to market and sales information analyses for agricultural machinery. Data costs vary with coverage and level of detail and are available in several formats.

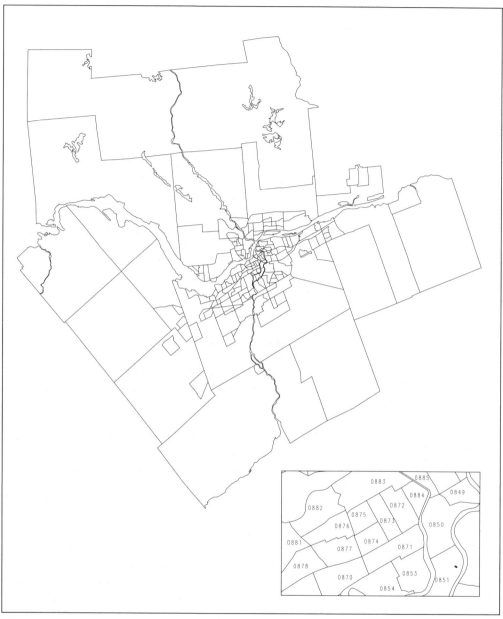

Figure 4.7 CARTLIB File of Census Tracts for the Ottawa-Hull Metropolitan Area. The insert shows census tract boundaries and identification codes for a portion of the map area at an enlarged scale. (Courtesy of Statistics Canada. Hull, Quebec.)

Area Master File Data

Statistics Canada creates and maintains digital street network files for over 300 municipalities, representing over half of the Canadian population. These data, termed Area Master Files (AMF), are used as a spatial data base to which census data can be related. City streets, railroad right-of-ways, rivers, municipal boundaries and other geographic features are represented in digital form (Figure 4.8). The data are designed to be used with census attribute data like population, household, income, and employment statistics.

The AMF data sets serve the same function in Canada as the GBF/DIME-Files and TIGER Files in the United States. However, the structure of the files is somewhat different. The AMF file is organized by the city block-face whereas the GBF/DIME-Files and TIGER Files are organized by city block. A block-face consists of one side of a street

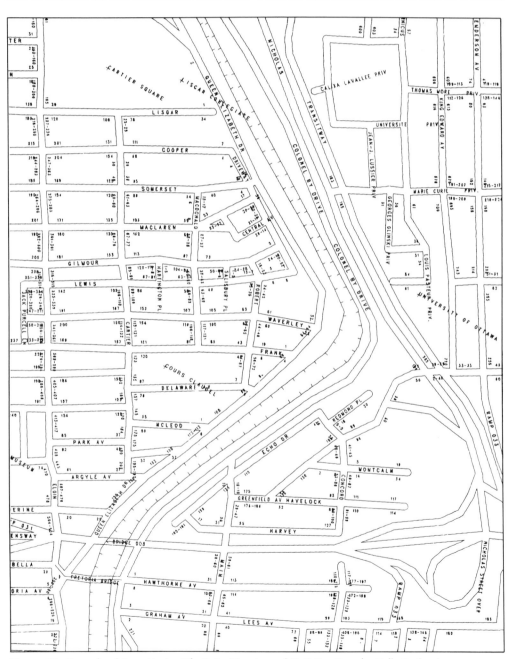

Figure 4.8 Example of Area Master File Data. (Courtesy of Statistics Canada. Hull, Quebec.)

between two successive intersections. A central point or centroid in the block-face serves as a label point to which attributes can be attached. In order to retrieve information for a specific area, the region of interest is digitized. Then the centroids falling within this region of interest are identified and used to retrieve the required census data. When the query area includes a large number of block-faces, this method of data retrieval provides a reasonable approximation (Statistics Canada 1972).

AMF data are most widely used for routing and district-delineation applications. Police and fire departments have implemented emergency vehicle dispatch systems based on Area Master Files. Municipalities, such as Burnaby B.C., Metropolitan Toronto, and Winnipeg, have used the AMF data to provide a geographically referenced street network with address ranges. In addition to dispatch applications, the retrieval of information by user-specified areas is used for planning. AMF data can be linked with postal codes, address lists, street names, and census data to provide special purpose map products and statistical summaries for specified areas. Because the same area coding is used, the spatial AMF data and census data layers like income, occupation, ethno-cultural, and housing can be used together. For example, maps of income, housing value, or population information can be produced. Similar integrated analyses are used to study purchasing power and market potential (Stanton 1987, Yan and Parker 1985).

Digital Elevation Data

Digital elevation data are a set of elevation measurements for locations distributed over the land surface. They are used to analyze the topography (i.e. the surface features) of an area. Several terms have developed that refer to digital elevation data and its derivatives. Digital Terrain Data (DTD), Digital Terrain Models (DTM), Digital Elevation Model (DEM), and Digital Terrain Elevation Data (DTED) are the more commonly used terms.

Digital elevation data are used in a wide range of engineering, planning, and military applications. They are used to calculate cut-and-fill requirements for earth works engineering, such as road construction or the area that would be flooded by a hydro-electric dam. They can be used to analyze and delineate areas that can be seen from a location in the terrain. Such intervisibility analyses can be used to plan route locations for roadways, to optimize the location of radar antennas or microwave towers, and to define viewsheds, for example. A more detailed discussion of DTM applications is presented in Chapter 7.

The methods used to capture and store elevation data can be grouped into four basic approaches: a regular grid, contours, profiles, and a Triangulated Irregular Network (TIN). These approaches are illustrated in Figure 4.9. Digital elevation data are generated from existing contour maps, by photogrammetric analysis of stereo aerial photographs, or more recently by automated analysis of stereo satellite data (as discussed in Chapter 3). Gugan and Dowman (1988) and Simard et. al. (1988) discuss the automated generation of elevation data from SPOT satellite imagery. Plates 8 and 9 illustrate the satellite imagery and the type of topographic mapping that can be produced in this way.

DTM data are most commonly provided in a grid format in which an elevation value is stored for each of a set of regularly spaced ground positions. Each data point represents the elevation of the grid cell in which it is located. One of the limitations of the raster form of representation is that the same density of elevation points is used for the entire coverage area. Ideally, the data points would be more closely spaced in complex terrain and sparsely distributed over more level areas. A number of methods have been developed to provide a variable

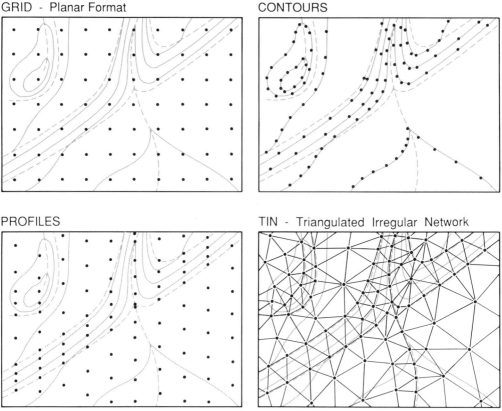

Figure 4.9   Four Basic Forms of Capturing and Storing Digital Elevation Data. The gray solid lines are contours and the dashed lines indicate distinct breaks in slope. The black points are the locations for which elevation values are recorded. (From Carter 1988a, courtesy of J.R. Carter and the American Society for Photogrammetry and Remote Sensing.)

point density. One method is to use a variable grid cell spacing to accommodate a variable density of points, with smaller cell sizes being used to capture the detail in more complex terrain.

Another approach has been to use irregularly spaced elevation points and represent the topography by a network of triangular facets. In this way, elevation data can be stored and manipulated using a vector representation. The Triangulated Irregular Network or TIN is produced from a set of irregularly spaced elevation points, as shown in Figure 4.9. A network of triangular facets is fit to these points. The coordinate positions and elevations of the three points forming the vertices of each triangular facet

are used to calculate such terrain parameters as the slope and aspect. A number of algorithms have been developed to generate the triangulated network, each of which may produce a slightly different solution for the same set of data points. Triangulations in which the triangles are most equilateral in shape tend to more accurately describe the surface. The advantage of a TIN compared with a gridded representation is that the TIN can use fewer points, capture the critical points that define discontinuities like ridge crests, and can be topologically encoded so that adjacency analyses are more easily done.

A topographic surface can also be represented by profiles showing the elevation of points along a series of parallel lines. When produced from a stereo airphoto model, elevation values should be recorded at all breaks in slope and at scattered points in level terrain. If the profiles are constructed from a topographic map, then elevation values can only be taken where the profile crosses a contour line.

The fourth approach is to digitize contour lines. Here the topographic surface is represented by series of elevation points taken along the individual contours. These may be digitized from a topographic map or produced from a stereo airphoto model. Although elevation data can be converted from one format to another, each time the data are converted some information is lost, reducing the detail of the topographic surface (Carter 1988a,b).

Availability of DTM Data

The first widely available DTM data in North America were developed for the United States by the Defense Mapping Agency. They were produced by scanning the contour overlays for 1:250,000 scale topographic maps. The data for the contour lines were interpolated to generate a regular grid of elevation values. A sampling interval of 3 arc-seconds of latitude and longitude, or approximately 90 m, between points was used. These data have an elevation accuracy of approximately 15 m RMS (root mean square error) in level terrain, 30 m RMS in moderate terrain, and 60 m RMS in steep terrain. The data are sold by the map sheet as 1° by 1° blocks and are available for the entire United States. The US Geological Survey plans to progressively upgrade the accuracy of this data set and is also producing a higher accuracy DTM file with a sampling interval of 30 m. The 30 m data are maintained in two data bases, one with a vertical accuracy of $\pm$ 7 m and the other with a vertical accuracy of $\pm$ 15 m but greater than $\pm$ 7 m. The data are available for about 30%

of the United States and are sold by the 7 ½ minute quad (the coverage of a 1:24,000 scale map sheet). Both digital elevation products are available from the National Cartographic Information Center (NCIC) of the US Geological Survey. The unit prices for these data decrease with the number of DTMs purchased. Prices for orders of six or more DTMs consist of a base charge of $90 US and $7 US for each additional unit.

In Canada, little digital elevation data are currently available. Only a few provinces, including Alberta, Nova Scotia, and Ontario, have begun production of DTM data.

DATA OUTPUT

Output is the procedure by which information from the GIS is presented in a form suitable to the user. Data are output in one of three formats: hardcopy, softcopy, or electronic. Hardcopy outputs are permanent means of display. The information is printed on paper, mylar, photographic film, or similar materials. Maps and tables are commonly output in this format. Softcopy output is the format as viewed on a computer monitor. It may be text or graphics in monochrome or colour. Softcopy outputs are used to allow operator interaction and to preview data before final output. The softcopy display is generally not used for final output because of its small size and the loss in quality when the screen image is photographed or electronically captured.

Hardcopy and softcopy output products are not just different media; they are used in fundamentally different ways. A softcopy output can be changed interactively, but the view is restricted by the size of the monitor. A larger map area can be seen but only at a coarser resolution. The hardcopy output takes longer to produce and requires more expensive equipment. However, it is a permanent record that is easily transported and displayed. A large map can be shown at whatever level of detail is required by making the physical size of the output larger. Individual map sheets can be assembled into as large a map as needed. A hardcopy output is easily written on and can be handled, folded, and viewed from various positions. For this reason, a GIS usually provides both softcopy and hardcopy output.

Output in electronic formats consists of computer-compatible files. They are used to transfer data to another computer system either for additional analysis or to produce a hardcopy output at a remote location.

HARDCOPY DEVICES

Hardcopy output hardware can be divided into two broad categories; vector and raster devices. The **pen plotter** is the most common type of vector-based equipment. It is basically a mechanical drafting device.

The paper or mylar material is laid flat on a drafting table. A penholder with one or more pens travels over the entire plotting area moving independently in the X and Y directions under control of the GIS. Separate commands move the pen down to begin drawing and lift it up at the end of a line. Some plotters handle large outputs by spooling the drawing material between two take-up reels, others use a single reel. A narrow drawing surface running the full width of the roll is used as the drawing platform. In this arrangement, the penholder need move only in the X direction since the Y movement is controlled by spooling the paper back and forth. An example of a pen plotter of this type is shown in Figure 4.10.

The software that drives the pen plotter generates a list of drawing commands from

Figure 4.10   Pen Plotter. (Courtesy Calcomp Canada Inc. Downsview, Ontario.)

the graphics data. The level of sophistication of this software can significantly improve the performance of the plotter by scheduling the drawing actions more efficiently. The level of detail that can be produced depends on the smallest step that can be registered by the motors controlling the pen position. Good cartographic work requires a minimum step size of about 0.025 mm. In addition to pens, these plotters can be equipped with scribing tools to prepare peel-coat masters for printing. Other plotter-type devices use a light beam or laser in place of the pen to draw directly onto a lithographic or diazo film (Bourrough 1986).

The first and simplest raster plotter was the **lineprinter**. It was designed to output alphanumeric characters. The image was distorted because the characters were rectangular and the raster cells they represented were square. Techniques to correct this distortion and to generate grey tone images by using overprinting were devised by ingenious programmers and frustrated users. The introduction of matrix printers (also called **dot matrix printers**) solved most of these problems by using a print head consisting of pins that can be independently fired. The matrix printer can be used to produce alphanumeric and graphics characters, square or rectangular. Matrix printers are inexpensive and versatile (they are the most common hardcopy output devices used on personal computers). They can be used to produce simple black-and-white maps easily and inexpensively.

Colour dot-matrix printers use a three-colour ribbon to generate colour output. Through over-printing, the three colour ribbon can produce six colours (i.e. six combinations of the three colours). A wider range of colours can be generated by using a cluster of dots to represent each picture element. Each dot in the cluster can still be only one of six colours, but the human eye will merge the combination of different coloured dots and perceive a single colour.

Different colours can be created by varying the combination of colours in a cluster of dots. However, the resolution of the output is then the size of the cluster of dots, not the size of a single dot.

A higher resolution and higher quality output is provided by **electrostatic plotters**, available as colour or black-and-white devices. Resolutions of 200 to 400 dots per inch are commonly available. These devices use a large number of fine needles placed in a line across the paper path. The line of needles represents a single row of the rasterized output image and each needle corresponds to a single grid cell. Those points that receive an electrostatic charge pick-up toner, thus creating an image. A multi-pass procedure is commonly used. The first pass applies black and a set of registration marks along the paper edge for correct alignment of subsequent passes. Red, green, and blue are then applied in separate passes.

Ink jet plotters generate a colour output by shooting different coloured ink drops at a paper or mylar plotting material. These plotters range in price from a few thousand dollars to tens of thousands of dollars depending on the resolution and range of colours produced. Inexpensive plotters have low resolutions on the order of 3 to 5 dots per mm. High resolution devices can have resolutions of 12 dots per mm and reproduce 4,000 colours. Earlier models were difficult to maintain because the ink jets and lines clogged easily. More recently plotters have been equipped with devices and solvents that reduce the frequency of clogging. The map shown in Plate 11 was produced on an ink jet plotter.

Thermal plotters cost about $13,000 US and are considerably less expensive than electrostatic plotters. They use a non-impact thermal transfer technology to apply yellow, magenta, and cyan colour onto paper or film output materials. Combinations of these colours produce red, green, blue, and black, a total of 7 colours for any one dot. By using clusters of dots, a wider range of colours can be produced. These devices typically have resolutions of 200 dots per inch and produce small format output (letter size to 11 in × 17 in).

Optical film writers can produce a smaller dot size and a greater range of colours for each dot. These devices use the data from an image file to modulate a light beam as it moves across a piece of black-and-white or colour photographic film. Spot sizes as fine as 12.5 microns (.0125 mm) are available that provide very high resolution photographic products. Typically, a brightness range of 256 levels is provided for each of the three primary colours, giving a theoretical range of 256 × 256 × 256 colours. The actual range of colours is considerably less as a result of limitations of the photographic film material. Photographic output devices vary widely in price from the tens of thousands to hundreds of thousands of dollars depending on the resolution and technology. They are used to generate high resolution images from such digital imagery as computer-generated graphics, satellite data, and digitized photographs. The satellite imagery shown in Plate 7 was produced using this type of device.

Often a quick hardcopy output of the monitor screen is needed to record a result or for use in presentations. **Screen copy devices** connect to the graphics display monitor and reproduce graphics and text. These devices use a variety of technologies, including impact printing, ink jet, thermal, and electrostatic plotting. They are generally less expensive ($4,000 to $7,000 US) and have lower resolution than those used for large format output (Croswell and Clark 1988).

Raster output devices can generate complex maps more rapidly than pen plotters. The output speed of a raster device is not dependent on the complexity of the image. The map material is passed through the raster plotter at the same rate no matter how complex the output graphics. A pen plotter must draw every line separately and so, the more complex the map, the longer it takes. A typical vegetation or soils thematic map sheet about 40 cm by 60 cm could take from 30 minutes to a couple of hours to produce using a pen plotter. A raster output device such as a colour electrostatic plotter would produce the same map in a matter of minutes. However, the pen plotter can produce a "drafted" line because it is in fact using a pen. It is also possible to attach scribing tools in order to plot map separates. In cases when these factors are critical, the pen plotter may be the preferred device.

SOFTCOPY DEVICES

The softcopy device most commonly used in a GIS is the television-type of computer monitor, the cathode ray tube or CRT. The image to be displayed is sent electronically from the computer to an electron gun in the CRT which directs a beam of electrons to the screen. The beam causes the phosphor-coated screen to emit light, the stronger the electron beam the brighter the light. A monochrome screen has a single electron gun and displays only one colour. A colour CRT has red, green, and blue light-emitting phosphors and three electron guns, one for each colour. By varying the intensity of the three guns, a full spectrum of colours is produced.

The light emitted from the screen fades quickly. To retain the image the screen must be refreshed, that is the electron beam must redraw the image on the screen. If the screen is not refreshed often enough the fading screen will appear to flicker. Refreshing the screen at a rate of 60 times a second is usually sufficient to maintain a clear image with no perceptible flicker. (The rate at which the screen is refreshed is termed the **scan rate**, **frame rate** or **refresh rate**.)

In a GIS, the screen image is generated using raster-scanning, a method that treats the screen as an array of discrete cells or pixels. The brightness of each pixel is individually controlled. The larger the number of pixels into which the screen is divided, the finer the detail that can be shown in the image and the higher the resolution of the display. The screen of a high resolution GIS monitor is commonly comprised of 1024 (rows) by 1280 (columns), 1024 × 1024, or 1024 × 768 pixels, i.e. on the order of 1 million pixels are used to represent the screen image. Other factors that contribute to the perceived sharpness of the monitor are the physical size and brightness of the screen pixels. A smaller screen pixel will give a sharper image. High resolution colour monitors typically can produce screen pixels as small as 0.3 mm (termed the **spot size** of the monitor).

The screen picture is generated by building the image from top to bottom, scanning each line of pixels in succession. During each scan the intensity of the beam is varied according to the brightness that each pixel is to be displayed. This scanning procedure is repeated continuously, 60 times a second. To redraw a high resolution screen 60 times a second requires a very fast data transfer rate. One method to reduce the transfer rate and the hardware cost is to redraw only half the screen every 1/60th of a second. The odd numbered scan lines are redrawn on the first pass and the even numbered lines on the second pass. In this way only half the amount of data needs to be transferred for each scan. This type of display is termed an **interlaced display**. In a **non-interlaced** display every line is redrawn on every pass. A higher scan rate and a non-interlaced display produce a brighter image which improves the apparent sharpness of the picture.

The image being displayed is stored in a computer memory, termed a **frame buffer**, as an array of numbers representing the brightness that each screen pixel is to

appear. These values are used to set the intensity of the electron beam as it scans across the screen. In the case of a colour system, the data in the frame buffer simultaneously control the intensity of the three electron guns. The displayed image is changed by writing new data to the frame buffer.

GIS display hardware vary in resolution and performance. Most systems employ a colour monitor to display graphics. Monochrome monitors are often used to display text information (like system commands) and simple graphics (such as statistical plots). In many systems a single high resolution colour monitor handles all softcopy output. The display hardware may be part of a computer terminal connected to a central computer or may be incorporated into a powerful workstation comprised of the graphics display with resident computing and data storage facilities. With the continued trend toward lower hardware prices, there is a steady increase in the use of workstations in GISes.

A detailed discussion of display technology is beyond the scope of this book. However the differences between the image processing and the computer graphics or vector graphics display approaches often causes confusion and is briefly described here. These two approaches to display design were developed to meet different sets of requirements. Their major components are illustrated in Figure 4.11. In both designs a frame buffer is used to store the image and drive the monitor, but they differ in the size and organization of the frame buffer, the method of generating the frame buffer data and the way these data are used to contol the screen image.

The image processing display was designed to handle photograph-like digital images. To produce a high quality picture from these data requires that the subtle variations in tone and colour be displayed. Adjacent screen pixels typically will be different in colour. To represent this high level of spatial variability the image data are stored in the

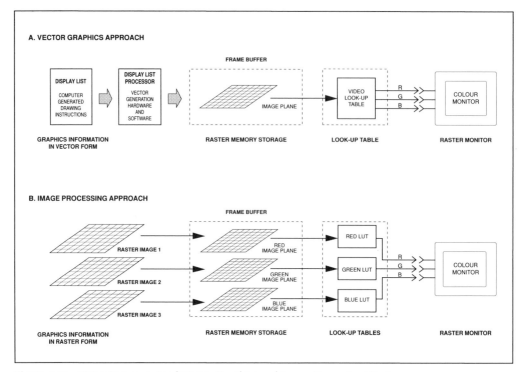

Figure 4.11   Major Components of Vector Graphics and Image Processing Displays.

computer system in a raster form, i.e. the image is divided into a large number of pixels. A colour image is stored as three separate raster data sets, one for each of the red, green, and blue colours. The images used for geographic analyses commonly have several times the number of pixels as can be stored in the frame buffer. This is easily accomodated by displaying either a portion of the image (a sub-area) or by displaying a sample of pixels from the image, e.g. selecting every 10th pixel in every 10th line produces an image with 1/100th the number of pixels.

To extract information from a digital image, such as satellite data, the image must be interpreted either visually or by auto-mated means. (Image interpretation methods are discussed in Chapter 3.) An interpreter can obtain more accurate information from an image that retains as much of the subtle variations in colour as possible. The image

processing approach excels at displaying fine gradations of colour and brightness.

In an image processing system the frame buffer is comprised of three separate sections termed **image planes**. Each image plane stores the data that controls one of the three electron guns. (A fourth image plane is often included to display graphics overlays on the image.) The image planes are normally designed with 8 bits of storage for each pixel. Eight bits represent the range 0 to 255, providing 256 different values (i.e. 2^8 values). The colour generated for each screen pixel is determined by the three values stored in the frame buffer for that pixel location, one value from each of the red, green, and blue image planes. Since each value can have 256 levels, the three colour guns can theoretically produce $256 \times 256 \times 256$ (about 16.7 million) different colours.

Look-up tables provide additional control of the colour produced for each screen pixel.

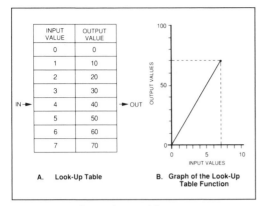

INPUT VALUE	OUTPUT VALUE
0	0
1	10
2	20
3	30
4	40
5	50
6	60
7	70

A. Look-Up Table

B. Graph of the Look-Up Table Function

Figure 4.12 The Look-Up Table Concept.

A **look-up table** or **LUT** is a set of values stored in computer memory. The LUT consists of a list of input values and corresponding output values. Figure 4.12 part A illustrates this concept. Using this simple look-up table, an input value of 4 would generate an output value of 40. As shown in Figure 4.11 part B, three LUTs are used, one for each image plane. The LUTs are positioned in the data path between the frame buffer and the monitor. In effect they process the image data ''on the fly'', as they are sent to the display.

The overall effect of a look-up table can be conveniently shown as a graph, as in Figure 4.12 part B. This LUT transforms the narrow range of input values (0 to 7) into a much wider range of output values (0 to 70). Its overall effect would be to make the image appear brighter. Look-up tables are used to alter the colour balance, contrast, and brightness of images. Mathematical functions, such as the addition of corresponding pixel values in two images, can also be implemented by means of look-up tables. Since a LUT is a very small data set, it can be loaded almost instantly and the image will be changed in the time it takes to refresh the screen. In this way LUTs can provide interactive image modification so that an image can be adjusted continuously to obtain the desired enhancement. This capability is particularly valuable to enhance satellite images for visual interpretation.

The image processing display approach is designed to handle very large raster data sets. In the same way that the red, green, and blue components of a colour image can be represented as three raster data sets for purposes of display, geographic information can also be stored and manipulated in the form of multiple raster data sets, as discussed in Chapter 6. Although image processing systems may display three raster data sets at a time, they are usually designed to perform computations simultaneously on more than three data sets and can handle any number of sequential operations. Although this discussion has focussed on the characteristics of the display, the specialized computational capabilities of an image processing system are perhaps more important in the context of a GIS. Image processing systems are specifically designed to perform calculations on large single or multiple raster data sets at very high speed. Some GIS functions, such as neighbourhood operations, are more efficiently processed in the raster domain of an image processing system than in the vector domain.

To recognize geographic features in an aerial photograph or a satellite picture it is important to retain the fine gradations of shading and colour. Since it is not known in advance which details will be needed to interpret the image, there is an advantage to retaining as much detail as possible. However in a map the individual features are already categorized. Colour, shading, line width, and other graphic details are chosen by the map designer to symbolize the geographic features. It is important to have a sufficient range of colours to represent these features, however if too many colours appear in a single map it becomes difficult to distinguish the individual categories. The fine gradations of colour that an image processing system can produce are generally not required. The display of maps and other thematic information in which the features are already categorized can be efficiently handled by a vector graphics (also termed **computer graphics**) type of display. (This assumes of course that the data are stored in vector form as well. Vector data storage is discussed in Chapter 6.)

In a vector graphics display the data used to create the image can be thought of as a set of drawing instructions, termed a **display list** or **vector list**. These instructions define where and how a feature is to be drawn. For example, the information needed to draw a straight line would consist of an operation code to indicate that a line is to be drawn, codes to define the start point and end point of the line, and codes for the line characteristics such as the line colour and thickness.

The drawing information is processed by special vector generation hardware and software, termed the **display list processor** or **graphics engine**, that generates the brightness values to produce the screen image and writes them to the frame buffer (see Figure 4.11 part A). Note that the graphics data are stored as drawing instructions but the frame buffer contains an image plane which stores the screen image in raster form and the monitor is a raster device, as in the image processing display. Unlike an image processing type of display, the vector graphics display has only one image plane. In a GIS an 8-bit image plane is normally used, so the values that represent the screen image can range from 0 to 255, i.e. 256 different values. Each number stored in the image plane represents the colour to be displayed at a specific pixel location on the monitor. Since the frame buffer can only accommodate numbers in the range 0 to 255, a maximum of 256 different colours can be produced in the screen image, more than enough to display even the most complex maps.

A special purpose look-up table, termed a **Video Look-Up Table**, generates the data to drive the colour monitor from the frame

INPUT VALUE	OUTPUT VALUE		
	RED	GREEN	BLUE
0	0	0	0
1	255	0	0
2	220	170	0
3	80	250	20
4	60	180	45
5	150	120	170
.	.	.	.

Figure 4.13 A Portion of a Video Look-Up Table.

buffer values. Figure 4.13 illustrates the concept of a video look-up table. Instead of generating one output value for each input value as in the LUTs discussed previously, the video look-up table generates three values, each in the range 0 to 255. The three values control the brightness levels of the red, green, and blue electron guns. Using the video look-up table in the Figure, an input value of 0, would produce output values of 0 for all three electron guns and the pixel would appear black on the monitor. A pixel with an input value of 1 would generate an output value of 255 for the red gun and 0 for the other two guns and the pixel would appear red. Similarly, an input value of 2 would produce a yellow screen colour.

Since each of the three output values in the video look-up table can have a range of 256 values, the combination of the three values gives a range of $256 \times 256 \times 256$ colours. However, unlike an image processing display, a vector graphics display cannot produce this very large range of colours at the same time, i.e. in a single image. This is because the numbers stored in the frame buffer can have only 256 different values. Each value can represent any one of the full range of colours, but only 256 separate colours can be specified in the video look-up table. As with other look-up tables, the colour representation can be quickly changed by modifying the values in the video look-up table.

A range of 256 colours is more than sufficient to display thematic information but is quite limiting for the display of photograph-like images. Remotely sensed data are an important source of geographic information and for many functions there is a considerable advantage to viewing and manipulating remotely sensed images in raster form and maps in vector form together. Plate 2 illustrates this type of integrated display for updating forestry maps. The forest cover map (vector format) is registered and overlayed on a Landsat TM satellite image which has been enhanced to show harvested and regenerating forest areas. The map can be updated by interactively drawing the new forest class boundaries on the screen and the updates can be directly transferred to the GIS data base.

A variety of techniques have been developed to display vector graphics on an image processing display and colour images on a vector graphics display. For example, to display a raster image on a vector graphics system, software routines can be used to transform the three raster input images into a single raster image where the pixel values are colour codes. These colour codes can then be loaded into the single image plane. The software also generates a video look-up table that converts the colour codes into appropriate screen colours. There is a considerable reduction in the possible range of colours, but images commonly contain only a fraction of this range. So by optimizing the colour selection high quality displays of photographs and satellite imagery can be achieved. However, converting three raster input images for display on a vector graphics system and then loading them into the display is considerably slower than displaying them on an image processing system.

The vector graphics approach is well-suited to the display of geographic features represented as points, lines, and areas. Maps of city streets, natural resources, administrative districts, and demographic data are familiar examples. There is no variation within the graphics elements. A red line remains the same colour red throughout its length. The image processing approach is well-suited to the display of data that represent phenomena with a high spatial variability, such as the changes in colour and brightness across a photograph. The phenomenon represented can just as easily be changes in water temperature over the surface of a lake, or incremental changes of distance (see for example the discussion of spread functions in Chapter 7).

Another way to view this comparison is that the image processing approach is designed to treat an image as a large grid of **measurements** taken at regularly spaced locations in an area. It is well-suited to situations where the information of interest is the variability in a measurement over space. Even when thematic information, such as a road, are represented the data are treated as a collection of individual pixels with the same value.

In the vector graphics approach the point, line, and area features are conceptual; they are **symbols** not measurements. The information of interest is the location of a particular type of road, not a set of measurements along the road. For this reason the road feature can be more efficiently represented as a single graphics element (Aronoff and Jones 1985).

In the past, the data in a GIS were stored either in raster or vector form and were displayed with image processing or vector graphics technology respectively. However there is considerable demand for the capability to display and manipulate raster and vector data sets together. It has been a major technical challenge to provide this type of integrated display functionality. GIS hardware and software are now available that provide some integrated raster and vector capabilities and there is rapid development of the technology to provide a more sophisticated level of integration (Aronoff et. al. 1987).

REFERENCES

Amos, L.L. and M.J. Chambers. 1988. Revising the Nation's Maps. Paper presented at the 1988 ACSM-ASPRS Annual Convention held in St. Louis, Missouri, March 14 1988.

Aronoff, S., R. Mosher, R.V. Maher. 1987. Operational Data Integration — Image Processing to Interface Vector GIS and Remotely Sensed Data. In *Proceedings of the GIS '87 Conference*. American Society for Photogrammetry and Remote Sensing. Falls Church, Virginia. pp.216–225.

Aronoff, S. and G. Jones. 1985. Using Raster Images to Integrate Data Sets. *IEEE Spectrum Magazine*, December Issue. Institute of Electrical and Electronics Engineers Inc. New York, New York. pp.45–52.

Batten, Lawrence C. 1988. Geographic Information Systems Research Utilizing 1:100,000 Digital Line Graph Data. In *Proceedings of the 1988 ACSM-ASPRS Annual Convention*. American Society of Photogrammetry and Remote Sensing. Falls Church, Virginia. Volume 5:184–189.

Blakeman, David A. 1987. Some Thoughts About GIS Data-Entry. In *Proceedings of GIS '87*. American Society of Photogrammetry and Remote Sensing. Falls Church, Virginia. pp.226–233.

Bourrough, P.A. 1986. *Principles of Geographic Information Systems for Land Resources Assessment*. Oxford University Press. New York, New York.

Carter, J. R. 1988a. Digital Representations of Topographic Surfaces. *Photogrammetric Engineering and Remote Sensing* 54(11):1577–1580.

Carter, J. R. 1988b. Digital Representations of Topographic Surfaces: An Overview. In *Proceedings of the 1988 ACSM-ASPRS Annual Convention*. American Society of Photogrammetry and Remote Sensing. Falls Church, Virginia. Volume 5:54–60.

Chrisman, N.R. 1987. Efficient Digitizing Through the Combination of Appropriate Hardware and Software for Error Detection and Editing. *International Journal of Geographical Information Systems* 1(3):265–277.

Croswell, P.L. and S.R. Clark. 1988. Trends in Automated Mapping and Geographic Information System Hardware. *Photogrammetric Engineering and Remote Sensing* 54(11):1571–1576.

Dangermond, J. 1988. A Review of Digital Data Commonly Available and some of the Practical Problems of Entering Them into a GIS. In *Proceedings of the 1988 ACSM-ASPRS Annual Convention*. American Society of Photogrammetry and Remote Sensing. Falls Church, Virginia. Volume 5:1–10.

Dulaney, R.A. 1987. A Geographic Information System for Large Area Analysis. In *Proceedings of GIS '87*. American Society of Photogrammetry and Remote Sensing. Falls Church, Virginia. pp.206–215.

Elassal, A.A. and V.M. Caruso. 1985. *Digital Elevation Models*. Geological Survey Circular 895-B. U.S. Geological Survey. Reston, Virginia.

Gugan, D.J. and I.J. Dowman. 1988. Topographic Mapping from SPOT Imagery. *Photogrammetric Engineering and Remote Sensing* 54(10):1409–1414.

Niemann Jr., B.J. and J.G. Sullivan. 1987. Results of the Dane County Land Records Project: Implications for Conservation Planning. In *Proceedings of the Eighth International Symposium on Computer-Assisted Cartography*. American Society of Photogrammetry and Remote Sensing. Falls Church, Virginia. pp.445–455.

NCIC. 1985. *Digital Cartographic and Geographic Data Products*. Bulletin produced by the National Cartographic Information Center of the US Geological Survey. Reston, Virginia.

Rhind, David and Stan Openshaw. 1987. The BBC Doomsday System: A Nation-Wide GIS for $4448. In *Proceedings of the Eighth International Symposium on Computer-Assisted Cartography*. American Society of Photogrammetry and Remote Sensing. Falls Church, Virginia. pp.595–603.

Rodrigue, M. and L. Thompson. 1983. The Availability and Use of Digital Topographic Data. In *Proceedings of the Sixth International Symposium on Automated Cartography*. University of Ottawa. Ottawa, Ontario. pp.580–587.

Simard, R., G. Rochon, and A. Leclerc. 1988. Mapping with SPOT Imagery and Integrated Data Sets. Invited paper presented at the 16th Congress of the International Society for Photogrammetry and Remote Sensing held July 1988 in Kyoto, Japan.

Stanton, P. 1987. *The Area Master File — Its Applications*. Unpublished Report. Geography Division. Statistics Canada. Ottawa, Ontario.

Statistics Canada. 1988. *1986 Census Products and Services*. Catalogue No. 99-103. Statistics Canada. Ottawa, Ontario.

Statistics Canada. 1972. *GRDSR: Facts by Small Areas*. Statistics Canada. Ottawa, Ontario.

Teng, Apollo T. 1985. Interfacing with GBF/DIME and AMF: A Comparison. In *Proceedings of the URISA '85 Conference*. Urban and Regional Information Systems Association. Washington, D.C. Volume 2:179–192.

Teng, Apollo T. 1983. Cartographic and Attribute Data Base Creation for Planning Analysis Through GBF/DIME and Census Data Processing. In *Proceedings of the Sixth International Symposium on Automated Cartography*. University of Ottawa. Ottawa, Ontario. pp.348–354.

Tomlinson Associates. 1984. Investigation of Digital Cartographic Status and Developments in Canada — Final Report (6 Volumes). Ottawa, Ontario.

Yan, J.Z. and J.P. Parker. 1985. A Framework for Coordinating the Development and Application of Street Network Files for Canada. In *Proceedings of the URISA '85 Conference*. Urban and Regional Information Systems Association. Washington, D.C. Volume 2: 132–143.

5. DATA QUALITY

INTRODUCTION

People routinely make judgments about data quality. A dentist friend explained to me that he would not automate his office records because he would not be able to recognize from the computer print-outs who had entered the information. He knew that some assistants were more perceptive and more accurate than others. From the handwriting, he could identify which assistant had recorded the data and he interpreted their entries accordingly. He was judging the quality of the data by the assistant who entered them. The quality of the data was encoded in the handwriting! Similarly, hikers learn from experience that on topographic maps the position of trails are less accurately shown than the position of roads. Their judgment of the relative quality of the trail and road information guides their use of the map data.

Knowing the quality of data is critical to judging the applications for which they are appropriate. When spatial analyses are done manually using map overlays, users quickly learn to shift the map slightly to align boundaries that should overlap. A map overlay may not be precisely registered but with these manual adjustments it can be shifted so that any local area can be registered closely enough for the work at hand. This type of informal assessment of data quality and allowance for inaccuracies break down when an automated GIS is used. Implicit assumptions about data quality must be made explicit so that they can be properly addressed. In a computer-based GIS, roads either meet or do not meet. The computer must be programmed to treat a line ending 1 mm short of the road as connected. Misalignment caused by positional error is one of several data quality issues to be taken into account in using and maintaining GIS data.

The cost of assessing data quality varies with the degree of rigour needed. The more rigorous the data quality testing, the more costly it becomes. This cost is not only a result of the expense of performing the test, but also of the delays caused in the production process to perform the tests and correct errors. For this reason, the level of testing should be balanced against the cost of the consequences of less accurate data or a less rigorously confirmed level of quality. Demanding higher levels of data quality than are actually needed quickly becomes a significant unnecessary expense when it is applied to an entire GIS data base. To illustrate the relationship between data quality assessment and consequences, consider the quality assessments provided for general purpose twine and climbing rope. People use general purpose twine for household needs without knowing about its tensile strength, deterioration over time, or other specifications. The consequences of misjudging the twine's strength are generally not serious, and when people are in doubt they use more twine. But a climber's life may depend on the properties of his safety rope. Climbing ropes undergo rigorous quality testing and are provided with detailed quality test results. Because the consequences of failure are serious, the need for a rigorous test of the rope's characteristics is considered to be worth the expense of testing its quality. In a similar way, the expense of testing and recording the quality of the data in a GIS should be matched to the consequences of its inappropriate use.

The data in a GIS may be used for a wider range of analyses than when the same data were in a non-digital form. Indeed this is one of the advantages cited for using a GIS, the capability to integrate diverse data sets that previously could not be analysed together. However, the data may be used in ways not foreseen by their producer and by users without the knowledge or experience to judge whether the application is appropriate.

An irate landowner in Wisconsin successfully sued the state for inappropriately showing the highwater mark around a lakeshore on a standard topographic map. In this case (discussed in Chapter 8), the user did not realize that this type of topographic map was not sufficiently accurate to show land parcel boundaries in the context of the elevation data. As a consequence, it appeared that a portion of the owner's land was below the highwater mark. According to the laws of the state, land below the highwater mark was the property of the state. Though the error was subsequently corrected, the owner successfully sued for damages because in the interim a reasonable interpretation of the map would have caused her title to the land to be in doubt (Epstein and Roitman 1987).

This is an example where a map of known quality, a US Geological Survey topographic map, was used to present data (the highwater level) of unknown quality. An incorrect assumption had been made about the quality of the combined dataset (the topographic map with the hand-drawn highwater mark). It had incorrectly been assumed to be as accurate as the topographic map itself. Producers of geographic information are increasingly being held responsible when reasonable reliance on their data results in financial losses or personal injury. In another case, the U.S. federal government was held responsible for inaccurately and negligently showing the location of a broadcasting tower on an aeronautical chart. This was shown to have been a contributing factor in a fatal plane crash (Epstein and Roitman 1987).

The quality of geographic data is often examined only after incorrect decisions have been made and financial losses or personal injury have occurred. Increasingly, producers of geographic information are being held liable when their products are found to contain errors, are poorly designed, or are used in ways and for purposes unintended by their designers (Epstein 1987a). Data quality standards, appropriatley defined, tested, and reported, can protect both the producer and user of geographic information. When data are provided in a standard format and at a defined and accepted level of quality, the producer is protected from liability in the case of inappropriate use. Such standards also protect the user from relying on inappropriate information.

A GIS provides the means for geographic information to be used for a broader range of applications and by users with a wider range of skill than ever before. In order for these data to be used in decision-making, their quality must be predictable and known.

A major contribution toward standardizing the definition, assessment, and reporting of GIS data quality has been made by the Data Set Quality Working Group of the National Committee for Digital Cartographic Data Standards. A summary of the data quality standards developed by this working group, as well as work on other aspects of cartographic data standards, is presented in the January 1988 issue of *The American Cartographer*.

This chapter provides an overview of the data quality factors that should be considered when using geographic information. The importance of each factor and the level of detail needed to report them will vary depending on the context in which the GIS is used. The standards of quality required of the data in a GIS data base and the methods used to measure quality must be explicitly defined before data entry begins. Ultimately, these data quality standards must serve the needs of the users, so the user community must be directly involved in specifying the data quality standards for the GIS data base and in dealing with practical constraints like budget, technical capabilities, and rate of production.

COMPONENTS OF DATA QUALITY

The characteristics that affect the usefulness of data can be divided into 9 components. These have been grouped into 3 categories: micro level components, macro level components, and usage components.

MICRO LEVEL COMPONENTS

Micro level components are data quality factors that pertain to the individual data elements. These components are usually evaluated by statistical testing of the data product against an independent source of higher quality information. They include positional accuracy, attribute accuracy, logical consistency, and resolution.

Positional Accuracy

Positional accuracy is the expected deviance in the geographic location of an object in the data set (e.g. on a map) from its true ground position. It is usually tested by selecting a specified sample of points in a prescribed manner and comparing the position coordinates with an independent and more accurate source of information. There are two components to positional accuracy: the bias and the precision. The bias refers to systematic discrepencies between the represented and true position. Ideally the bias should be zero, indicating no systematic tendency for the map position to differ from the true position. Bias is commonly measured by the mean or average positional error of the sample points.

Precision refers to the dispersion of the positional errors of the data elements. Precision is commonly estimated by calculating the standard deviation of the selected test points. A low standard deviation indicates that the dispersion of the positional errors is narrow, i.e. the error tends to be relatively small. The higher the precision of the measurement, the greater the confidence in

using the data. A measure of positional accuracy commonly used in surveying and photogrammetry is the **root mean square error** (RMS). It is calculated by determining the positional error of the test points, squaring the individual deviations and taking the square root of their sum. This measure does not distinguish between the bias and precision components of accuracy. There are many applications where the data can be more effectively used if this information is known. For example, a high precision but biased data set may be better used by subtracting out the bias. A discussion of the methods used to calculate positional accuracy can be found in surveying or photogrammetry texts. A concise review of positional accuracy standards for large scale topographic mapping can be found in Merchant (1987).

Attribute Accuracy

Attributes may be discrete or continuous variables. A discrete variable can take on only a finite number of values whereas a continuous variable can take on any number of values. Categories like land use class, vegetation type, or administrative area are discrete variables. Ratings are also discrete variables. They are, in effect, ordered categories where the order indicates the hierarchy of the attribute. For example, the severity of soil erosion might be ranked on a 4 point scale from 1 indicating low soil erosion levels to 4 indicating severe soil erosion. Tree heights might be given in height classes on a scale of 1 to 5 with 1 indicating a height less than 1 m and 5 indicating heights greater than 20 m. However, the range of conditions in category 1 is not necessarily equal to the range of conditions in category 2. Category 2 may be a height range of 1 m to 3 m. In some cases, ranges that don't occur might not even be given a rank at all. So a rank of 2.5 would not be valid. Variables like temperature, or average property value are continuous, the variable can take on any value so intermediate values are valid.

The method of assessing accuracy for continuous variables is similar to that discussed for positional accuracy. The assessment of the accuracy of discrete variables is the domain of classification accuracy assessment. The assessment of classification accuracy is a complex and somewhat controversial procedure. It has received considerable attention in the remote sensing literature. The difficulties in assessing classification accuracy arise because accuracy measurement is significantly affected by such factors as the number of classes, the shape and size of individual areas, the way test points are selected, and the classes that are confused with each other.

For example, wetlands along streams are typically long narrow areas. Though they are often important for planning purposes, these areas commonly make up less than 1% of the total map area. In a randomly selected sample of test points, these areas would probably not be chosen. A classification accuracy could be calculated from these test points, but if no wetland points were tested it would provide little information about the map accuracy of the wetland class.

Even if test points are selected separately for each class, thereby ensuring that the wetland areas would be assessed, the precision needed to accurately locate the test points in the field may exceed the positional accuracy of the data being tested. That is, the boundary may be mapped to within 10 m of the true position, but the wetland area may itself be less than 10 m across. It might not even be possible to locate the position in the field accurately enough to be sure that the correct spot is being sampled. A further difficulty is that sharp boundaries often do not actually exist even though they are mapped with clean sharp lines. A forest edge or wetland edge is usually a zone several meters or tens of meters in width.

As a result of factors such as these, the assessment of classification accuracy is not entirely objective. Despite these difficulties,

a number of useful approaches to classification accuracy have been developed. An appropriate, well-accepted, and well-documented accuracy test should be selected. The test results, test data, and the assessment method should all be reported. By reporting the test data used in the classification assessment, the user can re-interpret the test information for a specific application. Recent discussions of classification accuracy can be found in Aronoff (1982 a,b), Congalton and Mead (1983), Story and Congalton (1986), and Walsh et al. (1987).

Logical Consistency

Logical consistency refers to how well logical relations among data elements are maintained. For example, it would not be consistent to map some forest stand boundaries to the center of adjacent roads, and some to the road edge. They are normally all mapped to the road edge. Political and administrative boundaries defined by physical features should precisely overlay those features. The edge of a property that borders a lake should coincide with the lake boundary.

An unusual problem is encountered when mapping areas with reservoirs. The water level in a reservoir will fluctuate over the year. Different GIS data layers may show the reservoir boundary at different locations, depending on the date of the mapping. As a result, the reservoir boundaries may be accurately delineated, but logically inconsistent among data layers. In this case, the problem was solved by providing a standard outline for each reservoir. The representation of the reservoir on each data layer was then made to conform to the standard outline.

It is important to realize that two data sets may be correct to their specified level of positional accuracy and yet not be logically consistent. This is because the same boundary can be mapped in slightly different positions in two data sets, yet still meet the required level of positional accuracy. When the data sets are overlayed, this slight discrepency in position will create a thin unique area, termed a **sliver**, in the region between the two boundaries (as discussed in Chapter 7). Some GIS software are able to accommodate this type of positional discrepancy by assigning a band of uncertainty around a feature. Then two features that have overlapping bands of uncertainty would be treated as if they were coincident. (A boundary that is treated as a band of uncertainty is often termed a **fuzzy boundary**.)

There is no standard measure of logical consistency. Discrepencies in the position of coincident features can be measured. However, there may be many combinations of elements that should have a logically consistent relationship. Measuring the discrepancies of all possible overlay combinations would not be practical.

Logical consistency is best addressed before data are entered in the GIS data base. A map preparation stage is commonly used during which individual maps that are to be digitized are checked and, if necessary, re-drafted to correct errors and reconcile discrepancies like logical inconsistencies. (This process of reconciling data layers has been termed **conflation**. It is discussed in Chapter 7.)

Resolution

The resolution of a data set is the smallest discernable unit or the smallest unit represented. In the case of images, such as airphotos or satellite imagery, resolution refers to the smallest object that can be discerned, also termed **spatial resolution**. For camera systems this is usually reported in lines/mm. A resolution of 80 lines/mm is typical for aerial mapping camera systems. For digital scanning systems, such as satellite sensors, the spatial resolution is defined as the size of the picture elements or pixels of which the image is composed. A pixel is the area of the earth's surface represented by a single digital image value. A more detailed discussion of spatial resolution in imagery can be found in Chapter 3.

For thematic maps, such as soil maps, land use maps, and other categorized data, the resolution is the size of the smallest object that is represented, termed the **minimum mapping unit**. The decision of how small an object to include in a map is made during the map compilation process. Factors like the expected use of the map, legibility, source data accuracy, and drafting expense are considered in selecting the minimum mapping unit. The selection of the map resolution is fundamentally a judgment call in which both the presentation of the information and the storage of that information must be addressed.

In a GIS, the presentation of the information and the storage of the data are separate. The geographic data stored in the GIS data base can be presented at any scale. Labelling and other map details can be added as needed to suit the scale of the output. In this sense, the geographic data in a GIS do not really exist at a specific scale, and so the minimum mapping unit can be set very small, even for large coverage areas. The amount of detail to be shown on a map output can be selected when specifying the output. For a map of a large coverage area, the plotting of small features can be suppressed.

The ease with which the geographic data in a GIS can be used at any scale highlights the importance of accurate data quality information. Although the data do not have a specific scale, they were produced with levels of accuracy and resolution that make it appropriate to use them at only certain scales. Using a GIS, a 1:50,000 scale map for example, could be produced from data that were digitized from a 1:500,000 scale geology map. However, the map would not have the quality of geological mapping originally done at that scale. So the map should be used as if the data had the quality of the 1:500,000 scale map from which the data were taken, not the 1:50,000 scale at which it was plotted.

MACRO LEVEL COMPONENTS

Macro level components of data quality pertain to the data set as a whole. They are not generally amenable to testing but instead are evaluated by judgment (in the case of completeness) or by reporting information about the data, such as the acquisition date. Three macro level components are discussed: completeness, time, and lineage.

Completeness

There are several aspects to completeness as it pertains to data quality. They are grouped here into three categories: completeness of coverage, classification, and verification.

The completeness of coverage is the proportion of data available for the area of interest. A data set may not provide complete areal coverage of the area of interest or attribute data may not be available for some portion of the data set. Ideally, a data set would provide 100% coverage. However, many data sets are progressively updated. Agencies responsible for large mapping projects regularly publish a map index showing those maps currently available. When information is needed about the current status of a resource, the most current information may be the most suitable. In other cases, such as comparative analyses, it may be more important to have consistency within the data set. An older data set for the entire study area may be more appropriate than a patchwork of more recent data collected in different years.

The completeness of classification and the completeness of verification are important data quality factors that can determine the suitability of a data set for an application. These more subtle aspects of data completeness are more difficult to assess. Whereas percentage coverage can be measured, these more subtle characteristics are judged qualitatively and reported as a description rather than as a numerical value. It should not be considered that these necessarily qualitative assessments are any less important than the quantitative ones. In fact, the opposite is usually the case.

Completeness of classification is an assessment of how well the chosen classification is able to represent the data. The completeness of a classification may be evaluated with reference to a standard classification or on its own merits with reference to specific applications. Table 5.1 is an example of a classification that exhibits several types of incompleteness. For a classification to be complete it should be exhaustive, that is it should be possible to encode all data at the selected level of detail. In this example, the subdivisions of the *livestock* category are not exhaustive. If the *livestock* category *horses* occurs, it cannot be encoded at this level. It would have to be encoded at the second level as *livestock*.

The subdivisions of the *truck crop* category are exhaustive in that there is an appropriate category for any possible occurrence. However it may lack completeness of detail for a specific use. *Potato* would fall into the *Other* category. So, an estimate of potato crop area could not be obtained if the data had been encoded using this classification.

A more subtle completeness problem can arise when the definitions of the categories are overlapping, i.e. an observation could be assigned to more than one class. The way classes are defined will affect how consistently and accurately the features can be assigned to classes. An inappropriate classification can significantly bias the data. In Table 5.1, the *Forest* class has been subdivided into three classes. It is often difficult to define these forest classes precisely

Table 5.1 Sample Classification to Illustrate Concepts of Completeness.

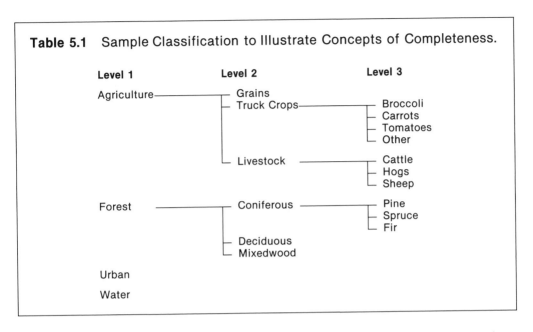

because there is usually a continuous gradation of forest types from purely deciduous through various mixtures of coniferous and deciduous species to pure coniferous stands. Depending on the definitions used, the reported area of each forest type will vary significantly. Also, distinctions among the classes are difficult to assess close to the cut-off point. The difference between 20% coniferous and 25% coniferous is difficult to determine visually on medium scale aerial photographs. This will tend to make the classification inconsistent.

Class definitions may also differ among map sheets as a result of the individual or the organization that produced them. Adjacent forest districts may use slightly different class definitions. As a result the maps may be accurate in terms of position and classification, but the boundaries from adjacent maps may not match if they were produced by different forest districts. By knowing the way the classes have been defined, the user can assess whether the classes are appropriate for the application at hand. Unfortunately, the information needed to assess the classification is usually buried in lengthy and complex internal documents, such as procedure manuals, that are not easily obtained or assessed by the user of the data.

Completeness of verification refers to the amount and distribution of field measurements or other independent sources of information that were used to develop the data. Geologists indicate this aspect of data quality by using solid lines to map rock types for which they have direct field evidence, such as boundaries they could actually see. Boundaries that were inferred but could not be verified are shown as dashed or dotted lines. This is the convention used in geology, but there is no standard method to report completeness of verificiation in GIS. Data sets are usually provided without this information and so the user would not know that different boundaries or classes were verified to different degrees. Completeness of verification may be indicated within the data set as an attribute of the geographic features. It may also be reported in the form of a separate map showing the location and type of verification data collected.

The assessment of completeness has usually been limited to reporting coverage.

Reporting of qualitative assessments of completeness, such as classification and verification, have been largely ignored. However, they are aspects of data quality that can be critical to the appropriate use of the data.

Time

Time is a critical factor in using many types of geographic information. Demographic information is usually very time sensitive. It can change significantly over a year. Land cover will change quickly in an area of rapid urbanization. Some data will be significantly biased depending on the time period over which they are collected. For example, in areas that produce multiple crops per year, the crop types grown in an area change with the seasons. The time period over which the land cover is mapped will then significantly affect the information obtained about crop areas and distribution.

The time aspect of data quality is most commonly reported as the date of the source material, such as the date of aerial photography. Topographic maps usually include updating information, and show updated features like roads and urban areas in a separate colour. On the new map the date of the original aerial photography and the date of the update information are both noted. Similar reporting should be included in the GIS data base.

For geographic information that changes relatively quickly over time, the date of acquisition may be a very important attribute. For example, a municipal GIS is usually organized so that the day-to-day transactions are used to update the data base directly. In this way the procedure used to construct a new road or register a change of land ownership includes transactions that directly update the data base. In natural resource applications, such as forestry and agriculture, geographic data are updated periodically. Forest inventory maps may be updated on a 5 to 10 year basis. Crop condition changes rapidly over the growing season and is commonly updated on a weekly basis.

Time is a frequently overlooked consideration when multiple data sets, collected independently, are used together. The boundaries of a floodplain taken from a land cover map may have been mapped much more recently than the topographic map from which the river boundaries were obtained. When the independent mappings are combined, the position of the river may be inconsistent with its floodplain. The position of the river may have changed significantly (see Logical Consistency discussed previously).

Lineage

The lineage of a data set is its history, the source data and processing steps used to produce it. The source data may include transaction records, field notes, airphotos, and other maps. The procedures may vary from sketch mapping to stereocompilation using photogrammetric instruments of high precision. A lineage report documents this information. For example, the lineage report for a topographic map would include the date of the aerial photography used, the photogrammetric methods used to map the contour lines and cultural features from the airphotos, the use of check points for photogrammetric control, and the methods used to generate the final map.

Each data source and processing method introduces a level of error into the information that is ultimately produced. In some cases, a knowledge of the lineage can be an important consideration in choosing a data set for a specific application. Lineage information, if it exists, is commonly in the form of procedure manuals or other internal documents but is not usually provided with the data set. Ideally, some indication of lineage should be included with the data set since the internal documents are rarely available and usually require considerable expertise to evaluate. Unfortunately, lineage information most often exists as the personal experience of a few staff members and is not readily available to most users.

USAGE COMPONENTS

The usage components of data quality are specific to the resources of the organization. The effect of data cost, for example, depends on the financial resources of the organization. A given data set may be too expensive for one organization and be considered inexpensive by another. Aronoff (1985) discusses how the same satellite data were inexpensive for an oil company to use for exploration but too expensive for a wildlife agency to use for habitat mapping. Another usage factor is accessibility. The accessibility of the data depends on imposed usage restrictions (e.g. the data may be restricted) and the human and computer resources of the organization.

Accessibility

Accessibility refers to the ease of obtaining and using data. The accessibility of a data set may be restricted because the data are privately held. Access to government-held information may be restricted for reasons of national security or to protect citizen rights. Census data are usually restricted in this way. Even when the right to use restricted data can be obtained, the time and effort needed to actually receive the information may reduce its overall suitability.

Direct and Indirect Costs

The direct cost of a data set purchased from another organization is usually well-known; it is the price paid for the data. However, when the data are generated within the organization, the true cost may be unknown. Assessing the true cost of these data is usually difficult or impossible because the services and equipment used in their production support other activities as well.

The indirect costs include all the time and materials used to make use of the data. When data are purchased from another organization, the indirect costs may actually be more significant than the direct ones. It may take longer for staff to handle data with which they are unfamiliar, or the data may not be compatible with the other data sets to be used. For example, the data may be in a non-digital format or in a digital format that cannot be directly input to the GIS on which it is to be used. Converting the data to a compatible format may simply involve running an existing conversion program. It might also be prohibitively expensive if non-digital data must be digitized or if special conversion software must be written. In these cases, the human and technical resources of the organization may largely determine whether the data are usable and how expensive it will be to handle the conversion.

SOURCES OF ERROR

There is error associated with all geographic information. Error is introduced at every step in the process of generating and using geographic information, from collection of the source data to the interpretation of the results of a completed analysis. The following discussion reviews the major types of errors that are introduced at each stage of geographic information processing. Some of the more common errors are listed in Table 5.2. **The objective in dealing with error should not be to eliminate it but to manage it.** Achieving the lowest possible level of error may not be the most cost-effective approach. There is a trade-off between reducing the level of error in the data base and the cost to create and maintain the data base. The level of error in a GIS needs to be managed so that data errors will not invalidate the information that the system is used to provide.

Table 5.2.   Common Sources of Error Encountered in Using a GIS.

Stage	Sources of Error
Data Collection	errors in field data collection errors in existing maps used as source data errors in the analysis of remotely sensed data
Data Input	inaccuracies in digitizing caused by operator and equipment inaccuracies inherent in the geographic feature (e.g. edges, such as forest edges, that do not occur as sharp boundaries)
Data Storage	insufficient numerical precision insufficient spatial precision
Data Manipulation	inappropriate class intervals boundary errors error propogation as multiple overlays are combined slivers caused by problems in polygon overlay procedures
Data Output	scaling inaccuracies error caused by inaccuracy of the output device error caused by instability of the medium
Use of Results	the information may be incorrectly understood the information may be inappropriately used

Data Collection Errors. Error exists in the original source materials that are entered into the GIS. These errors may be a result of inaccuracies in field measurements, inaccurate equipment, or incorrect recording procedures. Much of the data input to a GIS are generated using remote sensing techniques. There are inaccuracies in the photogrammetric methods used to draw maps and measure elevations. Airphoto or satellite image interpretations introduce a degree of error in the classification and in the delineation of boundaries.

Data Input. The data input devices used to enter geographic data all introduce positional errors. For example, digitizing tables are commonly accurate to fractions of a millimetre, but the accuracy varies over the digitizing surface. The center of a digitizing table commonly has a higher positional accuracy than the edges. The operator introduces error in the way the map is registered on the digitizing table, the boundaries are traced, and the accuracy with which the attributes and label information are entered. Error is introduced in the way spatial information is represented. Curved boundaries are approximated by a series of straightline segments. The smaller the segments used, the more closely the boundary is approximated. However, the smaller the line segments the more data points are generated and, consequently, the larger the data files produced. No matter how carefully boundaries and points are entered, some residual error will always remain. Errors in the position of natural boundaries are often introduced because the boundary does not in fact exist as a sharp line. A forest edge, though drawn as a definite line, usually exists as a zone that may be several meters or tens of meters wide.

Data Storage. When data are stored in digital form, they must be stored with a finite level of precision. A commonly used storage form in a vector-based GIS is the 32-bit real number format. This provides at most seven significant digits. All these digits may be needed. For example, UTM geographic coordinates use seven significant digits. A GIS data base that contains information with levels of detail ranging from fractions of a meter to full UTM coordinates would require greater precision. To retain the accuracy of such diverse data, more than seven significant digits would be needed. One solution is to store all values in double precision format, i.e. use 64 bits instead of 32 bits for each value. However, using double precision values increases the volume of data to be stored.

This type of limitation can be much more critical when the data are stored in raster form. Each value in a raster file represents a unit of the terrain, termed a pixel. The pixel size selected determines the positional accuracy with which the data will be stored. If the data are encoded using a pixel size of 10 m by 10 m, then even if the geographic position of a point is exactly known to a fraction of a meter it can only be represented to the nearest 10 m. There is a loss of accuracy here because the terrain has been regularly subdivided into discrete square areas, the pixels. In theory, higher levels of spatial resolution could be obtained by simply defining a larger number of pixels. In practise, the limitations of the computer hardware are quickly exceeded because the file size increases by the square of the resolution. Thus, increasing the resolution 10 times from 10 m pixels to 1 m pixels would increase the file size by about 100 times.

For both vector and raster-based systems, in order to obtain higher levels of precision there is a direct cost as a result of the increased storage and the decreased performance. As discussed in Chapter 6, the vector data model is better suited to storing high precision coordinates for discrete map elements. The raster data model tends to be better suited for representing measurements that vary continuously over an area.

Data Manipulation. Many GIS analysis procedures involve the combining of multiple overlays. As the number of overlays used in an analysis increase, the number of possible opportunities for error increase. The highest accuracy possible will not be better than the least accurate input overlay. Many manipulation errors arise from the representation of boundaries. As noted previously, the same boundary may be drawn slightly differently in two overlays. The more complex the shape of the boundary, the more of a problem this becomes. In an overlay operation, this mismatch will create inaccuracies in the results (Newcomer and Szajgin 1984).

There is also a level of inaccuracy inherent in the way classes are defined. Many continuous phenomena, such as vegetation and soils, are mapped as homogeneous map units with sharp boundaries. These maps are termed **choropleth** or **thematic** maps. However, in reality there is variability within each map unit. The map polygon labelled as a Pine stand may very well contain other species of trees in small numbers. When the data are compiled a decision is made that areas below a certain size (termed the **minimum mapping unit**) will not be recognized within an otherwise homogeneous map unit. While this may be quite acceptable for the applications envisioned by those creating the map, it may be unacceptable when applied to analyses they did not foresee. (See *classification accuracy* discussed previously.)

A soil materials map may show an area to be sandy soil. In an application like forestry, the presence of 15% clay soils in this map unit would not normally restrict the use of these data. However, in siting housing, the presence of clay inclusions is important. A house sited partly on clay and partly on sandy soil will tend to settle unevenly, cracking the walls and foundation. To avoid these settlement problems, an intensive soil survey would be needed. By understanding the accuracy of the class designations the need for a more detailed survey could be anticipated.

Data Output. At the data output stage, error can be introduced in the plotting of maps by the output device and by the shrinkage and swelling of the map material. As paper shrinks and swells, measurements taken from the map will be changed. On a small scale map, the millimeter changes can represent several meters at the ground resolution.

Use of Results. Error is also introduced when the reports generated by a GIS are incorrectly used. Results may be misinterpreted, accuracy levels ignored, and inappropriate analyses accepted. This source of error would seem to be independent of the GIS. But, in fact, the resulting errors in decision-making represent errors in the process of using geographic information. Unless critically evaluated, these errors are often attributed to the GIS facility.

A NOTE ABOUT ACCURACY

Accuracy is a term that is widely used and commonly misunderstood. **Accuracy** is the likelihood that a prediction will be correct. In the case of a map, the positional accuracy is the likelihood that the position of a point as determined from the map will be the "true" position, i.e. the position determined by more accurate information, such as by field survey. Classification accuracy is the probability that the class assigned to a location on the map is the class that would be found at that location in the field.

No map can be 100% accurate. There will always be some level of error. The degree of accuracy is often stated as a range or a value. A distance is measured as 150 m "plus or minus" 0.5 m or the classification accuracy is stated to be 90% correct. However, these statements are incomplete in an important way; they have not included the *probability* on which the accuracy assessment is based.

AN EXAMPLE

To illustrate the importance of probability in accuracy assessment, an example is presented in Figure 5.1. For simplicity, only four points are considered, although in practise tens of points would be used. Also for simplicity, the absolute distances between the mapped and true positions are used instead of separately evaluating the error in the X and Y directions and distinguishing positive and negative errors.

In Figure 5.1 part A the true coordinate positions of four points and their positions on the map are shown. The small differences between the true locations and map locations are the **positional errors** of the points. (Positional errors have been exaggerated in the illustration for clarity.) The accompanying table lists each point and its positional error (the distance between its true and map position).

The average error is 1.46. This is a measure of the errors for these four points. But what if other points are selected from the map? What would the positional error of these points be? A first guess might be that their error would be 1.46, on average. But what we really want to know is the likely error for an individual point, not an average. Looking at the four sample points we see that the positional errors ranged from 1.00 to 2.00. But there are only four test points for the entire map. Perhaps the test points happened to be relatively accurate ones because they were taken from an area that had been more accurately mapped. We would probably feel more confident in the error estimate if more points had been selected and the variation in error among the points was relatively small. A useful measure of the variability of error values is the standard deviation. The standard deviation value decreases as the errors for each point become smaller. It also decreases as the sample size increases. This property of the standard deviation incorporates our

A. TEST OF POSITIONAL ACCURACY

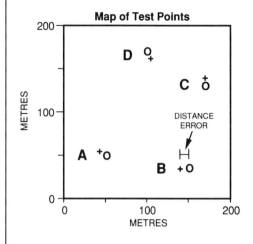

Map of Test Points

O - True Position
+ - Map Position

Positional Accuracy of Test Points

POINT	TRUE X	TRUE Y	MAP X	MAP Y	DISTANCE ERROR
A	50	50	49	51	1.41
B	150	35	148	35	2.00
C	170	130	170	131	1.00
D	100	170	101	169	1.41

Average Distance Error = 1.46
Standard Deviation of Distance Error = 0.36

B. ASSESSEMENT OF POSITIONAL ACCURACY USING THE NORMAL DISTRIBUTION

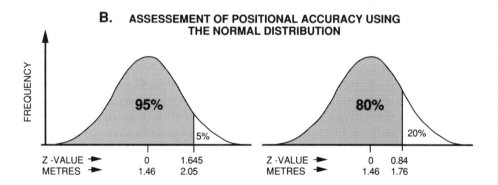

95%	5%	
Z-VALUE	0	1.645
METRES	1.46	2.05

80%	20%	
Z-VALUE	0	0.84
METRES	1.46	1.76

C. CALCULATION OF ACCURACY LEVELS

For: Mean of Distance Error = 1.46 m
Standard Deviation of Distance Error = 0.36 m

Confidence Level	Z-Value	Accuracy*
80%	0.84	1.76 m
85%	1.045	1.84 m
90%	1.28	1.92 m
95%	1.645	2.05 m

* Accuracy = Z-Value x SD + Mean
The accuracy is the maximum expected error at the selected
level of confidence.

Figure 5.1 Illustration of Accuracy and Probability.

increased confidence in the measurement of error when there are more data points.

The standard deviation is also useful in predicting the expected frequency with which a given level of error can be expected to occur. This is done by selecting an appropriate mathematical model. The strategy is to select a model that has been shown from past experience to be a good predictor of the distribution of errors for this type of map. A sample of test points can then be chosen from the map, checked, and the test results used to calibrate the model to the map. Then predictions can be made about the frequency with which any magnitude of error would be expected to occur. Of course *the correctness of our predictions would depend on whether a suitable model was being used*.

A frequently used model is the normal distribution, shown in Figure 5.1 part B. This model has been found to be a good predictor of the distribution of measurement errors like the positional errors being assessed here. It is also an easy model to use. The data needed to calibrate the model for a specific case are the mean and the standard deviation of the sample. In this case, the calibration values are the mean and standard deviation of the positional errors, i.e. the distance between the mapped and true position of the test points. Once the mean and standard deviation have been determined, the frequency with which an error of a specified magnitude will occur can be read from standard statistical tables. Anyone who studied basic statistics has completed numerous assignments using this distribution to model measurement errors of average heights of people, average lengths of leaves, and countless other sample problems.

The two graphs represent the expected frequency for each test result. The test in our case is the positional error of a point on the map. Each point on the curve represents the frequency (Y axis) with which a specific outcome (the value on the X axis) will occur. The area under each curve represents all the possible test outcomes. For convenience, the graphs are scaled so that the area under each curve sums to 1. To make the model easy to use, the X axis is divided into units of standard deviation termed Z-values. The Z-value of a sample measurement is calculated by dividing the measurement by the sample standard deviation.

In Figure 5.1 part B, 95% of the area under the curve is lower than the Z-value 1.645. This indicates that according to the model a test score with a Z-value of 1.645 or less would be expected to occur 95% of the time. The sample value corresponding to this point is easily calculated by multiplying the Z-value by the standard deviation and adding the mean sample value. (The mean is added because the model is calibrated to have a mean of zero.) As shown in the Figure, this works out to be 2.05. *If the model is a good predictor of error in this type of map*, then for this map the positional error for any point is expected to be 2.05 m or less, 95% of the time. Why 95% of the time? Because we selected a Z-value that would include 95% of the expected outcomes in our model, which is represented by 95% of the area under the curve. Conversely, the model predicts that 5% of the points would have errors greater than 2.05 m.

In selecting the 95% value, we are choosing a level of certainty or a level of confidence. We are "confident" that 95% of the time the observed errors will be 2.05 m or less. We are also accepting that 5% of the time errors exceeding 2.05 m will occur. **Our *choice* of the confidence level directly affects the value we obtain for the map accuracy**. The same analysis can be done using an 80% level of confidence and the predicted error would be 1.76 m or less. At the 85% level, the accuracy level would be 1.84 m and at 90% it would be 1.92 m (see Figure 5.1 part C). A higher accuracy level can be quoted by simply accepting a lower level of confidence.

So what then is the predicted accuracy of this map? The level of accuracy predicted for this map directly depends on the confidence level chosen. Without specifying a level of confidence, the stated accuracy value is actually quite meaningless. The accuracy of this map could be claimed to be 2.05, 1.92, 1.84 or 1.76 m, depending on the confidence level chosen. For this reason, **the proper reporting of any accuracy measure must include the level of confidence**. In this way the user of the data can judge whether the level of confidence is sufficient for the application.

ACCURACY TEST ASSUMPTIONS

The assumptions used in applying an accuracy test must be carefully considered in judging the reliability of an accuracy value. In the above example, it was assumed that the errors would be randomly distributed over the map. But what if one area was more difficult to field check and, as a result, was less accurately mapped than the rest of the map. A sample of points taken from this region might then have a lower level of accuracy than predicted by the accuracy assessment. This is because the errors would tend to be concentrated in one region of the map, and the assumption of randomly distributed errors would have been violated. In practice, minor violations of this model do not greatly affect the results, but other models may be more sensitive. Systematic errors such as these make accuracy values less accurate predictors.

Another assumption was that the sample points were representative of all the points on the map. In practice, "convenient" test points are often chosen to reduce the cost of testing. For example, test points may be chosen close to roads for easier access. As a result, the number, type, and distribution of test points may not in fact be representative.

It was also assumed that the model chosen was in fact a good predictor of the probability for the type of errors being assessed. The accuracy prediction made in the preceding example used the normal distribution to predict measurement accuracy. It was assumed that the map errors *really were* well predicted by the normal distribution. Experience has shown this to be a reasonable assumption. However, the models used for errors of measurement are usually not appropriate for classification errors. In the case of measurement errors, the error can theoretically be any real number because the variable is continuous.

In a classification, the variable (i.e. the class designation) is discrete. There are a finite number of classes, and these are the only allowable classes. As a result, the errors are restricted to labelling a location with an incorrect class designation. For this reason, classification errors in one class affect the accuracy of the class with which it is confused. Ideally errors in one class should have no affect on the accuracy of others (in statistics this property is termed **independence**). In the case of classification accuracy, an error in one class designation creates an error in the class with which it was confused, and so the errors are not independent of each other.

Classification accuracy will differ among the classes of a map, so ideally, each class should be tested separately. This can significantly increase the cost of the accuracy assessment and is usually not done. Instead, a single sample of test points can be selected from the map. How this sample is selected will affect which classes are tested and how intensively they are tested. A class that covers only a small map area is not likely to be tested at all!

CONCLUSION

Accuracy assessment can be an expensive procedure. Although it is valuable, its cost must be weighed against the benefits of the accuracy information. Less rigorous tests that are less expensive can be used for data sets where the consequences of errors are less critical. Accuracy assessments usually involve a comparison of values from the data set to be tested with values from an independent source of higher accuracy, such as field verification. Field verification of a statistically valid number of points may be more expensive than the application can justify.

Instead, less expensive approaches may be used. For example, indirect verification of test points may be done by interpreting airphotos instead of by field observations. The requirement for an independent data source might also be relaxed. The accuracy test points might be checked by re-interpreting the same airphotos used to generate the map being tested. In projects where the same mapping procedure is used repeatedly, accuracy assessments might be done on only a sample of maps of similar terrain conditions. The accuracy of all the maps in the group is then assumed to equal that of the sample maps. This is one of the least rigorous methods since considerable faith is placed in the consistency of the mapping procedure. Some maps would be assigned accuracy levels without any formal testing at all.

In the end, the specification of an accuracy level and the rigour with which it is assessed are judgment calls. They must take into account how the information is being used, the consequences of inaccuracies, and whether the accuracy measurements are indeed valid. Users often specify their accuracy needs on the basis of past experience (i.e. "the data have always been produced this way") or for reasons of consistency (i.e. "all classes on a map must have the same level of accuracy"). These approaches are often inappropriate.

Forest edges are usually "fuzzy" transition zones, not sharp boundaries. Requiring that forest boundaries be mapped with a positional accuracy of 1 m is probably not appropriate or even valid, because in the field the boundary could not be reliably positioned that precisely. On the other hand, it may be critical to measure the position of an underground cable more accurately than 1 m.

Requiring different levels of accuracy for different features in the same map or in the same data base is more cost effective than demanding that all features be represented at the same accuracy level. For this reason, the expenditure on accuracy assessment and data quality reporting in general must be matched to the consequences of errors. Higher levels of accuracy and more expensive accuracy assessments can be justified when the consequences of error are more costly.

The trade-offs in accuracy assessment costs, the mandate and budget of the producer of the data, and the willingness of the user to pay for data will all influence the assessment methods chosen. A rigorous accuracy assessment may not be justified for every data set in the GIS. But an accuracy rating of some form and a description of the method used to generate that rating should always be provided. Using data for which the level of accuracy or the data quality in general is unknown amounts to hoping for the best instead of recognizing the risk of the worst.

REFERENCES

Aronoff, S. 1985. Political Implications of Full Cost Recovery for Land Remote Sensing Systems. *Photogrammetric Engineering and Remote Sensing* 51(1):41–45.

Aronoff, S. 1982a. Classification Accuracy: A User Approach. *Photogrammetric Engineering and Remote Sensing* 48(8):1299–1307.

Aronoff, S. 1982b. The Map Accuracy Report: A User's View. *Photogrammetric Engineering and Remote Sensing* 48(8):1309–1312.

Chrisman, N.R. 1984. The Role of Quality Information in the Long-Term Functioning of a Geographic Information System. *Cartographica* 21(2):79–87.

Congalton, R.G. and R.A. Mead. 1983. A Quantitative Method to Test for Consistency and Correctness in Photo-Interpretation. *Photogrammetric Engineering and Remote Sensing* 49(1):69–74.

Croswell, P.L. 1987. Map Accuracy: What Is It, Who Needs It, and How Much is Enough. In *Proceedings of the URISA '87 Conference*. Urban and Regional Information Systems Association. Washington, D.C. Volume 2:48–62.

Epstein, E.F. and H. Roitman. 1987. Liability for Information. In *Proceedings of the URISA '87 Conference*. Urban and Regional Information Systems Association. Washington, D.C. Volume 4:115–125.

Epstein, E.F. 1987a. Litigation Over Information: The Use and Misuse of Maps. In *Proceedings of the International Geographic Information Systems Symposium: The Research Agenda*. Association of American Geographers. Washington, D.C.

Epstein, E.F. 1987b. Compatible Data for Land Decisions. In *Proceedings of the Conference on Compatible Data for Decisions*. National Governors' Association. Washington, D.C.

Fung, T. and E. LeDrew. 1988. The Determination of Optimal Threshold Levels for Change Detection Using Various Accuracy Indices. *Photogrammetric Engineering and Remote Sensing* 54(10):1449–1454.

Merchant, D.C. 1987. Spatial Accuracy Specifications for Large Scale Topographic Maps. *Photogrammetric Engineering and Remote Sensing* 53(7):958–961.

Newcomer, J.A. and J. Szajgin. 1984. Accumulation of Thematic Map Errors in Digital Overlay Analysis. *The American Cartographer* 11(1):58–62.

Story, M. and R.G. Congalton. 1986. Accuracy Assessment: A User's Perspective. *Photogrammetric Engineering and Remote Sensing* 52(3):397–399.

Vonderohe, A.P. and N.R. Chrisman. 1985. Tests to Establish the Quality of Digital Cartographic Data: Some Examples for the Dane County Land Records Project. In *Proceedings of AutoCarto 7*. American Society of Photogrammetry and Remote Sensing. Falls Church, Virginia. pp.552–559.

Walsh, S.J., D.R. Lightfoot, D.R. Butler. 1987. Recognition and Assessment of Error in Geographic Information Systems. *Photogrammetric Engineering and Remote Sensing* 53(10):1423–1430.

6. DATA MANAGEMENT

INTRODUCTION

For an organization to function effectively it requires accurate and timely information. Business enterprises are dependent on effective information handling to carry on their activities and maintain their competitiveness. Not surprisingly, it was the business community that first adopted computer-based data storage and retrieval technology. In the 1960s, computerized data bases were first used for materials management. Large engineering projects like the space program required a parts inventory of unprecedented volume and diversity. Efficient tracking and ordering of materials greatly reduced production expenses.

The other major data base application in the early 1960s was the Sabre airline reservation system developed by IBM for American Airlines. This system required a large communications network and for the first time addressed the problem of high-volume, simultaneous access to the data base. Since these early beginnings, the business community has invested heavily in data base technology to gather and maintain information. Indeed it was the demand of the business community that led to the commercial development of computerized information technology.

As the information system field developed during the late 1960s and 1970s, the concepts of the **data base** (i.e. the information to be stored) and the **data base management system** (the system used to manage the data base) were developed and refined. Today, sophisticated data base management systems are used to handle enormous data bases, such as national scale census information or global scale statistical data.

THE DATA BASE APPROACH

A data base is a collection of information about things and their relationships to each other. For example, a data base may consist of names and addresses. The names may themselves be categorized by other relationships, such as "client", "friend", or "family". The items to be stored in the data base may also be processes or concepts. Within a data base environment, the processes of erosion, water pollution, and agricultural development may be related to the item "rain forest clearcutting".

The objective in collecting and maintaining information in a data base is to relate facts and situations that were previously separate. This may simply require the retrieval of facts in the data base, such as the retrieval of the address associated with a person's name. Or it may require extensive data processing in which multiple relationships are evaluated, such as the analysis of a housing development or projecting the extent of rainforest clearcutting.

Early data base systems, like other computer software, were developed to provide a well-defined set of functions using a specified set of data. The data were stored as one or more computer files that were accessed by the special purpose data base software in whatever manner the designer believed to be most efficient. This **file processing** approach to data base management is illustrated in Figure 6.1 for a university administration application.

File processing is the most common approach to using a data base. However, it has some serious drawbacks. Since each application program must directly access each data file that it uses, the program must know how the data in each file are stored. This can create considerable redundancy

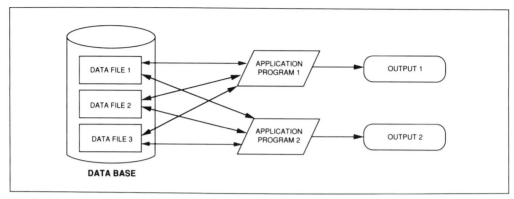

Figure 6.1   Sharing Data Files Among Applications in the File Processing Environment.

because the instructions to access a data file must be present in each application program. If modifications are made to the data file, these access instructions must be modified in each application program.

Another major problem arises when data are shared by different application programs and by different users. If data files can be accessed and modified by several programs and several users, then there must be some overall control over which users are given access to the data base and what modifications they are permitted to make. A lack of central control can seriously degrade the integrity (i.e. the quality) of the data base. The integrity of a data base is critical. Information of unpredictable quality can be worse than no information at all.

A data base management system (DBMS) is comprised of a set of programs that manipulate and maintain the data in a data base. They were developed to manage the sharing of data in an orderly manner and to ensure that the integrity of the data base is maintained. The DBMS concept was an important advance over the file processing approach. A DBMS acts as a central control over all interactions between the data base and the application programs, which in turn interact with the user (Figure 6.2). Application programs provide the functions that the user sees, such as inventory control transactions, order entry services, or geographic analysis functions. When these programs require access to the data base, the DBMS acts as the intermediary and supervisor.

One of the major benefits of a DBMS is that it provides **data independence**. That is, the application program does not need to

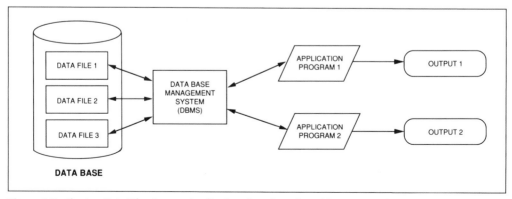

Figure 6.2   Sharing Data Files Among Applications in a Data Base Management System Environment.

know how the data is physically stored because all access to the data base is via the DBMS. The application program issues a command to the DBMS that retrieves and "re-packages" the data into the format needed by the application. When changes are made to a data file, the DBMS ensures that the data will still be correctly provided to the application programs that use them; modifications do not have to be made to the application programs. Similarly, if an application program requires different data or a different data format, the DBMS can handle the changes. In this way, the DBMS can greatly reduce the effort needed to maintain the application programs and the data base.

The services provided by the DBMS also simplify the development of new application programs. In fact, many data base systems incorporate a direct user interface. A wide range of data retrieval and manipulation operations can be executed using the DBMS services alone, without writing a lengthy program.

A DBMS is also used to tailor the style of information presentation to different users. Each style of presenting the data base is termed a **view.** Figure 6.3 illustrates this concept using a data base comprised of client names, addresses, purchases, and inventory information. This data base is presented to the account executive as if it were organized by sales accounts, whereas the same data base is presented to the inventory manager as if it were organized by item. By providing different views, the DBMS tailors the data base to each user group — a very valuable function — without storing multiple copies of all the data.

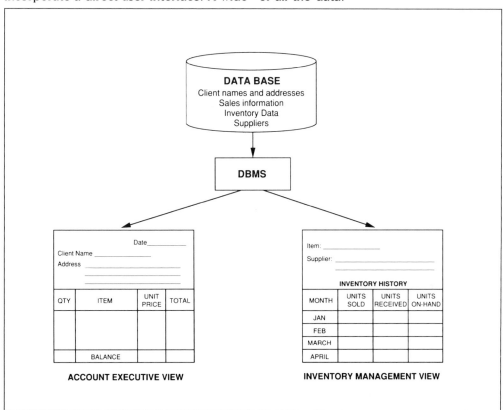

Figure 6.3 Presentation of Data Base Information Using Multiple Views.

ADVANTAGES OF THE DATA BASE APPROACH

The advantages of the data base approach over the file processing approach can be summarized as follows:

1. **Centralized Control**. A single DBMS under the control of one person or group can ensure that data quality standards are maintained, security restrictions are enforced, conflicting requirements are balanced, and the integrity of the data base is maintained.

2. **Data Can Be Shared Efficiently**. Using a DBMS, the information in a data base can be shared in a flexible yet controlled manner. The data handling services of the DBMS also facilitate the development of new applications of the existing data base.

3. **Data Independence**. Application programs are independent of the physical form in which the data are stored.

4. **Easier Implementation of New Data Base Applications**. New application programs and unique data base searches can be more easily implemented using the services provided by a DBMS.

5. **Direct User Access**. Data base systems now commonly provide a user interface so that non-programmers can perform sophisticated analyses. At the same time, the data base system provides the means to control data base access and operations, to maintain consistency, and to protect the integrity of the data base.

6. **Redundancy Can Be Controlled**. In a file processing environment, separate data files are used for each application and considerable data redundancy may result. There are sound reasons for maintaining multiple copies of some data. However, excessive data redundancy is expensive. In addition to the added data storage, an effective strategy must be provided to update the multiple copies of the data. A DBMS can be used to monitor and reduce the level of redundancy and where multiple copies of data are retained, the DBMS can manage the updating procedures.

7. **User Views.** A DBMS can provide a convenient user interface to create and maintain multiple user views.

DISADVANTAGES OF THE DATA BASE APPROACH

Not all applications of a data base can benefit from a DBMS. Some of the disadvantages of using a data base approach are:

1. **Cost**. The data base system software and any associated hardware can be expensive. As a minimum, they represent an additional acquisition and maintenance cost.

2. **Added Complexity**. A data base system is more complex than a file processing system. In theory, the more complex the system, the more susceptible it is to failure and the more difficult the recovery. In practise, full-featured DBMSes are provided with effective back-up and recovery systems.

3. **Centralized Risk**. In centralizing the location of data and reducing data redundancy, there is a greater theoretical risk of loss or corruption of data while running an application program. However, the backup and recovery procedures normally provided in a DBMS minimize these risks.

The cost-effectiveness of a DBMS depends on the applications to be supported. The first GISes used a file processing data base environment and many still do. However, the trend is increasingly towards the use of a DBMS, if not to manage all the data in a GIS at least to manage the non-spatial attribute components. Virtually all commercial GISes now incorporate some form of DBMS.

THREE CLASSIC DATA MODELS

The conceptual organization of a data base is termed the **data model**. It can be thought of as the style of describing and manipulating the data in a data base. There are three classic data models that are used to organize electronic data bases: the hierarchical, the network, and the relational models. They were first developed to handle the information needs of the business community and have been adapted to a wide range of other applications. These data models or their derivatives have also been adapted for use in the GIS environment.

RECORDS, FIELDS, AND KEYS

The organization of a data file can be described in terms of records, fields, and keys. In a computer-based data-storage system, a small group of related data items are stored together as a **record**. A record can be thought of as one row in a table, as shown for the student data in Figure 6.4. Here the first record contains the information for the student named Randy Thomas. The

record represents the information pertaining to a particular element or entity, the student. (An **entity** is an object, event, or concept. The terms *element* and *entity* are used interchangeably here.)

A record is divided into **fields**, each of which contains an item of data. A field defines where a particular type of data can be found in the record. In this example, the data fields are *Last Name*, *First Name*, *Year*, and *Major*. A record is retrieved from the data file by means of a **key**, i.e. a label comprised of one or more fields. The *Last Name* and *First Name* data fields could be designated as keys. Fields that are not designated as key fields are termed **attribute fields**. The software used to search the data base is designed to efficiently search for records that have a particular value of the key.

In general, the fewer the key fields, the more compact the data file, and the more quickly the data file can be searched. However, since the keys determine how the data can be accessed, the fewer the key fields, the more restricted the types of searches (also termed **queries**) that can be performed. For the student data file in Figure 6.4, if the first and last name fields were the keys then the records for a specified first and last

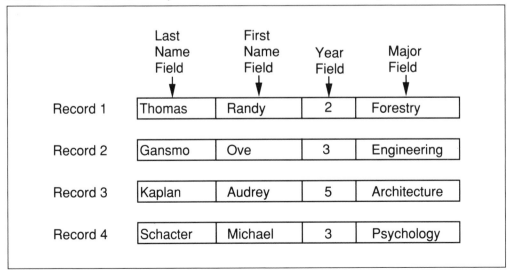

Figure 6.4 Organization of Information as Records in a Data File.

name could be easily retrieved. If the year data field was not a key, retrieval of the student records for a specified year would be more cumbersome or might not be possible at all (depending on the software).

THE HIERARCHICAL DATA MODEL

In the hierarchical data model, the data are organized in a tree structure as shown in Figure 6.5 part A. The relations among the five entities (*University, Department, Students, Professors*, and *Courses*) are defined by the organization of the hierarchy. The organization is encoded in the data records for each entity, as shown in Figure 6.5 part B. The field names are shown in the top half of each box and a sample data record is shown in the lower half. There is one field that is designated as the key field. It is used to organize the hierarchy. In Figure 6.5 part B, the hierarchy is represented by the arrows connecting the key field in each data record (the key fields are stippled).

The top of the hierarchy is termed the **root**. It is comprised of one entity, in this case a *University*, the University of California. The root may be represented by a record containing a single data field (as shown here), or by a record containing many fields. Except for the root, every element has one higher level element related to it, termed its **parent**, and one or more subordinate elements, termed **children**. An element can have only one parent but can have multiple children.

In the hierarchical data model, every relation is a many-to-one relation or a one-to-one relation. The many departments belong to one university, there are many students in each department, and so on. In the figure the *many* side of the relation has an arrow head, the *one* side does not.

In a hierarchical data model, information is retrieved by traversing the tree structure. Retrieval of all the students or all the professors in a specific department is a very efficient search because there is a direct link between student and department entities and between professor and department entities.

However, to find all the courses offered by a specific department requires a two

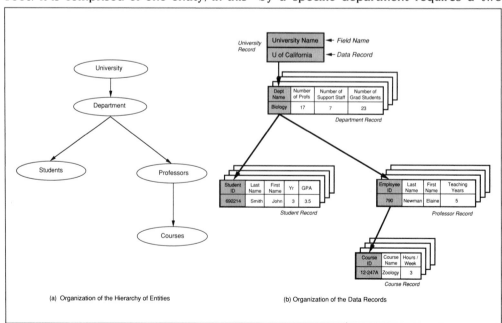

(a) Organization of the Hierarchy of Entities   (b) Organization of the Data Records

Figure 6.5　Organization of a Data Base Using the Hierarchical Data Model.

stage search. First, the records for all the professors teaching in that department would be retrieved and then the courses that each of those professors taught would be retrieved. This is a less efficient type of retrieval because an intermediate entity, the professors, must be retrieved. However, because the links between the key fields are encoded in the data structure, this type of retrieval can still be quite efficient if it does not involve too many intermediate levels. The search would be more efficient if a course could be directly related to a department as well as to a professor. However, in the hierarchical model an entity can have only one parent, so the *Course* entity is not permitted to have both the *Department* and *Professor* entities as parents.

Another limitation of the hierarchical model is that searches cannot be done on the attribute fields. In this example, the retrieval of all second year students could not be done because the *Year* data field is not a key. For this search to be possible, the data base would have to be restructured or special linkages, such as pointers, would have to be used to modify the data base organization. (A **pointer** is a code that indicates a location in a file, such as the location in a file where the attributes of a geographic feature are stored.) Because the relations between entities are encoded in the data base, it is difficult to modify.

Hierarchical systems are easy to understand and they are easy to update. They can provide high speed access to large data sets. In part, the high speed access is achieved by encoding the data relationships into the data base itself. They work well when the structure of the hierarchy is optimized for the searches to be performed. However, this requires that the complete range of queries be known in advance. For applications similar to bibliographic data bases and airline reservation systems, the types of searches are very predictable, and so they can be tightly specified.

Major disadvantages of the hierarchical model are that the data relationships are difficult to modify and queries are restricted to traversing the existing hierarchy. For applications like environmental assessment or geographic information analysis, the data searches are often exploratory and cannot be predicted in advance. The inflexibility of the hierarchical data model makes it too restrictive for this type of application. Another limitation is that multiple parents are not allowed. There are many applications where an element needs to be represented as a member of multiple groups. The network model addresses some of these restrictions.

THE NETWORK DATA MODEL

The network data model overcomes some of the inflexibility of the hierarchical model. In the network data model, an entity can have multiple parent as well as multiple child relations and no root is required. As a result, data records can be directly searched without traversing the entire hierarchy above that record. Figure 6.6 shows the university data base organized as a network model. The *Course* entity can now have two parents and is related to both the *Department* and *Professor* entities. A search of all courses in a specified department can now be done more directly than in the hierarchical example.

The *Student-Course* relation is a many-to-many relation. That is, each student can be enrolled in many courses and each course can have many students. While the network model does not allow many-to-many relations, this relation can be handled indirectly by using an intermediate relation, often termed an **intersection record**. As shown in the Figure, the intersection records represent the *Student-Course* combinations, i.e. the registration of students in courses. Each *Student-Course* combination is unique. One *Course* entity can have many *Registration*

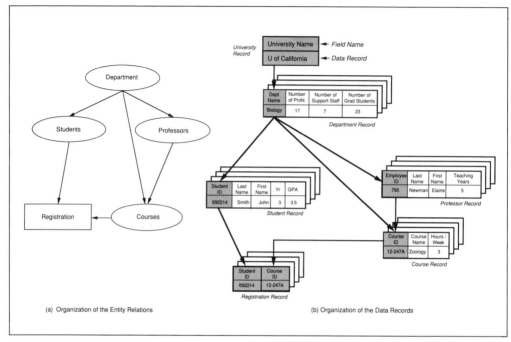

Figure 6.6   Organization of a Data Base Using the Network Data Model.

entities, and one *Student* entity can have many *Registration* entities, so both of these relations are one-to-many and are permitted. Intersection records can also be used in the hierarchical model, although the other restrictions of that model make the implementation somewhat more complex.

Network models tend to have less redundant data storage than the corresponding hierarchical model. However, more extensive linkage information must be stored, adding to the size and complexity of the data files. In a complex data base, the linkage information can be substantial and the time needed to update the linkages when changes are made can be significant.

When the data structure to be represented is in fact a simple hierachy, there is no real difference in the expressive power of the two models. However, where a more complex real-world data structure must be represented, the network model can accommodate the added complexity. As with the hierarchical model, the relations among data

elements are encoded in the data base. This provides high speed retrieval, but the data relationships are difficult to modify. Thus the principal disadvantages of the network model are that it is more complex than the hierarchical and not as flexible as the relational model.

THE RELATIONAL DATA MODEL

Figure 6.7 illustrates the university data base organized using the relational data model. In the relational data model there is no hierarchy of data fields within a record; every data field can be used as a key. The data are stored as a collection of values in the form of simple records, termed **tuples**. Each tuple represents a fact, i.e. a set of permanently related values. The tuples are grouped together in two-dimensional tables, with each table usually stored as a separate file. The table as a whole represents the relationships among all the attributes it contains, and so it is often termed a **relation**.

1. Course Information

Professor ID	Course Dept	Course Name	Course Hours	Course ID
790	Biology	Zoology	3	12-247A
745	Chemistry	Organic	4	14-200A
807	Chemistry	Organic	4	14-200B
642	Chemistry	Biochem	5	14-280A
689	English	Medieval	3	17-340A

2. Registration Information

Course ID	Student ID
12-247A	692214
14-200B	692214
17-340A	692214
17-340A	728437
14-200B	728437
14-280A	728437
14-200B	745870

3. Student Information

Student ID	Last Name	First Name	Yr	GPA	Dept
692214	Smith	John	3	3.5	Biology
728437	Green	John	2	2.4	English
745870	Thomas	Randy	4	3.7	Physics

4. Department Information

Dept Name	Number of Professors	Number of Support Staff	Number of Graduate Students
Biology	17	7	23
Chemistry	10	8	7
English	11	3	20
French	5	1	15
Physics	6	3	8

5. Professor Information

Professor ID	Last Name	First Name	Teaching Years	Dept
745	Brown	Al	5	Chemistry
790	Newman	Elaine	5	Biology
807	Ross	Grant	4	Chemistry
642	Geist	Val	8	Biology
689	Colwell	Bob	8	English

Figure 6.7 Organization of a Data Base Using the Relational Data Model.

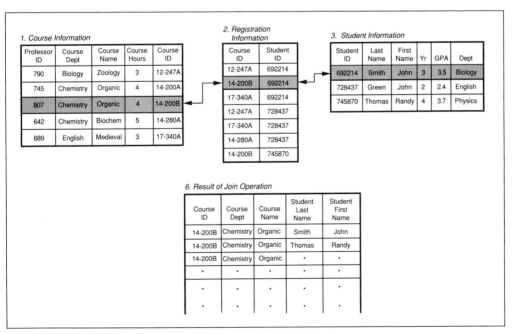

Figure 6.8 The Relational Join Operation. The data from three relational tables are used to generate a student list for a course (Table 6). Common data fields (indicated by the arrows) are used to interrelate the records in different tables.

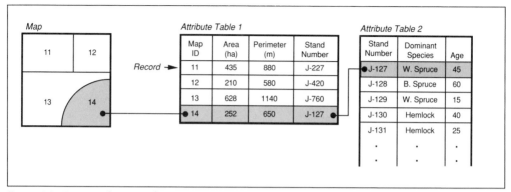

Figure 6.9   Storage of GIS Attribute Information in a Relational Data Base.

Using the relational model, a search can be made of any single table using any of the attribute fields, singly or together. For example, the Student Information Table can be searched for all students in year 4. Just as easily, the table could be searched for all students with the last name Johnson.

Searches of related attributes that are stored in different tables can be done by linking two or more tables using any attribute they share in common. This procedure is termed a **join** operation. The shared attribute need not itself be part of the relation being analyzed. Figure 6.8 illustrates how the data base could be searched to generate a student list for a specific course. To retrieve this information, tables 1, 2, and 3 are joined by means of the Course-ID and the Student-ID attributes. In effect, table 1 is joined to table 2 by means of the Course-ID attribute that they have in common. Tables 2 and 3 are joined by means of the Student-ID attribute that they have in common. A new table, table 6, can be created from this join operation. Notice that only a few of the fields are needed for the table. By including only the data fields required, redundant data storage is reduced. In fact, table 6 need not be stored at all. Instead it can be treated as a ''virtual table'' that is defined, can be queried, but is not actually created. This ''logical join'' operation gives the relational model tremendous flexibility. It is able to accommodate diverse queries for which it was not specifically designed.

As can be seen from table 6, there is a certain amount of redundancy in a relational table. The *Course-ID ,Course Department*, and *Course Name* information is repeated. However, each tuple (row) is unique. Since a tuple represents a fact, there should never be two identical tuples because there is no need to store the same fact twice. Theoretically, the attributes could all be stored in a single data table; however, there is a performance trade-off. It takes longer to search data stored in several tables than to search the same data stored in one table. But as the number of data tables is reduced, the redundancy of data storage tends to increase. So the number, size, and organization of the data tables directly affect the amount of data storage required and the speed with which queries can be done.

Despite these performance trade-offs, there are no restrictions on the types of queries that can be done so long as there are the necessary common data fields. This is the major advantage of the relational model over the network and hierarchical models. This flexibility has made the relational model the one most commonly used to store attribute information in a GIS. Figure 6.9 illustrates a typical GIS application. Here the forest stand map is linked to an Attribute Table 1 by means of an ID code. This table contains the area and perimeter measurements. The stand number provides a link to Attribute Table 2 that contains the dominant species and age information.

The major advantages of the relational model over the hierarchical and network models can be summarized as follows (Bowers 1988, Date 1983):

1. The relational model is more flexible than the other models. The way the data values exist in the relational tables does not in any way restrict the kinds of processing that can be done. In the hierarchical and network models, manipulation of the data is restricted by the structure built into the data model.

2. The relational model has a sound theoretical base in mathematical theory. The opportunity exists to use the mathematics of relations as the basis for data processing procedures instead of programming. However, most implemented systems provide a programming language interface, not a mathematical one.

3. The organization of the relational model is simple to understand and, therefore, is a good vehicle to communicate data base ideas.

4. The same data base can generally be represented with less redundancy using the relational model than the other two models.

The major disadvantages of the relational model are:

1. It is more difficult to implement.

2. It tends to have slower performance. The absence of physical links or pointers (as used in the hierarchical and network models) requires that manipulation of the data be based on matching values in the relational tables. This is a much more time-consuming operation. As a result, a relational data base system tends to be significantly slower than the corresponding hierarchical or network data base system.

The relational model can accommodate a very wide range of relations and is easily modified. Although it is not necessary to know in advance the types of queries to be performed, such knowledge can be used to design the relational tables so as to optimize performance. The simplicity of the relational model and its flexibility make it an attractive choice. However, the absence of pointers and linkages embedded in the data records, which gives the relational model its flexibility, also reduces its speed of operation.

QUERY LANGUAGES FOR THE THREE CLASSIC DATA MODELS

The types of queries supported by both the hierarchical and network data base models are defined when the data base is constructed. Physical linkages embedded in the data records are used to traverse the data base. As a result, in order to search the data base the user is required to know the hierarchy in which the data have been stored. Languages that require the user to know the hierarchy are termed **procedural query languages.** Often there is no sound basis for deciding in advance which queries require the highest level of performance. This makes it difficult to optimize these data base system designs.

In the relational data base, more flexibility is achieved by abolishing the hierarchy of attributes. Any attribute can be used as a key to retrieve information, and the data in separate tables can be related using any attribute field that they share in common. Unlike in the hierarchical and network models, in a relational model the relations are not explicitly encoded in the data base.

Since the relational model does not restrict the range of queries the user does not need to know the structure of the data base to construct a query. A query language that is not dependant on the structure of the data base is termed a **non-procedural language.** The query language SQL (for

Standard *Query Language*) developed by IBM is a widely used example of such a non-procedural language. Non-procedural query languages have become very popular because they are easy to learn yet powerful. They have made information data bases much more accessible to users with little or no computer training. Although originally developed for relational data base systems, non-procedural query languages are now commonly available for network systems as well (Larson 1987).

THE NATURE OF GEOGRAPHIC DATA

The map is perhaps the most familiar form in which geographic data are represented. A map consists of a group of points, lines, and areas that are positioned with reference to a common coordinate system. It is usually represented in two dimensions so that it is easily portrayed on a flat sheet of paper. The map legend links the non-spatial attributes, such as place names, symbols, and colours, to the spatial data i.e. the locations of the map elements.

The map itself serves to both store the data and to present the data to the user. It is a relatively inexpensive means of storing a considerable amount of spatial information. However, its double service as a medium of storage and presentation has a number of limitations. For the map to be legible, the amount of data represented in the map must be limited. Similarly, the form of presentation the map provides is constrained by the need to carry considerable information. The map is relatively expensive to draft manually, and so a map is commonly designed to serve many different uses. This may require that compromises be made in the scale, level of information detail, and other design considerations. Maps also become rather cumbersome to use when there are large numbers of them and when the information from several maps must be analyzed together.

In a computer-based GIS, the storage and presentation of geographic data are separate. The data may be stored at a high level of detail and then plotted at a more general level and at a different scale. In a GIS, the plotted map becomes one of many forms of presenting the data. It becomes, in effect, a view of a geographic data base. The same data may be viewed as many different types of maps. Each can be customized for a specific use because it is relatively inexpensive to plot a map by computer. In addition to maps, the data may be presented in the form of tables, or even as text descriptions.

In a computer-based GIS, geographic data are represented as points, lines, and areas, as with maps. However, for efficient computer implemetation, these elements are organized somewhat differently than the organization of a paper map. Geographic data have a set of characteristics that make them distinctly different from the more familiar lists and tables of data used in the information systems developed for business applications.

The information for a geographic feature has four major components: its geographic position, its attributes, its spatial relationships, and time. More simply, the four components are: where it is, what it is, what is its relationship to other spatial features, and when did the condition or feature exist.

GEOGRAPHIC POSITION

Geographic data are fundamentally a form of spatial data. Each feature has a location that must be specified in a unique way. The locational definitions can be quite complex because geographic phenomena tend to occur in irregular complex patterns, such as a sinuous shoreline or a web of transportation routes.

For geographic data, locations are recorded in terms of a coordinate system like the Latitude/Longitude, UTM (Universal Transverse Mercator), or State Plane coordinate systems. In some cases the coordinates of one system can be mathematically transformed into the coordinates of the other. But in many cases this is not possible. For example, property boundaries were at one time described with reference to local features like fences or trees. That type of reference system cannot be mathematically transformed into a geographic coordinate system. Similarly, a geographic coordinate cannot be calculated from a street address.

A GIS requires that a common coordinate system be used for all the datasets that will be used together. For a small study area the coordinate system can be any convenient grid. For larger coverage areas, one of the nationally or internationally accepted coordinate systems is generally used. The UTM coordinate system is a convenient one for scales of 1:500,000 or larger.

Geographic data may be stored at different levels of positional accuracy. To some extent the locational data are always imprecise at some level of detail. Some data may be accurate to within a few centimeters, while other data may only be accurate to 10 m.

ATTRIBUTES

The second characteristic of geographic data are their attributes, i.e. "what it is". For example, the feature might be a forest stand. Its attributes might include the species composition, average tree height, the crown closure, and the date it was last logged. These attributes are often termed **non-spatial attributes** in that they do not in themselves represent locational information. There is a level of inaccuracy inherent in non-spatial attribute data as there is for spatial data. A commercial district may not be 100% commercial and a stand of pine trees is generally not 100% pine. Often this type of inaccuracy is not addressed by GIS users, but for many types of analyses it is important to recognize and take into account this imprecision.

SPATIAL RELATIONSHIP

The third characteristic of geographic data are the spatial relationships among the geographic features. These relationships are generally very numerous, may be complex, and are important. For example, it is not only important to know the location of the fire and the fire hydrants, but also how close those fire hydrants are to the fire. The effects of an oil spill depend on what it is near, as well as on where it is located. These relationships are intuitive to the person looking at the area or at a map. However, for a computer-based GIS, relationships must be expressed in a computer-usable manner. In practical terms, it is not possible to store information about all the possible spatial relationships. Instead, only some of the spatial relationships are explicitly defined in the GIS, and the remainder is either calculated as needed or is not available. The trade-offs made in choosing the spatial data model directly affect the performance characteristics of the system.

TIME

Geographic information is referenced to a point in time or a period of time. Knowing the time when geographic data were collected can be critical to using those data appropriately. An area may be covered by trees one year and have been clear-cut the next. An urban area may be zoned residential for twenty years and then be re-zoned commercial. Agricultural crops are grown in certain seasons. In some regions of the world, different crops are grown in the same area at different times of year. So, it would be important to know the time of year when the crop area data were collected.

Historical information may also be a valuable component of the GIS data base. Knowing the previous condition of a geographic location may be very useful. For example, knowing the forest that previously grew on a clear-cut site that has been harvested can be useful in deciding how best to replant an area. Knowing that a site was once a waste dump will restrict its future use.

The representation of time in a GIS is an added level of complexity that is difficult to handle. As a result, the time factor has generally not been addressed in a sophisticated manner.

Taken together the four characteristics, geographic position, attributes, spatial relationship, and time, make geographic data uniquely difficult to handle. It is too complex to record all the information for geographic entities. As with other data base systems, a data model is used to represent the information considered to be most relevant to the applications at hand. If the model is appropriately designed, the GIS will mimic the behavior of the real world accurately

enough to provide useful information. The data base system of a GIS provides the means to organize the spatial and non-spatial attribute data for efficient storage, retrieval, and analysis. In the following sections, the common data models used to handle the spatial data in a GIS are discussed.

SPATIAL DATA MODELS

There are two fundamental approaches to the representation of the spatial component of geographic information: the vector model and the raster model. In the vector model, objects or conditions in the real world are represented by the points and lines that define their boundaries, much as if they were being drawn on a map. The position of each object is defined by its placement

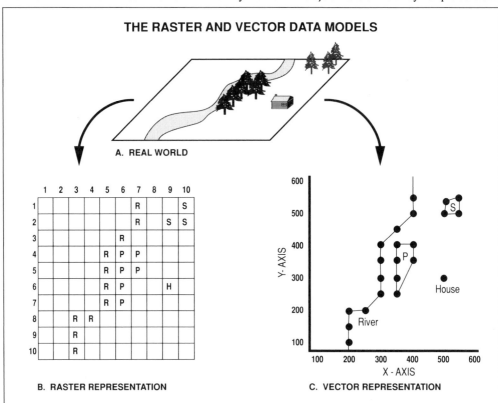

Figure 6.10 Comparison of the Raster and Vector Models. The landscape in A is shown in a raster representation (B) and in a vector representation (C). The pine forest stand (P) and spruce forest stand (S) are area features. The river (R) is a line feature, and the house (H) is a point feature.

in a map space that is organized by a coordinate reference system, as shown in Figure 6.10 part C. Every position in the map space has a unique coordinate value. Points, lines, and polygons are used to represent irregularly distributed geographic objects or conditions in the real world. (A polygon is an area bounded by a closed loop of straight-line segments.) A line may represent a road, a polygon may represent a forest stand, and so on. The spatial entities in the vector model correspond more or less to the spatial entities that they represent in the real world.

In the raster model, the space is regularly subdivided into cells (usually square in shape), as shown in the Figure 6.10 part B. The location of geographic objects or conditions is defined by the row and column position of the cells they occupy. The area that each cell represents defines the spatial resolution available. Because positions are defined by the cell row and cell column numbers, the position of geographic features is only recorded to the nearest cell. For example, if the area is divided into 10 m × 10 m cells, then the position of an object can only be recorded to the nearest 10 m × 10 m area. The value stored for each cell indicates the type of object or condition that is found at that location. Thus in the raster approach, the space is populated by a large number of regularly distributed cells, each of which can have a different value. The spatial units are the cells, each of which corresponds to an area at a specific location, such as an area on the earth's surface. The cell values report a condition at a location and that condition pertains to the entire cell. Unlike those of the vector model, the units of the raster model do not correspond to the spatial entities they represent in the real world. The spatial entitities or units in the raster data model are not the objects we conceptualize; they are the individual cells. For example, a road does not exist as a distinct raster entity; the cells representing the road are the entities. Thus, a road is represented by a group of cells with the condition *road*. The road is not itself recognized as a single entity.

In both models, the spatial information is represented using homogeneous units. **In the raster approach, the homogeneous units are the cells**. (The area within a cell is not subdivided and the cell attribute applies to every location within the cell.) A very large number of relatively small cells, all the same size, are used. Raster data files commonly contain millions of cells and the position of each unit is rigidly defined. **In the vector approach, the homogeneous units are the points, lines, and polygons**. Relative to the raster approach, these homogeneous units are relatively few in number and variable in size. In a vector file, the elements might number in the tens of thousands, but not in the millions as commonly occurs in a raster file. The positions of these homogeneous vector units are defined using a nearly continuous range of coordinate values. This method provides a much more flexible and usually more precise coordinate position than the row and column positioning used in the raster approach.

The different approaches have their advantages and disadvantages. The major trade-offs are summarized in Table 6.1. Each approach tends to work best in situations where the spatial information is to be treated in a manner that closely matches the data model. Where the geographic information of interest is the spatial variability of a phenomenon, the raster representation is generally better suited. The subtle colour variations from point to point in a digital image are well represented by very large numbers of cells each assigned a set of values to represent the red, green, and blue intensity at that cell position. Similarly, the shape of a surface, its topography, is well represented by a set of evenly spaced elevation measurements. Where the information of interest is the distribution of objects in

---

Table 6.1   Comparison of Raster and Vector Data Models.

RASTER MODEL	VECTOR MODEL
Advantages:	**Advantages:**
1. It is a simple data structure.	1. It provides a more compact data structure than the raster model.
2. Overlay operations are easily and efficiently implemented.	2. It provides efficient encoding of topology, and, as a result, more efficient implementation of operations that require topological information, such as network analysis.
3. High spatial variability is efficiently represented in a raster format.	
4. The raster format is more or less required for efficient manipulation and enhancement of digital images.	3. The vector model is better suited to supporting graphics that closely approximate hand-drawn maps.
Disadvantages:	**Disadvantages:**
1. The raster data structure is less compact. Data compression techniques can often overcome this problem.	1. It is a more complex data structure than a simple raster.
2. Topological relationships are more difficult to represent.	2. Overlay operations are more difficult to implement.
3. The output of graphics is less aesthetically pleasing because boundaries tend to have a blocky appearance rather than the smooth lines of hand-drawn maps. This can be overcome by using a very large number of cells, but may result in unacceptably large files.	3. The representation of high spatial variability is inefficient.
	4. Manipulation and enhancement of digital images cannot be effectively done in the vector domain.

---

space or the conditions that apply to an area feature (such as the soil unit or forest stand of a thematic map), then the vector approach tends to be better suited. The raster and vector approaches are examined in greater detail in the following sections.

THE RASTER DATA MODEL

In its simplest form, the raster data model consists of a regular grid of square or rectangular cells. The location of each cell or **pixel** (for picture element) is defined by its row and column numbers. The value assigned to the cell indicates the value of the attribute it represents. As shown in Figure 6.10, a point (the house) is represented by a single cell, a line (the river) by several cells with the same value forming a linear grouping, and an area (the forest stand) by a clump of cells

all having the same value. More complex implementations have used regular shapes other than a square, such as triangles and hexagons. However, the square cell is easily handled as a numerical array by the more common programming languages. The raster data model is also easily interfaced to the hardware devices commonly used for the input and output of spatial data. For this reason, the first GISes were written in the Fortran programming language and were raster-based.

Each cell in a raster file is assigned only one value. So, different attributes are stored in separate files. The soil types and forest cover for an area would be stored as separate soil and forest data files. Operations on multiple raster files involve the retrieval and processing of the data from corresponding cell positions in the different data files.

Conceptually, the process is like stacking the files as shown in Figure 6.11 and using the vertical stack of cell values to analyze each cell location. For example, in order to find all the cells with a *Pine* forest cover and a *Sandy* soil type, each cell in the soil file and each corresponding cell in the forest file would be retrieved and evaluated. All those cells that were coded as *Pine* forest and also as *Sandy* soil would be identified and could be output to a new data file. This procedure, termed *overlay analysis*, is discussed in Chapter 7.

In the raster data model, each cell represents an area of the land surface. Since the attribute of each cell is stored as a unique value, the total number of values to be stored is the product of the number of columns times the number of rows. The smaller the area of land that each cell represents, the higher the resolution of the data, and the larger the file needed to store the data. The size of the file increases rapidly with resolution. If a cell represents a 250 m × 250 m area on the ground, then a distance of 1 km would be represented by 4 cells. A 1 km × 1 km area would be represented by 16 cells. If a higher resolution was used and the cells represented 100 m × 100 m areas, then a distance of 1 km would be represented by 10 cells and the same 1 km × 1 km area would require a total of 100 cells. Since the file size is related to the area of coverage, it increases by the square of the increase in resolution. For these reasons, raster files tend to be

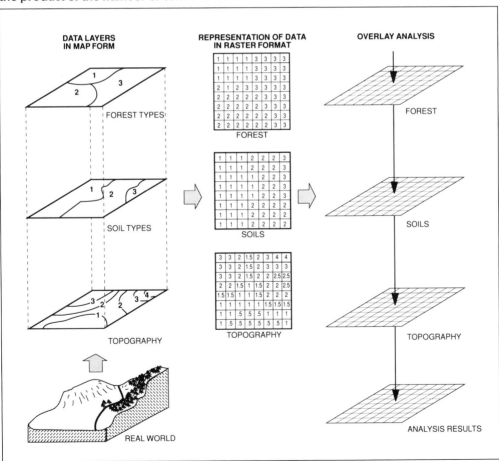

Figure 6.11 Overlay Analysis Using Raster Data Files.

relatively large. It is common for a raster file to be comprised of several million cells. However, many of the cells may contain the same value as neighbouring cells. Where there is considerable redundancy of this type, significant reductions in the size of the raster file can be achieved by using various methods of data compression, such as *run-length encoding* and *quadtrees*. (The representation of data in a more compact form is termed **data compression**.)

Run-Length Encoding

If the data are highly variable from cell to cell, as with digital terrain data or a photographic image, then the large number of cells serve to capture the high spatial variability. If the number of values were reduced, some of the spatial information would be lost. However, in many cases the spatial variability is not high and the information can be represented with less redundancy and without loss of detail. This often occurs when the data are thematic. Since cells representing areas of the same class have the same value, the pattern of values tends to be spatially clumped. The quantity of data needed to capture a clumped pattern of spatial variability can be considerably reduced by using data structures that code these repeated values more compactly than the simple raster data structure. Two common methods of compact raster data storage are run-length encoding and quadtrees.

In **run-length encoding**, adjacent cells along a row that have the same value are treated as a group termed a **run**. Instead of repeatedly storing the same value for each cell, the value is stored once, together with information about the size and location of the run. Several run-length encoding strategies have been developed, two of which are illustrated in Figure 6.12. In **standard run-length encoding** the value of the attribute, the number of cells in the run, and the row number are recorded. In this example, the

100 cell values have been reduced to a file of 54 values, see part B.

The second data compression technique shown in part C is termed **value point encoding**. Here the cells are assigned position numbers starting in the upper left corner, proceeding from left to right and from top to bottom. The position number for the end of each run is stored in the point column. The value for each cell in the run is in the value column. Using value point encoding, only 32 entries were needed to encode the same data. However, larger values (requiring more digits and hence more storage) occur in the point column.

The degree of compression obtained using these methods depends on the complexity of the map. These forms of data compression become less efficient as the number of edges or transitions increase. The greatest degree of compression is achieved when there are only a few classes and they occur in large clumps. As the spatial variability increases, i.e. when there are many different classes distributed in small clumps, then the compressed formats could actually require *more* storage space than the full raster file.

Run-length encoding can significantly reduce the storage needed for a raster data file, but the trade-off between cell size and file size still remains. A coarse grid (large cell size) gives smaller data files that use less storage space and are faster to process. However, boundaries can be positioned only as accurately as the size of a cell. A finer grid (smaller cell size) provides more accurate positioning but greatly increases the size of the file and increases processing times when using the file. The quadtree structure addresses both the resolution as well as the redundancy issue.

Quadtrees

The **quadtree data model** provides a more compact raster representation by using a

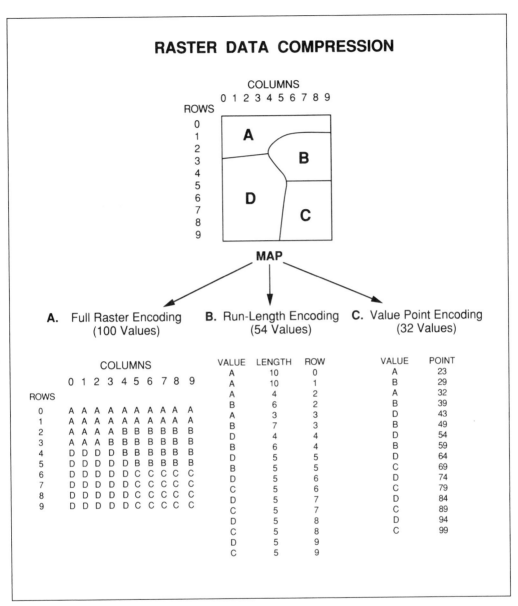

Figure 6.12 Run-Length Encoding of Raster Data.

variable-sized grid cell. Instead of dividing an area into cells of one size, finer subdivisions are used in those areas with finer detail. In this way, a higher level of resolution is provided only where it is needed.

For a thematic map, the fine grid is only needed in the vicinity of lines, points, and polygon boundaries. A large area of a single class would be just as accurately encoded with one large cell as with many small cells because they all have the same attribute value. Using the quadtree data structure, a coarse resolution (large cells) is used to encode large homogeneous areas. A finer resolution (small cells) is used for areas of high spatial variability, i.e. with many relatively small features.

Conceptually, the construction of a quadtree can be thought of as a process of regularly subdividing a map. If the entire map is assigned the same class, e.g. is covered by the same forest type, then the process stops and the map would be represented as one cell representing a very large area. (In the conventional raster model, the map would be represented by a large number of small cells each with the same value.) If there is more than one class present, then the map is subdivided into four equal-sized quadrants. The same test is repeated for each quadrant. Every quadrant that contains more than one class is again subdivided into four, whereas homogeneous quadrants are not subdivided.

The result is a quadtree representation as shown in Figure 6.13. Notice that more cells and smaller cells are created at feature boundaries. The dividing process is limited to a chosen maximum number of iterations. This in effect establishes the minimum cell size that can be represented. So, the resolution is limited by cell size. However, a very small minimum cell size can be used without creating an enormous data file because most of the area will not actually be represented at the finest resolution.

Figure 6.14 provides a more detailed look at the quadtree structure. Part A in the Figure shows a land use map for an urban area and part B shows its representation as a quadtree. The schematic diagram (part C) illustrates the components of the quadtree. The **root** is defined as the point from which all other branches expand. A **leaf** is a point from which there is no further branching. All other points in the tree are termed **nodes**. Several systems have been developed to assign identification or key numbers to the quadtree nodes. The Morton matrix number is perhaps the most widely known because it is convenient for computer implementation (see Peuquet 1984). The numbering scheme shown in part D of the Figure is that developed by Abel and Smith (Abel and Smith 1983). This type of quadtree is termed a linear quadtree because the number assigned to each cell is an ordered list of its parent nodes. For example, cell 212 is contained in cell 21, which is contained in cell 2.

A.   Area as Represented on a map

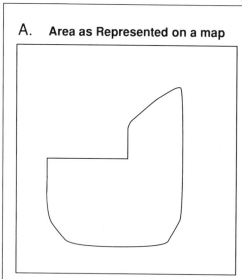

B.   Quadtree Representation

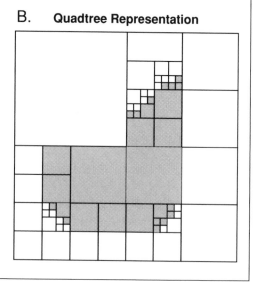

Figure 6.13   Variable Division of Space Using the Quadtree Data Model.

THE QUADTREE REPRESENTATION

A. Land-Use Map

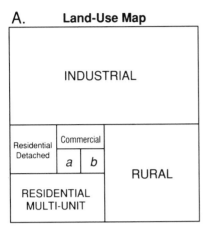

a. *Community Services*
b. *Recreation*

B. Quadtree Representation

C. Schematic Representation of the Quadtree

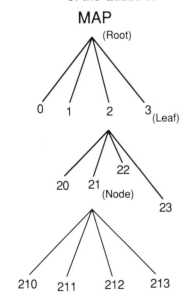

D. Table of Attributes

Quadtree Levels			ATTRIBUTES
1	2	3	
0			Industrial
1			Industrial
2			Residential
	20		Residential, Detached
	21		Services
		210	Commercial
		211	Commercial
		212	Community Services
		213	Recreation
	22		Residential, Multi-Unit
	23		Residential, Multi-Unit
3			Rural

Figure 6.14 The Organization of Data Using the Quadtree Model.

The numbering scheme of a linear quadtree has several useful properties. The neighbours of any point can be efficiently identified from its key number. For example, the neighbours of cell 212 will be the other 21x cells (210, 211, and 213) and the 2x cells (20, 21, 22, and 23). The physical structure of the computer file is also organized according to the numbering scheme. As a result, cells that are close together on the map are close together in the file. For operations that use data for a neighborhood, this storage organization provides efficient data retrieval, which in turn improves system performance. Quadtrees are particularly efficient for identifying the nearest neighbour of a selected point and for identifying the area (polygon) in which a point is located (termed a **point-in-polygon** search).

Another property of quadtrees is that the attribute coding makes it relatively easy to generalize the data to any level of detail. If a lower level of detail is needed, a lower level of branching is specified. In part D of Figure 6.14, the attribute table lists the class of every cell, including the parent cells of the finer divisions. To produce a map with a minimum cell size at the second level of subdivision, the attribute for cell 21 (Services) is used instead of those for cells 210, 211, 212, and 213.

As with other data models, there are trade-offs in using quadtrees. One of the major disadvantages of quadtrees is the time it takes to create and modify them. Compared with a simple raster structure, the more complex quadtree data structure requires more processing time to generate the quadtree with its indexes and tables. For complex areas, the processing time to generate a quadtree can be significant and whenever the map is changed, the quadtree structure must be modified — also a more complex task than modifying a raster file. Depending on the software and the spatial complexity of the data, the updating procedure can be made relatively efficient. The overhead tasks needed to maintain the quadtree can be optimized. However, they cannot be eliminated. The significance of this overhead depends heavily on the type of data, the application, and the design of the software.

Quadtrees can provide more efficient storage of data but only if the data are fairly homogeneous. The fewer the classes and the larger the clumps, the greater the degree of data compression and the more efficient the quadtree structure. But if the clumps are small and there are many different classes, then quadtree encoding can actually produce a file *larger* than a raster file for the same data.

In summary, the use of quadtrees can make certain spatial analysis functions (such as set operations and point-in-polygon searches) more efficient than when a conventional raster data structure is used. Quadtrees tend to have their greatest advantage when the data are relatively homogeneous, and do not require frequent updating, and when fast execution of certain types of functions are needed. Their advantages diminish as the map becomes more complex, as the data have to be updated more frequently, and when analyses less suited to quadtrees are done (Waugh 1986).

THE VECTOR DATA MODEL

The vector data model provides for the precise positioning of features in space. The approach used in the vector model is to precisely specify the position of the points, lines, and polygons used to represent features of interest. The map area is assumed to be a continuous coordinate space where a position can be defined as precisely as desired. The vector model assumes that position coordinates are mathematically exact. In fact, the level of precision is limited by the number of bits used to represent a single value within the computer, although it is a very fine resolution compared with the cell sizes generally used in raster systems.

The location of features on the earth's surface are referenced to map positions using an XY coordinate system (termed a **Cartesian coordinate system**). Geographic features are commonly recorded on two-dimensional maps as points, lines, and areas. The vector model uses a similar approach. A point feature is recorded as a single XY coordinate pair, a line as a series of XY coordinates, and an area as a closed loop of XY coordinate pairs that define the boundary of the area. (An area bounded by a closed loop of straight-line segments is termed, a **polygon**.)

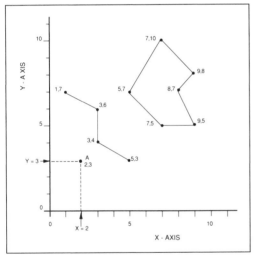

Figure 6.15 Representing Points, Lines, and Polygons as XY Coordinate Strings.

Figure 6.15 illustrates how geographic features are digitally encoded using XY coordinates. The position of point A is represented by the single coordinate pair 2,3, and the line is represented by an ordered list of coordinate pairs 1,7; 3,6; 3,4; 5,3. The area is represented by an ordered list of coordinate pairs that begins and ends at the same position, thereby forming a closed loop. The coordinates for the area feature are: 7,10; 9,8; 8,7; 9,5; 7,5; 5,7; 7,10. In this example, the coordinates are in arbitrary units. However, in a GIS, positions are usually stored using a standard geographic coordinate system like UTM, State Plane, or Latitude and Longitude.

A number of systems have been devised to organize the storage of point, line, and polygon coordinate data. The early systems were designed to meet the needs of automated mapping where the principal objective was to store the positions of the points, lines, and polygons, as well as the drawing instructions to plot them (such as line weight, colour, and pattern). Little information was needed about the geographic features that these graphic elements represented.

These systems were later developed to provide for storage of geographic attributes and recognition of the graphic elements that represented a particular geographic feature (such as a group of line segments that represented a particular roadway). However, the data were stored as a more or less unorganized collection of elements. An example of this type of vector model is the **spaghetti model**.

The Spaghetti Data Model

The spaghetti data model is illustrated in Figure 6.16. In this model the paper map is translated line-for-line into a list of XY coordinates. A point is encoded as a single XY coordinate pair and a line as a string of XY coordinate pairs. An area is represented by a polygon and is recorded as a closed loop of XY coordinates that define its boundary. The common boundary between adjacent polygons must be recorded twice, once for each polygon. A file of spatial data constructed in this manner is essentially a collection of coordinate strings with no inherent structure — hence the term **spaghetti model**.

The structure of this model is very simple and easy to understand. The data model is really the map expressed in Cartesian coordinates. The data file of XY coordinates is actually the **data structure**, the form in which the spatial data are stored in the computer. Although all the spatial features are recorded, the spatial relationships between these features are not encoded. For example, information is not recorded about the

THE "SPAGHETTI" DATA MODEL

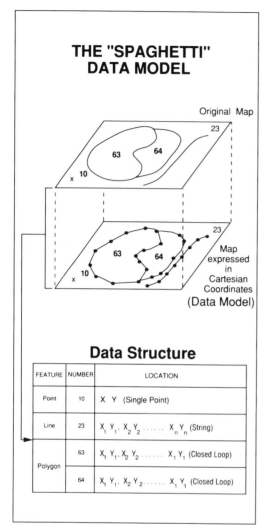

Original Map

Map expressed in Cartesian Coordinates (Data Model)

Data Structure

FEATURE	NUMBER	LOCATION
Point	10	X Y (Single Point)
Line	23	$X_1 Y_1, X_2 Y_2 \ldots\ldots X_n Y_n$ (String)
Polygon	63	$X_1 Y_1, X_2 Y_2 \ldots\ldots X_1 Y_1$ (Closed Loop)
	64	$X_1 Y_1, X_2 Y_2 \ldots\ldots X_1 Y_1$ (Closed Loop)

Figure 6.16 The Spaghetti Data Model. (Adapted from Dangermond 1982).

features adjacent to each polygon. This information would have to be generated by searching all the features in the data file and calculating whether or not they were adjacent.

The spaghetti model is very inefficient for most types of spatial analyses since any spatial relationships must be derived by computation. However, it is an efficient model for digitally reproducing maps because information extraneous to the plotting process, such as spatial relationships, are not stored (Peuquet 1984).

The Topological Model

The topological model is the most widely used method of encoding spatial relationships in a GIS. **Topology** is the mathematical method used to define spatial relationships. Figure 6.17 illustrates a map and its associated data tables. This particular form of topological model is termed the **Arc-Node** data model. The basic logical entity is the **arc**, a series of points that start and end at a node. A **node** is an intersection point where two or more arcs meet. A node can also occur at the end of a "dangling" arc, i.e. an arc that is not connected to another arc, such as the end of a dead-end street. Isolated nodes, not connected to arcs, represent points. A polygon is comprised of a closed chain of arcs that represents the boundaries of the area.

In Figure 6.17, the topology is recorded in three data tables, one for each type of spatial element, and the coordinate data are stored in a fourth table. In a GIS, polygons and points are often stored in one type of data layer and lines are stored in a separate data layer, in which case a separate set of topology and coordinate tables would be associated with each data layer. For purposes of illustration, points, lines, and polygons are shown here together in the same data layer.

The **Polygon Topology Table** shows the arcs that make up the boundaries of each polygon. For example, polygon *A* is bounded by arcs *a1, a3*, and *a5*. (By convention, the arcs that make up a polygon are defined moving in a clockwise direction.) Polygons can have islands within them. Polygon *C* is an island in polygon *B*. This is indicated in the arc list for polygon *B* by a zero preceding the list of arcs that make up the island. In this case, there is only one arc (*a7*) in polygon *C*. The point in polygon *B* is also treated as a polygon, polygon *D*, which is comprised of the single arc *a6*. A point can be considered a polygon with no area. In

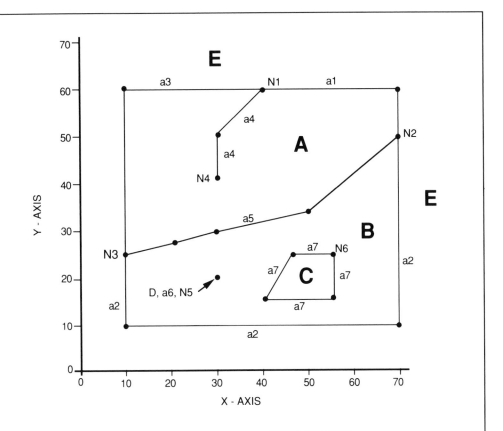

SPATIAL DATA ENCODING

POLYGON TOPOLOGY	
POLYGON	ARCS
A	a1, a5, a3
B	a2, a5, 0, a6, 0, a7
C	a7
D	a6
E	area outside
	map coverage

NODE TOPOLOGY	
NODE	ARCS
N1	a1, a3, a4
N2	a1, a2, a5
N3	a2, a3, a5
N4	a4
N5	a6
N6	a7

ARC TOPOLOGY				
ARC	START NODE	END NODE	LEFT POLYGON	RIGHT POLYGON
a1	N1	N2	E	A
a2	N2	N3	E	B
a3	N3	N1	E	A
a4	N4	N1	A	A
a5	N3	N2	A	B
a6	N5	N5	B	B
a7	N6	N6	B	C

ARC COORDINATE DATA			
ARC	START X, Y	INTERMEDIATE X, Y	END X, Y
a1	40, 60	70, 60	70, 50
a2	70, 50	70,10; 10,10	10, 25
a3	10, 25	10,60	40, 60
a4	40, 60	30,50	30, 40
a5	10, 25	20,27; 30,30; 50,32	70, 50
a6	30, 20		30, 20
a7	55, 27	55,15; 40,15; 45,27	55, 27

Figure 6.17 The Topological Data Model.

this example, it is also treated as an isolated node and as an arc comprised of a single point. (Details such as these may be treated differently, depending on the software used.) In order to complete the spatial definitions, there must be a way to refer to the area that is outside the map boundary. This outside area is designated as polygon E, for which the arcs are not explicitly defined.

In the **Node Topology Table**, each node is defined by the arcs to which it belongs. For example, node $N1$ is an endpoint for arcs $a1$, $a3$, and $a4$. Node $N5$ is a single point that is also defined as arc $a6$ and as polygon D. The **Arc Topology Table** defines the relationship of the nodes and polygons to the arcs. The end points are distinguished by designating one node as the *start* or *from* node and the other as the *end* or *to* node. For example, arc $a5$ starts at node $N3$ and ends at node $N2$. Moving from $N3$ to $N2$, the polygon to the left is Polygon A and the polygon on the right is polygon B.

From the topology alone, i.e. the three topology tables, analyses of the relative position of the map elements can be done. For example, all polygons adjacent to polygon B can be found by searching the Arc Topology Table. Every polygon paired with B in this table is adjacent to it because they have a common arc. For example, polygon A and B are paired in the entry for arc $a5$. Therefore polygon A is adjacent to polygon B.

The topology tables can be used to find all features contained within a polygon by searching the polygon topology table for arc lists that contain a zero. The arcs following each zero are then searched in the Arc Topology Table to identify the elements. Polygon B is seen to have two contained features, one defined by arc $a6$ and the other by arc $a7$. From the Arc and Node Topology Tables, arc $a6$ is seen to be a single point (it has the same right and left polygon and has only one node). Arc $a7$ is an island polygon (it has a different polygon

on its right and left and one node). Spatial queries of this type can be processed much more quickly using the topology tables than they can be done by calculation from the coordinate data (as required for non-topological data models, such as the spaghetti model).

To relate the map features to "real world" positions, the XY coordinates are needed. These are stored in the **Arc Coordinate Data Table**. Each arc is represented by one or more straight-line segments defined by a series of coordinates. The more complex the shape, the more coordinates are needed to represent it as a series of straight-line segments. Arc $a1$ makes one sharp turn and can be represented by its end points and a single intermediate point. To represent the curved shape of arc $a5$, several intermediate points must be encoded. The coordinates of the nodes can be obtained from this table by reference to the Arc Topology Table in which the node numbers for the start and end points are identified.

Attribute data are commonly stored in the form of relational tables in which one data field contains an identification code for the spatial entity. This is illustrated in Figure 6.9 (discussed previously). A relational data base is easily adapted to handle large quantities of attribute data and provides a very flexible approach to data retrieval.

A topologically structured data model is well-suited to such spatial operations as contiguity and connectivity analyses (discussed in Chapter 7). **Contiguity** is the spatial relation of adjacency, i.e. elements that touch each other are adjacent. Contiguity analysis is applied to a wide range of applications. A biologist might be interested in the habitats that occur next to each other. A city planner might be interested in zoning conflicts, such as industrial zones bordering recreational areas.

Connectivity refers to interconnected pathways or networks that transport something. The streets of a city, the cables of a telephone system, and the streams and rivers in a landscape are examples of transportation networks. Connectivity functions are used to find optimum routings through a network, such as the most efficient delivery route or the fastest travel route. A network analysis can also be used to optimize transportation scheduling, such as in bus route planning. Network functions can also be used to predict loading at critical points in a network, such as the water flow at a bridge crossing that will result from heavy storm runoff.

Trade-Offs of the Topological Model

One of the advantages of a topological structure is that spatial analyses can be done without using the coordinate data. Many spatial analyses, such as contiguity and connectivity analyses, can be done largely, if not entirely, using the topological data alone. This avoids the time-consuming calculations needed to derive spatial relationships from the geographic coordinates. When spatial data are stored using a non-topological model, extensive calculations are needed to derive the topological information when they are needed. As a result, many spatial operations are performed much more efficiently by a topologically-based GIS.

Creating the topological structure does, however, impose a cost. When a new map is entered or an existing map is changed, the topology must be updated. This updating procedure can be relatively time-consuming (on the order of minutes to an hour or two), depending on the size of the map, efficiency of the software, and speed of the hardware. Many systems optimize the updating procedure by saving all the changes from an editing session and then updating the topology at a later time as a batch job (i.e. a process that can be run without operator interaction).

Systems that do not have a topological structure can use a simpler internal data structure but require more complex algorithms to analyze spatial relationships. More powerful hardware can be used to handle the more complex spatial analysis operations, or slower performance for these operations might be acceptable. The trend has been strongly towards the inclusion of topology in GIS data bases. Virtually all full-featured, vector-based GISes now use a topological data model.

The Triangulated Irregular Network (TIN)

The Triangular Irregular Network or TIN is a vector-based topological data model that is used to represent terrain data. A TIN represents the terrain surface as a set of interconnected triangular facets. For each of the three vertices, the XY coordinate (geographic location) and the Z coordinate (elevation) values are encoded.

The structure of a TIN is illustrated in Figure 6.18. Each triangle or facet is designated by a letter and is defined by three nodes designated by numerals. The coordinate data and topology for the TIN are stored in a set of tables. The **Node Table** lists each triangle and the nodes which define it. The **Edge Table** lists the three triangles adjacent to each facet. Triangles that border the boundary of the TIN show only two adjacent facets. The **XY Coordinate Table** and **Z Coordinate Table** store the node coordinate values. TIN algorithms differ in the way they generate the network of triangles, and so the resulting solutions will be slightly different. In practise, triangulations in which the triangles are most equilateral in shape tend to most accurately represent the surface. Figure 6.19 shows a section of the US Geological Survey 1:24,000 La Honda Quadrangle and a TIN for the same area.

Using a TIN model, terrain parameters like slope and aspect are calculated for each facet and stored as an attribute of the facet

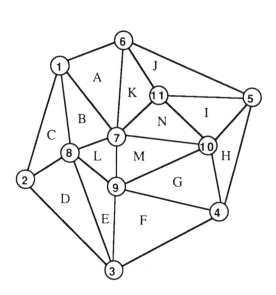

X-Y COORDINATES	
node	coordinates
1	x1, y1
2	x2, y2
3	x3, y3
.	
.	
11	x11, y11

Z COORDINATES	
node	coordinate
1	z1
2	z2
3	z3
.	
.	
11	z11

EDGES	
Δ	adjacent Δ
A	B,K
B	A,C,L
C	B,D
D	C,E
E	D,F,L
F	E,G
G	F,H,M
H	G,I
I	H,J,N
J	I,K
K	A,J,N
L	B,E,M
M	G,L,N
N	I,K,M

NODES	
Δ	node
A	1,6,7
B	1,7,8
C	1,2,8
D	2,3,8
E	3,8,9
F	3,4,9
G	4,9,10
H	4,5,10
I	5,10,11
J	5,6,11
K	6,7,11
L	7,8,9
M	7,9,10
N	7,10,11

Figure 6.18   The Structure of a TIN. The TIN is a topological data model. The data are stored in a set of tables that retain the coordinate values as well as the spatial relations of the facets, as explained in the text. (Adapted from an illustration by ESRI. Redlands, California.)

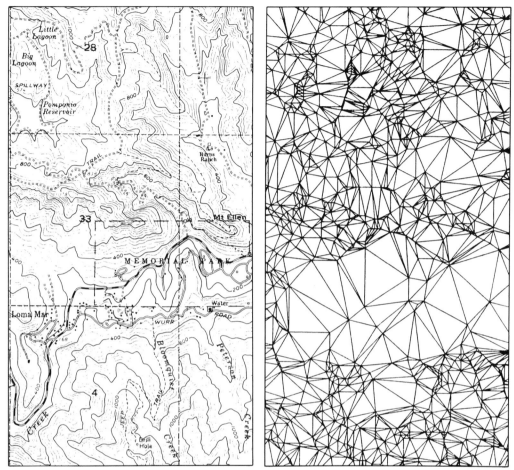

Figure 6.19 A Topographic Map and the Corresponding TIN Representation. The map on the left is a section of the US Geological Survey 1:24,000 La Honda, California Quadrangle. It represents a ground distance about 2.4 km across. The TIN on the right was generated from 6% of the values (about 700 points) from the digital elevation model for the area. (From R.J. Pike et. al. 1987, courtesy of the US Geological Survey and the American Society for Photogrammetry and Remote Sensing.)

in the same way as attributes are stored for polygons. These values can then be queried using the same types of data base operations. Figure 6.20 shows examples of the slope and aspect information that can be generated using TIN data.

Terrain data in digital form are usually provided as a raster data set. The regular grid of values are produced by photogrammetric analysis of airphoto or satellite stereopairs. Where the elevation values are an irregular set of data points, such as elevation contours, or spot elevation measurements, a regular grid of points can be generated by estimating values for the cells that do not contain data points. The process of estimating the value for missing points is termed **interpolation**. The values derived in this way are not necessarily the true values; they are a mathematical "best guess" based on the known values.

A major limitation of the raster representation of elevation data has been the use of a uniform cell size. If a small cell size is chosen, a large data file is produced and areas with low spatial variability, such as gentle terrain,

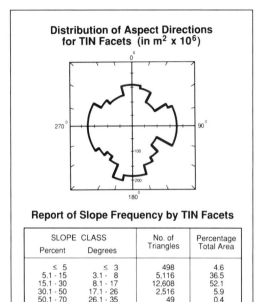

Distribution of Aspect Directions for TIN Facets (in m² x 10⁶)

Report of Slope Frequency by TIN Facets

SLOPE CLASS		No. of Triangles	Percentage Total Area
Percent	Degrees		
≤ 5	≤ 3	498	4.6
5.1 - 15	3.1 - 8	5,116	36.5
15.1 - 30	8.1 - 17	12,608	52.1
30.1 - 50	17.1 - 26	2,516	5.9
50.1 - 70	26.1 - 35	49	0.4
>70	>35	60	0.5

Figure 6.20 Examples of Slope and Aspect Reports Generated from the TIN Attribute Data. (Adapted from R.J. Pike et. al. 1987).

use more cells than are actually needed. Conversely, if a large cell size is used, the data file is smaller but the grid may not be sufficiently fine to capture the detail in highly variable terrain. Raster formats with variable-sized grids have been developed that allow a predetermined range of cell sizes to be used within a single terrain model. However, this does not completely solve the problem of precisely representing terrain break-points, such as ridges.

One of the advantages of a TIN is that extra information is encoded for areas of complex relief without requiring large amounts of data to be collected from areas of simple relief. Because the size of each facet is variable, smaller triangles and therefore a more detailed representation can be provided where there is a higher density of data points. Break-point features in the terrain, such as ridge lines, valley bottoms, highpoints, and saddle-shaped passes, can be accurately encoded by using a higher

density of elevation points. As a result, these features can be more precisely encoded in a TIN than in a grid cell representation where these sharp features may be smoothed.

The disadvantage of the TIN compared with a raster representation is that significantly more processing is needed to generate the TIN file itself. However, once the TIN file is produced, the more compact representation can be more efficiently processed. Image processing hardware and software can efficiently handle very large raster data files at high speed, but vector-based GIS hardware is not designed to process large raster files. The various triangulation algorithms used to generate TINs differ in the results and types of errors they produce, and editing is often needed to remove unwanted artifacts near the edges of the model. However, TIN models can represent a surface more accurately and with far less storage than a raster model. The organization of the TIN and its smaller data volume have also made it attractive for interactive editing and manipulation procedures, such as interactive surface modelling.

DATA BASES FOR GIS: MANAGING SPATIAL AND ATTRIBUTE DATA TOGETHER

Over the past three decades, a considerable body of theory and practical experience has been developed in handling non-spatial data. The development of data base systems to handle spatial information with their associated attribute data has been much more recent, mainly within the past five years. Spatial information is an inherently more complex type of data to store and to manipulate than non-spatial data.

Most spatial data are still being stored in the form of paper maps, imagery, tables, or text descriptions and are analyzed using manual techniques. However, there is a rapid increase in the amount of geographic information that is now being collected and stored in digital form suitable for computer-based retrieval and analysis. Many of these data sets are digital representations of conventional maps. But computer-based processing has enabled a greater range of geographic data sets, many of which are not usually represented in map form, to be accommodated within the same spatial data base.

ORGANIZING GEOGRAPHIC INFORMATION WITHIN A DBMS

Digital mapping was historically viewed as a set of tools for automating the drafting function. It was a task generally confined to a single unit within an organization, such as a cartography section or a municipal public works department. The digital mapping function was not viewed as an integral part of the data management system of the organization.

Early GISes and automated cartography systems used data files directly without using a data base management system (DBMS). This file processing approach to data management (discussed previously) is still used in many GISes today. In an automated cartography environment, spatial data handling tends to be greatly simplified. The work is generally performed by a single operator using a stand-alone system. The data files usually contain a standardized set of information for a standard size area, such as a map sheet. Rapid response time is critical in a high volume automated mapping environment, and so vendors have tended to favour the file processing approach, which does not suffer from the overheads imposed by a DBMS. Since automated cartography systems and GISes perform similar geometric manipulations, it is often expected that they should provide comparable levels of performance. However, the additional complexities of handling the data components other than the geometric ones and providing a multi-user environment, makes this expectation difficult to satisfy.

In a GIS, geographic information is not approached as a drafting task but as a data base application. The advantages of organizing data using a DBMS were discussed earlier. They include minimizing redundancy of data storage, providing central control of data access, manipulation, integrity, and security of the data base, and making application programs independent of the form in which the data are stored.

Following are some of the ways that DBMS concepts are applied in the context of a GIS.

1. The views of the data are independent of the way the data are stored. So for example, instead of storing different maps, the data that describe the geographic elements (i.e. the spatial and attribute information) are stored with minimum redundancy, and then maps or other types of output are generated as needed in the form best suited for a specific analysis.

2. Automated updating of interrelated data files is provided. As changes occur within an organization, the single transaction that registers that change, such as the sale of a land parcel, can be used to update all the data files that are affected. Then all users immediately have their views of the data base updated.

3. The relationships among all the spatial and attribute information are explicitly defined. Keys are used to relate the attribute information to the corresponding spatial features, and topology is used to relate all the spatial elements to each other.

4. The central control of the DBMS provides better control of the integrity of the data base by means of security and consistency checking to prevent misuse or degradation of the information as it is managed.

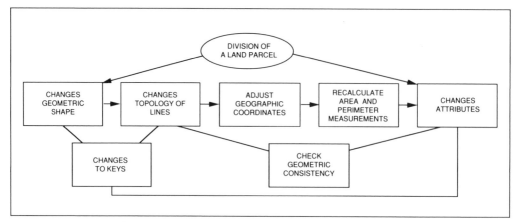

Figure 6.21   The Cascading Effect of Changes in a GIS Data Base. (Adapted from an illustration courtesy of ESRI. Redlands, California.)

LIMITATIONS OF GENERAL PURPOSE DBMSES FOR GIS APPLICATIONS

The data model most widely accepted for handling non-spatial attribute data in GIS applications has been the relational model. The organization of attribute data into a series of tables that can be used individually or together is simple to understand and provides efficient data storage. Relational data bases use non-procedural query languages (discussed previously) that are easily learned by GIS users and provide very flexible analysis capabilities. The relational model is thus well-suited to handling the storage of the non-spatial attribute data and is well-adapted to the unpredictable nature of GIS analyses. Using a relational DBMS to store the spatial data and adapting the DBMS query language to provide spatial analysis functions has been a much more difficult problem to handle.

When a relational data base is used to manage tabular data, the types of changes or transactions most commonly done involve adding, deleting, and changing a record (i.e. altering the contents of a field). For example, in a property data base, if a new property is added, a new record is added. If the owner changes, the name field of the corresponding property record is altered.

Spatial information is more complex and the transactions performed on them are more complicated. Instead of a change causing a single data record to be altered, a modification to the spatial information commonly involves the simultaneous updating of large numbers of records in multiple files. This is illustrated in Figure 6.21 for a transaction to divide a land parcel. The transaction changes the geometric shape from one parcel to two smaller parcels. As a result, the topology of the lines bounding the parcel must be altered, which in turn requires that the coordinates be changed and the area and perimeter measurements for the parcels be recalculated. The attributes of the land parcels (such as ownership, property ID, assessed values, and so on) will be updated, and the keys that link the attributes to the spatial data will also have to be modified.

The functions needed to handle geographic data are not done well in the tabular data base environment of a standard relational DBMS. Some of the major difficulties are listed below:

1. The spatial data records used in a GIS are variable length records which are needed to store variable numbers of

coordinate points, whereas general purpose data base systems are designed to handle fixed length records. Also, a rather complex topology that is interrelated with the spatial coordinate data must be correctly maintained. To provide these data base functions, additional software is needed to extend the capabilities of the relational DBMS (see for example Waugh and Healey 1987, Charlwood et. al. 1987).

2. Manipulation of geographic data involves spatial concepts, such as proximity, connectedness, containment, and overlay, that are not easily accommodated by general purpose data base query languages.

3. A GIS requires sophisticated graphics capabilities that are not normally supported by a general purpose DBMS.

4. Geographic information is complex. The representation of a single geographic feature requires multiple records in multiple files. It may involve geodetic networks, feature coordinates, topology, measurements of spatial features, keys to the non-spatial attribute data, and the non-spatial attributes themselves.

5. The highly interrelated nature of GIS data records requires a more sophisticated security system than the record locking approach used by general purpose DBMSes. To ensure the integrity of the geographic data base, the security system must protect the integrity of the multiple files in which the spatial data are stored. A change in one record can create multiple errors in multiple files.

The urgent need for operational spatial data handling in computer-based GIS applications has outstripped the ability of the research community to develop the needed spatial data base theory and practical spatial data base systems. As a result, the design of GIS data base management systems still depends heavily on *ad hoc* techniques to bridge the gaps in our theoretical framework.

PRACTICAL APPROACHES USED TO IMPLEMENT A GIS

Over the past decade, various practical approaches were taken to provide data management services for a GIS. They may be broadly grouped into the following four, somewhat overlapping, strategies (see Figure 6.22):

1. Develop a proprietary system providing the individual data management services required by the different application modules. This is the file processing approach.

2. Develop a hybrid system using a commercially available DBMS (usually a relational one) for storage of the non-spatial attributes. Develop separate software to manage the storage and analysis of the spatial data, using the services of the relational DBMS to access the attribute data.

3. Use an existing DBMS, usually a relational one, as the core of the GIS. Then develop extensions to the system where needed. Although the spatial and attribute data may be managed by the DBMS, a significant amount of software is generally added to the DBMS to provide the spatial functions and graphics display used in geographic analysis.

4. Start from scratch and develop a spatial data base capable of handling the spatial and non-spatial data in an integrated fashion.

The first category includes most existing GISes. The widely distributed Map Analysis Package developed by Tomlin (1983) is one example of many GISes that use this approach. Each data set is stored as a

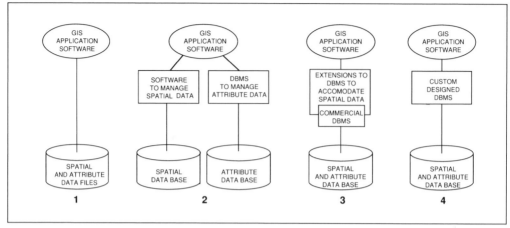

Figure 6.22   Four Approaches to GIS System Design.

separate file. The user can invoke a separate analysis functions to manipulate one or more of these data files. Results are produced in the form of new data files that can be output using suitable routines.

The major commercial GISes use one of the remaining three approaches. The ARC/INFO system from the Environmental Systems Research Institute (ESRI) is an example of the hybrid approach. The non-spatial attribute data are stored in the INFO data base management system. (ESRI has also implemented their GIS using the ORACLE and INGRES relational DBMS.) The commercial DBMS provides both storage and manipulation functions for the non-spatial attribute data. The ARC system (manufactured by ESRI) provides for the storage and manipulation of the spatial data. Spatial analysis functions are provided using the "toolbox" approach. In ARC, individual modules are provided for such functions as data entry, editing, network analysis, and so on. These modules are then used sequentially to perform the required geographic analyses (Morehouse 1985).

The GIS from Kork Systems also uses a hybrid approach. The non-spatial attribute data are maintained in a relational DBMS and the spatial data are maintained in an object-oriented data base system, called PANDA. (An **object-oriented data base** treats a set of attributes pertaining to a feature as a single unit, termed an object. An object may be composed of other objects, thus allowing interrelated features to be represented as a hierarchically structured set of objects.) PANDA uses a quadtree structure for data storage, and the data base system is organized on the network model. Spatial and non-spatial data can be manipulated together using the SQL query language to which have been added special commands to handle spatial functions (Ingram and Phillips 1987).

System 9, developed by Wild Heerbrugg and Prime Computers, is an example of a GIS built around an existing DBMS. In this case, a relational DBMS operating in the UNIX environment was used. Both the spatial coordinate and topological data as well as the non-spatial attribute data are stored using the relational data model. The relational DBMS supports variable length fields and extensions were added to the SQL query language to handle spatial referencing (such as the points and lines that comprise a geographic feature) and spatial query functions (e.g. overlay, connectivity, and neighbourhood operations) (Charlwood et. al. 1987).

The GEOVIEW system under development at the University of Edinburgh is being used to research applications of the relational data base model to store spatial and non-spatial attribute data. The system handles both raster and vector format spatial data. The raster data are encoded either as point data, data blocks, or using quadtrees. The vector data are encoded using a topological model. All data are stored using relational tables and queries are done using an extended SQL language. The first implementation was built on the ORACLE data base system (Waugh and Healey 1987).

A number of GISes under development at research facilities have incorporated artificial intelligence (AI) techniques so that the system itself can create new information about objects as the system is used (i.e. to "learn" about objects from experience). A knowledge-based GIS, KBGIS-II, has been developed at the University of California at Santa Barbara to explore these possibilities (Smith et. al. 1987). The major system functions provided are query-answering, learning, editing, and training. Data storage uses the quadtree structure and routines are included to optimize the execution of complex searches. The system does not yet provide the cartographic capabilities and polygon processing functions available in a full-featured commercial GIS. However, it has provided an interactive system that can use AI methods to modify its knowledge base, can handle large multi-layered data sets, and integrate both image and digital cartographic data.

The MAPS system being developed at Carnegie-Mellon University (McKeowan and Lai 1987, McKeowan 1987) has taken a somewhat different approach than most other GISes. Instead of trying to develop one or perhaps two representations of spatial data that are sufficiently general for all types of spatial information, the MAPS system supports several different representations in a manner that is transparent to the user. Data in a number of vector and raster forms can be accommodated and data base

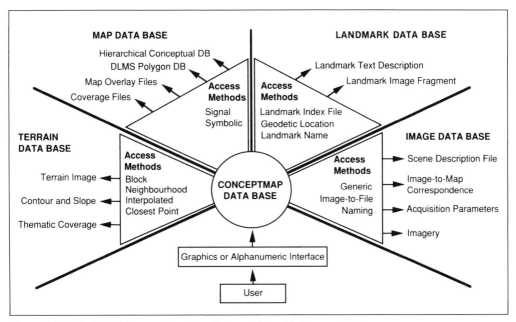

Figure 6.23 The MAPS System Overview. (Adapted from McKeowan 1987.)

queries are optimized to use the representation best-suited to execute a particular function.

The system is used to store and manipulate a data base of some 200 high-resolution airphoto images, digital terrain data, and a variety of map datasets for the Washington, D.C. area. MAPS has also been used as a component of several expert systems such as an automated road finder and follower and a system to interpret airport scenes. The system offers the types of functions needed for change detection, automated image interpretation, and report generation tasks.

Five data bases are maintained within the MAPS system: a digital terrain data base, a map data base, a landmark data base, an image data base, and the CONCEPTMAP data base (see Figure 6.23). Together they provide for the storage and integrated analysis of such diverse data sets as digitized aerial photography, digital satellite imagery, digital terrain data, topologically encoded map data, symbols, and non-spatial attributes. CONCEPTMAP contains the information about spatial entities and manipulation procedures that are used to interact directly with a user or with an applications program. In this way, MAPS provides a single uniform access to all spatial data, independent of their internal structure.

CONCLUSION

A DBMS provides a number of functions critical to the effective operation of a GIS. However, the difficulties of adapting existing DBMSes to handle spatial data have required that hybrid or modified DBMSes be developed. These developments are leading to improved methods of representing and manipulating spatial and non-spatial data within a single environment. Continued research is needed to develop improved spatial DBMSes that can meet the performance requirements of operational GISes.

There is a strong demand for systems that can integrate diverse raster and vector format data sets without requiring the operator to first convert the data to a common format. Researchers are developing new methods to provide these capabilities.

REFERENCES

Abel, D.J. and J.L. Smith. 1983. A Data Structure and Algorithm Based on a Linear Key for a Rectangle Retrieval Problem. *Computer Graphics, Vision, and Image Processing* 24:4–14.

Aronoff, S., R. Mosher, and R.V. Maher. 1987. Operational Data Integration — Image Processing to Interface Vector GIS and Remotely Sensed Data. In *Proceedings of the GIS '87 Symposium*. American Society of Photogrammetry and Remote Sensing. Falls Church, Virginia. pp.216–225.

Aronson, P. 1987. Attribute Handling for Geographic Information Systems. In *Proceedings of Autocarto 8*. American Society of Photogrammetry and Remote Sensing. Falls Church, Virginia. pp.346–355.

Bourrough, P.A. 1986. *Principles of Geographic Information Systems for Land Assessment*. Oxford University Press. Oxford, U.K. 193p.

Bowers, D.S. 1988. From *Data to Database*. Van Nostrand Reinhold (UK) Co. Ltd. Berkshire, England.

Cebrian, J.A., J.E. Mower, and D.M. Mark. 1985. Analysis and Display of Digital Elevation Models within a Quadtree-Based Geographic Information System. In *Proceedings of Autocarto 7*. American Society of Photogrammetry and Remote Sensing. Falls Church, Virginia. pp.55–64.

Charlwood, G., G. Moon, and J. Tulip. 1987. Developing a DBMS for Geographic Information: A Review. In *Proceedings of Autocarto 8*. American Society of Photogrammetry and Remote Sensing. Falls Church, Virginia. pp.302–315.

Dangermond, J. 1982. A Classification of Software Components Commonly Used in Geographic Information Systems. In *Proceedings of the US–Australia Workshop on the Design and Implementation of Computer-Based Geographic Information Systems*. Honolulu, Hawaii. pp. 70–91.

Date, C.J. 1981. *An Introduction to Database Systems, Volume I*. Third Edition. Addison-Wesley Publishing Company. Reading, Massachusetts.

Date, C.J. 1983. *An Introduction to Database Systems, Volume II*. Addison-Wesley Publishing Company. Reading, Massachusetts.

Frank, A.U. 1988. Requirements for a Database Management System for GIS. *Photogrammetric Engineering and Remote Sensing* 54(11):1557–1564.

Ingram, K. and W. Phillips. 1987. Geographic Information Processing Using A SQL-Based Query Language. In *Proceedings of AutoCarto 8*. American Society of Photogrammetry and Remote Sensing. Falls Church, Virginia. pp.326–335.

Larson, J.A. 1987. *Database Management*. IEEE Computer Society Press. Washington, D.C.

McKenna, D.G. 1987. The Inward Spiral Method: An Improved TIN Generation Technique and Data Structure for Land Planning Applications. In *Proceedings of AutoCarto 8*. American Society of Photogrammetry and Remote Sensing. Falls Church, Virginia. pp.670–679.

McKeown, D.M., and R.C.T. Lai. 1987. Integrating Multiple Data Representations for Spatial Databases. In *Proceedings of AutoCarto 8*. American Society of Photogrammetry and Remote Sensing. Falls Church, Virginia. pp.754–763.

McKeown, D.M. 1987. The Role of Artificial Intelligence in the Integration of Remotely Sensed Data with Geographic Information Systems. *IEEE Transactions on Geoscience and Remote Sensing* GE-25(3):330–348.

Morehouse, S. 1985. ARC/INFO: A Geo-Relational Model for Spatial Information. In *Proceedings of AutoCarto 7*. American Society of Photogrammetry and Remote Sensing. Falls Church, Virginia. pp.388–397.

Peuquet, D. 1984. A Conceptual Framework and Comparison of Spatial Data Models. *Cartographica* 21(4):66–113.

Peuquet, D. 1983. The Application of Artificial Intelligence Techniques to Very Large Geographic Databases. In *Proceedings of Autocarto 6*. University of Ottawa. Ottawa, Ontario. pp.419–420.

Pike, R.J., G.P. Thelin, and W. Acevado. 1987. A Topographic Base for GIS from Automated TINs and Image Processed DEMs. In *Proceedings of the GIS '87 Symposium*. American Society of Photogrammetry and Remote Sensing. Falls Church, Virginia. pp.340–351.

Samet, H., C. Shaffer, R.C. Nelson, Y. Huang, K. Fujimara, and A. Rosenfeld. 1986. Recent Developments in Quadtree-Based Geographic Information Systems. In *Proceedings of the Second International Symposium on Spatial Data Handling*. International Geographical Union. Williamsville, New York. pp.15–32.

Smith, T., D. Peuquet, S. Menon, and P. Agarwal. 1987. KBGIS-II: A Knowledge-Based Geographic Information System. *International Journal of Geographical Information Systems* 1(2):149–172.

Tomlin, C.D. 1983. *Digital Cartographic Modelling Techniques in Environmental Management*. Unpublished Doctoral Dissertation. School of Forestry and Environmental Studies, Yale University. New Haven, Connecticut.

Ullman, J.D. 1982. *Principals of Database Systems*. Second Edition. Computer Science Press. Rockville, Maryland.

Waugh, T.C. 1986. A Response to Recent Papers and Articles on the Use of Quadtrees for Geographic Information Systems. In *Proceedings of the Second International Symposium on Spatial Data Handling*. International Geographical Union. Williamsville, New York. pp.33–37.

Waugh, T.C. and R.G. Healey. 1987. A Relational Data Base Approach to Geographical Data Handling. *International Journal of Geographical Information Systems* 1(2):101–118.

Wiederhold, G. 1984. Databases. *Computer* (October 1984):211–223.

7. GIS ANALYSIS FUNCTIONS

INTRODUCTION

What distinguishes a GIS from other types of information systems are its spatial analysis functions. These functions use the spatial and non-spatial attribute data in the GIS data base to answer questions about the real world.

The GIS data base is a model of the real world that can be used to mimic certain aspects of reality. To mimic behaviour, a model must represent certain entities (i.e. things) and the relationships among them (i.e. the rules that govern how they interact). The entities might be the names of individuals and a list of properties. The relationships might include *ownership, mortgagee*, and *mortgagor*. Together these entities and relationships provide a model for land ownership. A model may be represented in words, in mathematical equations, or as a set of spatial relations displayed as a map or stored in the computer hardware and software of a GIS.

It is important to recognize that models are designed to mimic only selected aspects of reality. In general, the more factors that a model takes into account, the more complex it becomes, and the more expensive it is to use and maintain. A more complex model may or may not provide "better" answers; it depends on the questions to be addressed.

Models are used when it is more convenient or it is not possible to collect the information directly. It is more convenient to read road distances off a map than to go out and measure them, and it is not possible to measure the height a forest will reach in 100 years time. The value in using a model is that it can be tested and manipulated more conveniently, at a faster (or slower) rate, and less expensively than the conditions it mimics. In many cases, the model is used to repeatedly perform analyses that test alternative scenarios, such as the effectiveness of an emergency evacuation plan in response to different kinds of events. A model is used to answer questions about what exists now or existed at some point in the past. Perhaps most importantly, it can be used to predict what will happen or has happened in another location or at another point in time. A GIS provides these capabilities by means of its analysis functions.

An important GIS application is predicting the consequences of proposed activities. They may involve large areas, e.g. when a reservoir is formed behind a dam, or they may involve relatively small areas, such as the effect on traffic flow of closing a city street. This ability to model what will occur provides the opportunity to select the "best" alternative. But what is "best"? A GIS can use spatial analysis functions to provide part of the answer to "What is best?". The part it cannot answer is the human value judgments that define the wants, goals, and values of the organization or society that are using the information. These judgments define the allowable trade-offs among alternatives. A failure to consider the relevant value judgments in a GIS analysis can make the procedure a theoretical exercise instead of a practical tool. Or worse yet, the analysis results may be misleading.

To develop the best answers from the information available requires a systematic framing of the questions to be addressed. There is often a strong tendency to begin the analysis with only a general idea of the questions to be answered and the data needed. This is a particularly common tendency whenever a new computer-based system, such as a GIS, is introduced.

In order to find useful answers, one must ask the right questions. Working backwards through an analysis is often a good approach. That is, begin at the end. Assuming that good answers have been produced, what questions would they have answered, what concerns would they have addressed, and what data, including judgments, would have been used in the analysis that produced those answers. This approach to the design of an analysis ensures that the effort is focussed on answering the appropriate questions. The following framework and examples illustrate this approach.

The answers provided by a GIS can be categorized into three types as illustrated in Figure 7.1:

1. a presentation of the current data, i.e. the data in the data base such as a map of the city streets,

2. a pattern in the current data, such as all houses valued at over $100,000, and;

3. a prediction of what the data could be at a different time or place. For example, predicting the services that would be lost in the event of an earthquake. This type of analysis might be used to develop emergency response plans.

The types of questions to be answered can also be characterized by three categories:

1. What are the data?, i.e. what is the information currently stored in the data base. For example, what is the name and address of the owner of a specified property?

2. What is the pattern in the data? This type of question is a search for entities that possess a specified set of characteristics. For example, plotting a map of all lots with houses valued at over $100,000 would be defining a pattern in the data that may not be obvious when all the data are viewed together — the pattern for this type of house.

3. What could the data be? This type of question implies that a predictive model will be used. The model may be as simple as predicting that a field will produce the same crop next year as this year. It may be as complex as predicting the change in stream flow after a forest has been removed from a watershed.

The functions used to produce these answers can similarly be categorized by the types of answers they provide:

1. storage and retrieval functions,

2. constrained query functions, and

3. modelling functions.

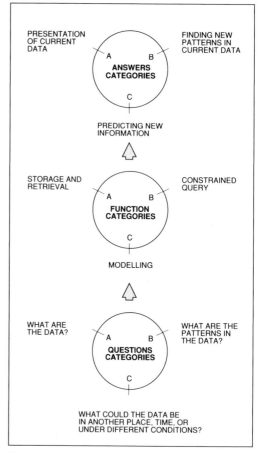

Figure 7.1 Categorizing Questions, Functions, and Answers in a GIS Analysis.

These categories of questions, functions, and answers are not mutually exclusive. A given answer, function, and question will have aspects of each category. This is represented conceptually by their position in the circles. The following examples illustrate this idea. The letter representing each example is shown in the Figure.

A. Retrieving the street map for an area primarily involves the retrieval of existing information. The map exists in the GIS and need only be recalled and output.

B. Retrieving those lots with houses valued over $100,000 is an example of the second category of use. Here the value of each lot must be considered and only those satisfying the constraint are accepted.

C. Determining the optimum routing of a powerline is an example of modelling. Multiple layers of information are used together to weigh different alternatives in order to optimize the design.

The strategy for undertaking a specific GIS analysis will depend on the answers that are to be provided. By defining the most important answers to be provided, an appropriate set of questions and analysis methods can be specified.

The remainder of this chapter presents the different functions used in a GIS to manipulate and analyze geographic information. The art and science of using a GIS is to know how to combine the analysis functions available on a particular system to provide the required information using the available data. The purpose of first presenting this framework of answers, questions, and functions is to highlight the fact that individual analysis functions must be used in the context of a complete analysis strategy. One of the unwanted characteristics of computer technology is the ability to produce incorrect information at a rapid rate, and with all

the apparent authenticity of the "real McCoy." The quality of the information produced from a GIS depends on the intelligent use of a systematic analysis approach.

ORGANIZING GEOGRAPHIC DATA FOR ANALYSIS

Geographic information are organized within a GIS so as to optimize the convenience and efficiency with which they can be used. The form of organization chosen will be influenced by the types of data to be used, the types of analyses to be performed, and the methods used to encode the data. Methods used to represent geographic data within a GIS were discussed in Chapter 6. In this section, the logical organization of the data is addressed.

On a paper map, geographic information is usually organized as a set of themes, such as roads, streams, land cover types, and political boundaries. They are often thought of as map layers, and each layer may actually have been plotted separately in the process of compiling the final map. To cover a large area, such as a country, it may be necessary to use several map sheets, and so a convenient system is used to divide the coverage area into individual map sheets. The level of detail used to present the geographic information is chosen according to the information needs that were specified and the limitations of the storage medium. In the case of a paper map, the map itself is both the means of storing the geographic information, as well as the form of presentation. Symbols, colours, line widths, and other map elements are selected to suit the need for visual analysis of the map data. The double service of the paper map as both a storage and presentation medium requires that trade-offs be made between the amount and accuracy of information that can be shown and the need for the map to be legible.

In a computer-based GIS, these organizational considerations are handled somewhat differently. Since the data storage and presentation (or output) are separate, the level of detail at which the geographic information can be stored is limited by the storage capacity of the hardware and the method used by the software to represent the data. The legibility of an output product like a map can be controlled by selecting the scale, amount, and level of detail of information, and symbols at the time when the map is to be plotted. (One can present information at a less detailed level than it was stored but cannot present more detailed information than exists in the data base. For this reason, information need only be entered once, at the finest level of detail that will be required. However, in the case of paper maps, separate maps are needed to show information at different levels of detail and different scales.)

Large coverage areas are subdivided into smaller units for more efficient storage, in a manner similar to the map sheet concept. Each unit is commonly stored as a separate set of data files. Unlike the paper map, a GIS

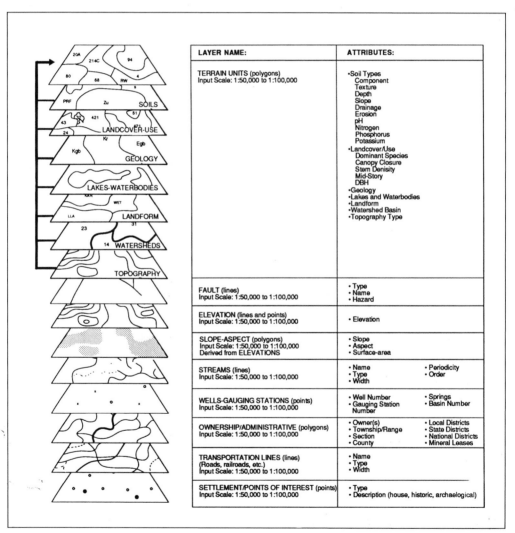

LAYER NAME:	ATTRIBUTES:	

TERRAIN UNITS (polygons)
Input Scale: 1:50,000 to 1:100,000
	•Soil Types
	  Component
	  Texture
	  Depth
	  Slope
	  Drainage
	  Erosion
	  pH
	  Nitrogen
	  Phosphorus
	  Potassium
	•Landcover/Use
	  Dominant Species
	  Canopy Closure
	  Stem Density
	  Mid-Story
	  DBH
	•Geology
	•Lakes and Waterbodies
	•Landform
	•Watershed Basin
	•Topography Type	

FAULT (lines) Input Scale: 1:50,000 to 1:100,000	• Type • Name • Hazard	
ELEVATION (lines and points) Input Scale: 1:50,000 to 1:100,000	• Elevation	
SLOPE-ASPECT (polygons) Input Scale: 1:50,000 to 1:100,000 Derived from ELEVATIONS	• Slope • Aspect • Surface-area	
STREAMS (lines) Input Scale: 1:50,000 to 1:100,000	• Name • Type • Width	• Periodicity • Order
WELLS-GAUGING STATIONS (points) Input Scale: 1:50,000 to 1:100,000	• Well Number • Gauging Station Number	• Springs • Basin Number
OWNERSHIP/ADMINISTRATIVE (polygons) Input Scale: 1:50,000 to 1:100,000	• Owner(s) • Township/Range • Section • County	• Local Districts • State Districts • National Districts • Mineral Leases
TRANSPORTATION LINES (lines) (Roads, railroads, etc.) Input Scale: 1:50,000 to 1:100,000	• Name • Type • Width	
SETTLEMENT/POINTS OF INTEREST (points) Input Scale: 1:50,000 to 1:100,000	• Type • Description (house, historic, archaeological)	

Figure 7.2   GIS Data Layers Commonly Used in Natural Resource GIS Applications. (Courtesy of ESRI, Redlands, California.)

can provide sophisticated functions to ensure that adjacent units match precisely along their borders. In addition, many systems hide these subdivisions from the user, presenting a seamless coverage of the entire area as if it were a single very large map. The different types of thematic information, represented as different map layers in paper maps, are treated as different data layers in a GIS.

DATA LAYERS

A **data layer** consists of a set of logically related geographic features and their attributes. The features to be grouped in a single data layer are chosen for the convenience of the users. The organizing principal may be to group similar feature types. For example, the data may be organized thematically, i.e. by the type of geographic features

they represent. For example, roads and railways might be combined as a single transportation data layer and streams and lakes as a hydrology data layer. Figure 7.2 illustrates the organization of data layers for a natural resource application, and Figure 7.3 illustrates an urban application. The organization of the data layers will also depend on the restrictions imposed by the GIS software used. It may be necessary or more convenient to store point, line, and area features in separate data layers.

PARTITIONING THE COVERAGE AREA

When a GIS must handle large amounts of spatial data, the coverage area may be subdivided into smaller units termed **tiles**, as shown in Figure 7.4. The allowable shapes and sizes of tiles depend on the restrictions

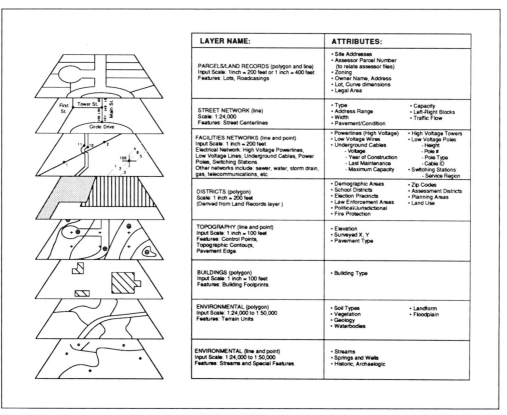

Figure 7.3 GIS Data Layers Commonly Used in Urban GIS Applications. (Courtesy of ESRI, Redlands, California.)

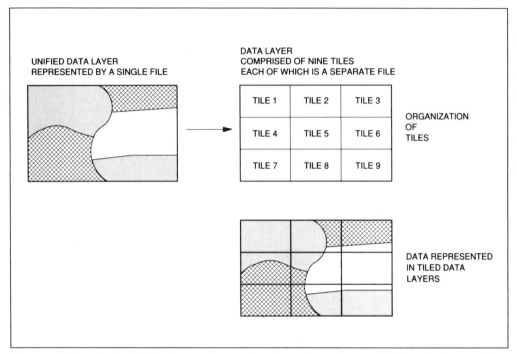

Figure 7.4 Unified and Tiled Data Layers. Geographic information can be organized as a single unified data layer, or may be subdivided into separate units termed tiles. By using tiles, a large coverage can be stored as a set of small files.

of the software. In general, tile boundaries should be chosen that will be stable for the life of the data base and that will enhance the performance and use of the system. A grid defined by latitude and longitude or UTM coordinates is commonly used. Tiles may also be organized by interest areas. Each tile might represent a different ranger district in a national forest. Searches are faster when the areas to be retrieved correspond to the tile structure used. If users most frequently access data by 7.5' quadrangle map sheets, then tiles representing one or even multiples of map sheets might be suitable.

In some systems the user must directly create and manage the tiles as separate coverage areas, re-assembling the files for adjacent areas when needed. A more sophisticated approach is to provide special purpose software to automatically create and manage the tiling so that the data are re-assembled as needed without operator interaction. The automated management of tiling is a data base operation that provides a service analogous to a map library. (For this reason the software is often termed **map library software**.) In addition to managing the partitioning and re-assembling of data layers, map library functions may include checks on data consistency, control of access and updating, and storage of repeatedly used map output formats.

The way that a map library is implemented will affect the overall performance of the system, the ease with which users can access and analyze their data, and the maintenance of the data base. Once the map library has been set up it is usually difficult to change. For this reason, the design of a map library should be given careful consideration by experienced personnel. There should be a systematic design stage, followed by a pilot implementation and evaluation. Interviews can be used to define the coverage areas and the types and level of detail of data required by the different user groups. Other important factors in developing the design are the frequency with which the data will be used and the volume of data to be stored and accessed. If there are different user groups that have competing needs, it may be necessary to provide them with separate map libraries in order to optimize system performance.

The partitioning of data layers into tiles can increase system performance by providing efficient retrieval of a subset of data. However when an entire coverage area must be retrieved, tiling will slow the storage and retrieval operations because additional steps are needed to re-assemble the data. For smaller data volumes, the overhead of partitioning the data layers may not be justified and the other data base functions provided by map library software, such as consistency checks and access control, may not be needed. However, for large data volumes, there is no choice but to use tiles. At some point the size of the file needed to store a data layer will exceed the file size limit of the system and the area will have to be subdivided into units that can be stored in smaller files.

A CLASSIFICATION OF GIS ANALYSIS FUNCTIONS

The development of GIS techniques has provided a constantly growing number of ever more sophisticated analysis functions. A description of even the most common functions would quickly overwhelm the uninitiated. The approach used here is to group the functions into four major categories, each with several subdivisions. Figure 7.5 presents a classification of GIS analysis functions. The first level of classification contains four groups: 1. Maintenance and Analysis of the Spatial Data, 2. Maintenance and Analysis of the Attribute Data, 3. Integrated Analysis of Spatial and

Attribute Data, and 4. Output Formatting. Each major group is further subdivided into types of functions. The distinctions among these categories are somewhat artificial and not clear-cut, but they do provide a useful framework.

The way that a GIS function is implemented depends on such factors as the data model (e.g. raster versus vector), the hardware, and performance criteria (e.g. how fast it must run, what options must be provided). These details are important and require considerable expertise to properly evaluate. However, this level of detail is not needed to understand the types of analysis functions that a GIS can provide, how they are used, and why they are valuable. The remainder of this chapter discusses GIS analysis and manipulation functions in this context.

MAINTENANCE AND ANALYSIS OF THE SPATIAL DATA

Maintenance and analysis functions are used to transform spatial data files, edit them, and assess their accuracy. They are primarily concerned with the spatial data and require little if any reference to the associated non-spatial attribute information. The approaches used to provide these functions differ among GISes. All GISes require the capability to transform source data into the data structure used within the system and to edit those files once they have been created. In addition, the data may need to be transformed so that different data layers for the same area are properly registered to each other or to a selected geographic coordinate system.

It may also be necessary to assemble files for adjacent areas into a single file (termed **mosaicing**). In order to improve storage efficiency, it may be desirable to reduce the quantity of data used to store the information. **Coordinate thinning** is a procedure that reduces the number of coordinate pairs

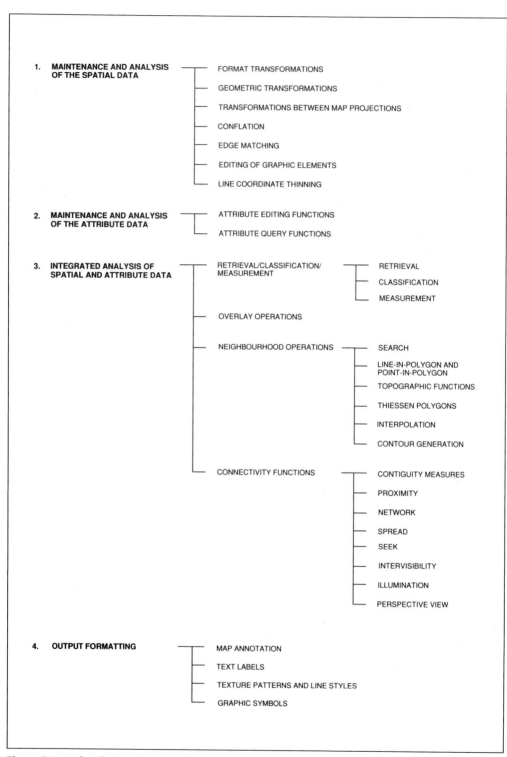

Figure 7.5   A Classification of GIS Analysis Functions.

used to define boundaries. It may also be necessary to reconcile lines that represent the same boundary in different data layers but do not coincide. A particular GIS may include all of these functions or depend on the user to supply data that have been suitably pre-processed.

FORMAT TRANSFORMATIONS

Data may be supplied to a GIS in the form of lists of points that were generated from a digitizer. They may be input as a digital file of gridded elevation values or as base maps in Digital Line Graph (DLG) format. These files must be transformed into the data structure and file format used internally by the GIS. The transformation procedure may be fast and straightforward in cases where little additional processing is needed. A raster file that is input to a raster-based GIS may require virtually no re-formatting. The internal files may only differ from the original data by the addition of some information (often termed **header information** or simply a **header**) to identify the name, origin, size, and other parameters used by the system.

In the case of topologically structured vector-based systems, it is usually necessary to "build" or create the topology from the coordinate data. This procedure is critical, for without it the topology of the overlay is not available to the system. It is also relatively time-consuming, taking minutes to hours depending on the number of map elements (i.e. the number of points, lines, and polygons) and the capabilities of the hardware and software.

The format transformation procedure can become a very costly and time-consuming operation when the data are not collected in a form well-suited to the GIS. For example, map information is often digitized for auto-mated drafting applications in a format that is not topologically structured. When these files are later used as input to a GIS, they are often difficult or impossible to transform into the topological structure needed by the GIS. Such problems as polygons that don't close and lines that don't meet may require considerable editing. It may, in fact, be less expensive to re-digitize the entire map. For this reason, the cost of format transformation can significantly affect the cost of using digital data sets and will influence the source data selected for input to the GIS.

GEOMETRIC TRANSFORMATION

Geometric transformations are used to assign ground coordinates to a map or data layer within the GIS or to adjust one data layer so it can be correctly overlayed on another of the same area. The procedure used to accomplish this correction is termed **registration** (i.e. the different data layers are registered to a common coordinate system or to one data layer that is used as a standard). Data layers for the same area must all be registered so that the same location in each overlay has the same map coordinates. The data may not have been precisely registered during data entry because the digitizing was inaccurate, there were inaccuracies in the source maps, or different map projections were used.

Two approaches are used in registration: the adjustment of absolute positions and the adjustment of relative position. The term **relative position** refers to the location of features relative to other features. The term **absolute position** refers to the location of features in relation to a geographic coordinate system.

Registration by Relative Position

In this procedure, one data layer, termed the **slave**, is registered to a second data layer, termed the **master**. The relative position of the slave data layer is adjusted by choosing features that can be easily and precisely identified on both the data layers to be registered. A road intersection, the confluence of two streams, or a small island are some examples of features that can be used for registration.

The locations of these features are then input to the GIS, a procedure that is usually done graphically and interactively. One commonly used procedure is to position a cursor on a display of the data layers or to use the cursor of a digitizer to identify the registration points. After the corresponding points have been entered for both data layers, the GIS then calculates a mathematical function to transform the coordinates of the slave into coordinates that more closely fit the master data layer. The operator is then usually presented with statistics indicating the quality of the registration, i.e. how well the data layers match. If the operator chooses to proceed, then the data layer is processed and new coordinate values are assigned to the features in the slave data layer.

This type of registration operation is often termed **rubber sheeting**. (The procedure is analogous to stretching one data layer, as if it were a rubber sheet, to fit another.) It is an empirical solution that makes few assumptions about the coordinate systems being used in the two data layers. The accuracy of this registration is predicted from the position errors of the registration points. It is a somewhat biased prediction because the transformation was calculated using these same points. Other areas of the map may not be as accurately registered. (In some systems, a set of points that was not used to calculate the transformation function can be used for accuracy assessment.) If the registration points can be accurately identified on both maps and the points are well distributed throughout the map, then this method of assessing registration accuracy does provide a reasonable prediction of the registration for the data layer.

Registration by Absolute Position

The other approach used to register data layers is to correct the absolute position of each data layer. In this case, each data layer is separately registered to the same geographic coordinate system (such as UTM coordinates). They should then be registered to each other. When data layers are registered by relative position, position errors in the master data layer are propogated to the slave data layers. The advantage of registering each data layer by absolute position is that this type of error propogation does not occur. Also, the accuracy of positions as represented on a data layer (i.e. the digital map) can be directly assessed with reference to ground coordinates. The disadvantage is that the small position errors that occur in each data layer will be independent, and so boundaries that should precisely overlay may be slightly misaligned. These discrepancies can be reconciled using an additional processing step termed *conflation* (discussed below).

TRANSFORMATIONS BETWEEN GEOMETRIC PROJECTIONS

The earth has a spherical shape. In order to uniquely reference locations on the earth's surface, the system of latitude and longitude coordinates was developed. As shown in Figure 7.6, this system is based on the angles formed by a line drawn from the center of the earth to a point on the surface. (A more detailed explanation is given in the next section.)

The spherical surface of the earth can be easily represented on a spherical map, such as a globe, using this coordinate system. However, a globe is not as convenient to use as a flat 2-dimensional map. A map projection is a mathematical transformation that is used to represent a spherical surface on a flat map (Figure 7.7). The transformation assigns to each location on the spherical surface a unique location on the 2-dimensional map. However, this transformation cannot be done without some distortion. Map projections differ in the degree of distortion that

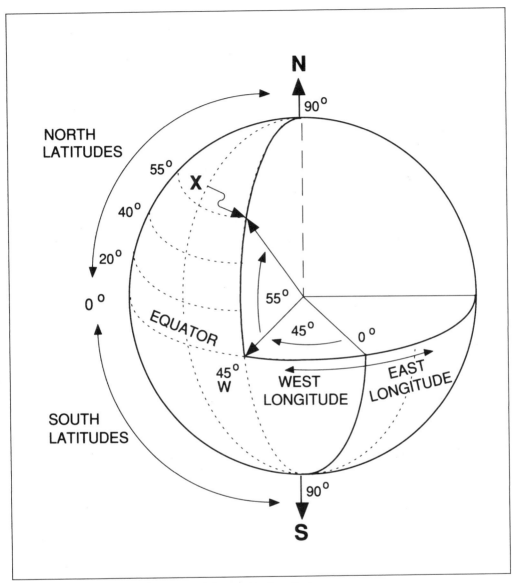

Figure 7.6 Latitude/Longitude System of Geographic Coordinates.

is introduced in the representation of area, shape, distance, and direction. The trade-offs in the degree and types of distortions that will be accepted should be considered in selecting a map projection. However, a more important selection criterion may be to use the projection commonly accepted for the discipline or application. Wherever possible, geographic information should be stored in the GIS so that it can be output in a form that is familiar to the user, i.e. a map should use the commonly accepted map projection for the discipline.

The data layers to be used together in a GIS should all be represented using the same coordinate system. A GIS commonly

Figure 7.7 Map Projections. A map projection is used to represent a spherical surface on a 2-dimensional map. (Adapted from maps provided by S. Prashker. Carleton University, Ottawa, Ontario.)

supports several projections and has software to transform data from one projection to another. The map projections most commonly used for mapping at scales of 1:500,000 or larger in North America is the UTM (Universal Transverse Mercator) projection. In the United States, the State Plane coordinate system is also used for regional scale mapping. For maps of continental extent, the Albers, Lambert's Azimuthal, and Polyconic projections are commonly used. Map projections tend to be standardized in a given field either by conscious decision or by tradition. It is important to ensure that the GIS can use the map projections in which the input data will be provided and map outputs are to be produced. Where more than one projection is to be used, then appropriate projection transformations should be provided as well.

The Latitude/Longitude System

Figure 7.6 illustrates the principles of the latitude/longitude system of geographic coordinates. Lines of longitude are drawn from the north pole to the south pole. The line of longitude passing through the Greenwich Observatory in England has the value of 0°. Moving west, the value of any line of longitude is the horizontal angle formed between the line drawn from that point to the center of the earth and a line drawn from the center of the earth to a point along the 0° line of longitude. Since these values are west of 0° longitude, they are termed **west longitude** values. In the Figure, point X is located at 45° west longitude. Similarly, lines of longitude east of 0° longitude are termed **east longitude** values. The two sets of longitude values meet at 180° longitude on the opposite side of the earth from 0°.

The lines of latitude are drawn perpendicular to the lines of longitude. In this case the lines of latitude are referenced to the equator designated as 0° latitude. The latitude of any point is defined as the vertical angle formed by a line drawn from the point to the earth's center and a line from the equator to the earth's center. Latitudes in the northern hemisphere are termed **north latitudes**, those in the southern hemisphere are **south latitudes**. In Figure 7.6, point X is located at 55° north latitude and 45° west longitude.

CONFLATION

Conflation is the procedure of reconciling the positions of corresponding features in different data layers. For example, consider two forest cover maps for the same area mapped in different years. Ideally, when these two maps are digitized and registered, features that had the same geographic position should precisely coincide when the two maps are overlayed. In practise, they may not precisely overlay because small errors were introduced during the input operation, the source maps were slightly different, the position of such features as streams had actually shifted slightly over the intervening years, or for other reasons.

Conflation functions are used to reconcile these differences so that the corresponding features overlay precisely. This is important when data from several data layers are used in an analysis. If the boundaries are slightly in error, "new" polygons, often termed *slivers*, may be created that do not represent information about the area but are rather inaccuracies in the mapping.

A manual procedure to minimize these errors is to re-draw the source maps using feature boundaries in one of the overlays as a standard or template. The representation of roads, streams, and lakes are commonly reconciled in this way. The template is drawn on a map that becomes the basemap which is then used to guide the redrafting of these features on all the maps. Templates are also used to standardize the shape of features with geographic positions that change but that are more conveniently handled as having a fixed location over time. The position of the shoreline of a water reservoir will change over the year as water accumulates and is then released. If each data layer uses the position of the shoreline at the time of mapping, there will be discrepancies in its position on different data layers. To standardize the shoreline position, a standard template can be used to represent the shoreline at the same location on each data layer instead of using the actual shoreline position at the time of mapping. All mapping is then extended to the shoreline of the template. In this way, the representation of the reservoir is made to be consistent for all data layers.

Computerized techniques can be used to perform a similar function. The procedure usually requires the operator to identify corresponding features in the two maps or data layers that should have the same location. As more points are identified, one or both maps are adjusted. In some cases this procedure can be interactive so that the best match can be developed by adding points where discrepancies exist. Computerized conflation is a relatively recent development. It is an important function in applications where integrated analyses of data sets from diverse sources will be required. However, it is generally much less expensive to reconcile maps manually in the map preparation stage, before they are digitized. Reconciling maps that are already in the GIS data base incurs the added expenses of the computer system, the more highly trained and higher-paid operator, and the need to re-process the edited files.

EDGE MATCHING

Edge matching, illustrated in Figure 7.8, is a procedure to adjust the position of features that extend across map sheet boundaries. The coverage area stored in a GIS is usually larger than a single map sheet. Data entered from separate map sheets are usually organized within the GIS so as to present the data to the user in the form of a continuous geographic coverage. Theoretically the data from adjacent map sheets should precisely meet at the map edges. However, in practise, features that extend across map boundaries often do not align perfectly. Minor errors can be caused by such factors as errors in the original mapping, differences in the dates of mapping, paper shrinkage in the

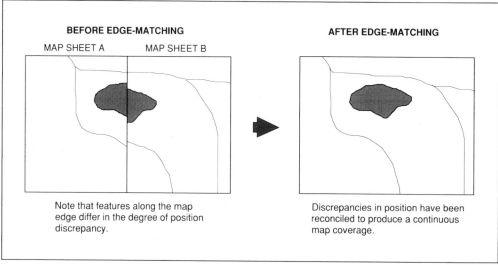

BEFORE EDGE-MATCHING

MAP SHEET A   MAP SHEET B

AFTER EDGE-MATCHING

Note that features along the map edge differ in the degree of position discrepancy.

Discrepancies in position have been reconciled to produce a continuous map coverage.

Figure 7.8 Edge Matching. Edge matching is used to reconcile the position of features that extend onto an adjacent map but are not correctly aligned at the map boundary.

source maps, or errors in the digitizing process. The discrepancies may be difficult to correct because the features along the map edge may not be shifted to the same degree or in the same direction. Some GISes provide software to reconcile these differences by making adjustments to the position of features in one or both maps. Edge matching must be done for the geographic information from several adjacent maps to be represented as a single continuous data layer.

Software differ considerably in the degree of automation provided. Beard and Chrisman (1986) review the approaches used in edge matching and provide an example of a highly automated procedure.

EDITING FUNCTIONS

Editing functions are used to add, delete, and change the geographic position of features. The sophistication of the editing software can greatly affect the speed and accuracy with which these essential functions can be done.

Slivers or **splinters** are thin polygons that are often created during digitizing and overlay operations. In Figure 7.9 polygons **A** and **B** share a common boundary. If the boundary between these polygons was mistakenly digitized twice a sliver would be created as shown. Polygons **C** and **D** represent features in two different data layers that share a common boundary. For example, a forest stand and an adjacent agricultural field will share a common boundary. If the boundary is not coded with precisely the same geographic coordinates in the two data layers, then, as shown in the Figure, a sliver may be created if they are used together in an overlay operation. Some software packages provide automated detection and correction of slivers or splinters. These functions can substantially reduce the editing time needed to remove slivers, but some features are correctly represented as long thin shapes. As a result, some operator supervision is required to check that only the slivers are removed, not the valid features.

Often a digitized line is short by a few millimeters and does not quite reach the feature to which it is connected. Automated **line-snapping** can correct this type of error by connecting lines to a node if they end within a specified distance. Other editing aids will search for such inconsistencies as lines that are left dangling (i.e. are not connected to another element) or polygons that do not close. In some cases, the program will aid the operator by moving the cursor to each inconsistency and waiting for the operator to correct the problem.

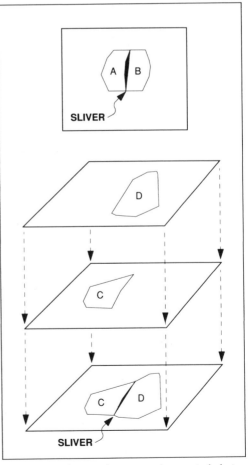

Figure 7.9 Slivers. Slivers may be created during digitizing and overlay operations.

LINE COORDINATE THINNING

This function is used to reduce the quantity of coordinate data that must be stored by the GIS. Often, more coordinates are entered than are actually needed to define a line or a polygon. This usually occurs during digitizing or scanning operations. Coordinate thinning, by reducing the number of coordinate points, reduces the size of the data file, thereby reducing the volume of data to be stored and processed in the GIS. Reducing the data volume will tend to improve system performance.

The thinning function reviews all the coordinate data in a file, identifies and then removes unnecessary coordinates. The degree of coordinate thinning is controlled by the operator. As the level of thinning is increased, fewer coordinates are stored. This results in a less detailed but more compact representation of lines (see Figure 7.10). In practise, the number of coordinate points can usually be significantly reduced without a perceived loss of detail.

MAINTENANCE AND ANALYSIS OF NON-SPATIAL ATTRIBUTE DATA

This group of functions is used to edit, check, and analyze the non-spatial attribute data. Many GIS analyses can be performed using these attribute functions alone. For example, in a vector-based GIS, the area and perimeter of polygons are commonly stored in the attribute file, along with the class, and other characteristics. To produce a table of areas for all polygons of a certain class, the data can be retrieved from the attribute file, without reference to the spatial data.

In a simple raster-based GIS, non-spatial attribute data may be embedded in the spatial data file. For example, a legend, the latitude and longitude coordinates of the corners of the data layer, and a title for the data layer might be attached to the beginning or end of the file.

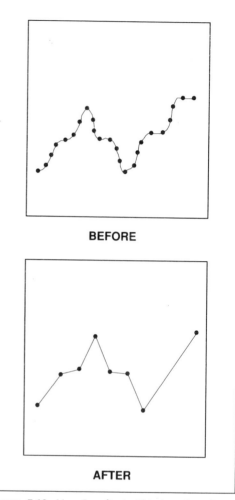

BEFORE

AFTER

Figure 7.10 Line Coordinate Thinning. Coordinate thinning reduces the number of coordinate pairs used to store a line segment within the GIS.

In more sophisticated systems, the attributes are stored separately from the spatial data, often in a separate data base system. The greater complexity of this type of GIS also can provide greater analysis flexibility and the power and capacity to handle large data sets.

ATTRIBUTE EDITING FUNCTIONS

Editing functions allow the attributes to be retrieved, examined, and changed. New attributes can be added or old ones deleted. In some systems, it is difficult to add new categories of attributes once the data base has been defined. This is usually related to restrictions of the data base software. Where the attributes are not well-known in advance, as is the case with many natural resource applications, such restrictions may be an important system limitation. It is often valuable to import data from other sources. For example, in municipal applications, files containing ownership data keyed to addresses might be added to the attribute file for a street map so that the owner's name and address becomes an attribute of the spatial data. Many systems provide a function to match corresponding records in the two attribute data sets using a common data field, in this case the address. This capability is termed **file matching** or **address matching**.

ATTRIBUTE QUERY FUNCTIONS

Query functions retrieve records in the attribute data base according to conditions specified by the operator. Depending on the software, query functions can be restricted to very simple retrievals, such as finding the class assigned to a selected polygon. A full-featured GIS will have a data base that supports more complex queries that involve selective searches of all the attributes for one or more data layers and the generation of a report that tabulates the results.

Figure 7.11 illustrates the use of an attribute query to generate a report of forest area by dominant species. The attributes for the forest cover data layer are stored as two tables. They can be queried together, as shown, to generate a report of the total area of forest more than 30 years old. The stand number, a data field common to both tables, is used as the link between the tables. (The procedure of retrieving information from the two tables by means of a common data field is termed a **relational-join** operation.) The

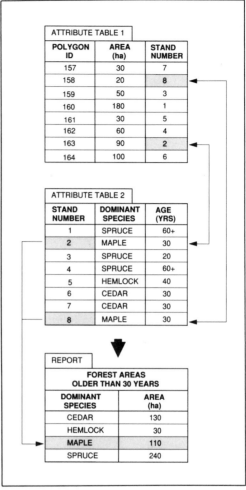

Figure 7.11 An Attribute Query Used to Produce a Forest Cover Summary Report. The figure illustrates the use of an attribute query to generate a report of forest cover by species for stands older than 30 years. The data are retrieved from two attributes tables by using a common data field, the stand number, to identify corresponding records in the two tables. This procedure is termed a relational-join.

information is summarized by dominant species in the report. Note that to generate the report, it was only necessary to query the attribute data files, the spatial data were not needed.

Attribute analyses can be very powerful and efficient because complex spatial operations are not used. A full-featured data base provides commands for an operator to perform a wide range of queries and to output the results in the desired report format. In many operational applications, the attribute entry and retrieval functions are the principal day-to-day operations. Such questions as how large an area was planted to a specific crop, who is the owner of a specified property, and what is the property value, can be answered by operations on the attribute data base alone.

INTEGRATED ANALYSIS OF SPATIAL AND ATTRIBUTE DATA

The power of a GIS lies in its ability to analyze spatial and attribute data together. It is these capabilities that most distinguish a GIS from automated mapping and computer-aided drafting systems. The range of analysis procedures in this group of functions is very large. They have been subdivided into four categories; retrieval/classification/measurement, overlay, neighbourhood, and connectivity or network functions.

A specific GIS may implement a function as a single computer task or by linking several tasks together. The way a specific function is implemented affects the ease and flexibility of operation, the storage requirements (e.g. for intermediate files), and the level of performance. Though important for system evaluations and comparisons, these considerations are too detailed to be considered here. For the purpose of understanding how GIS analysis and manipulation functions are used, the method of implementation is of secondary consideration. The emphasis in this section is on the conceptual procedures used to generate different types of information from the GIS data base.

RETRIEVAL, CLASSIFICATION, AND MEASUREMENT FUNCTIONS

In this set of functions, spatial and attribute data are retrieved, but only the attribute data are modified or created. No changes are made to the location of spatial elements and no new spatial elements are created.

Retrieval Operations

Retrieval operations on the spatial and attribute data involve the selective search, manipulation, and output of data without the need to modify the geographic location of features or to create new spatial entities. These operations work with the spatial elements as they were entered in the data base. The production of a city map showing buildings classified by their age is an example of this type of operation (see Figure 7.12).

Classification and Generalization

The sets of elements (the land parcels with buildings of a certain age range) retrieved in the previous example (Figure 7.12) could be assigned class names such as "pre-1900", "1900–1930", "1931–1950", and "post-1950". These class names could be stored as attributes of the buildings in the data base. This new class designation could then be used to select these buildings for further analyses.

This procedure of identifying a set of features as belonging to a group is termed **classification**. Some form of classification function is provided in every GIS. In the case of a single data layer, classification may involve the assignment of a class name to each polygon as an attribute. The classification may be for land cover, and so the class names might be *forest land, agricultural land, urban areas*, and so on. In this case, the classification process involves looking at the attributes for a single data layer and assigning an additional attribute, the new class name. In a raster-based GIS, numerical values are often used to indicate classes. A cell might be assigned the value 1 to indicate agricultural land, 2 for forest land, and so on. The classification process would then involve assigning numerical values to the cells (sometimes termed **recoding**) and writing these new values into a new data layer.

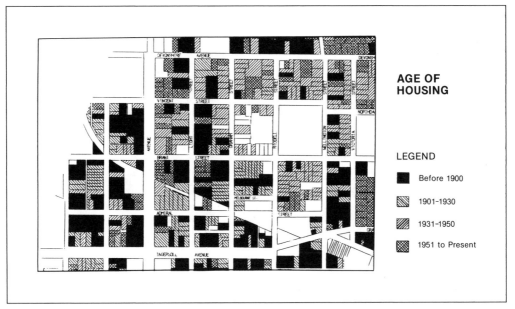

Figure 7.12 Map of an Urban Area Showing Buildings Classified by Age. (Courtesy of Oxford County, Ontario.)

Classification is important because it defines patterns. One of the important functions of a GIS is to assist in recognizing new patterns. These patterns might be areas of the city with the highest crime rate, areas of forest land suitable for timber harvest, or areas of agricultural land most likely to be converted to residential development.

Classification is done using single data layers, as illustrated in the building age classification, as well as with multiple data layers as part of an overlay operation (overlay operations are discussed later in this chapter). For example, a desirable site for a cottage might be a forested area, with well-drained soils, a southern exposure, and a non-agricultural land use zone. Each of these factors might be represented as separate data layers in the GIS. An overlay analysis could be used to identify the areas meeting these criteria and to assign them the class name "good cottage areas".

Generalization, also called **map dissolve**, is the process of making a classification less detailed by combining classes (see Figure 7.13). Generalization is often used to

reduce the level of classification detail to make an underlying pattern more apparent.

Measurement Functions

Every GIS provides some measurement functions. Spatial measurements include distances between points, lengths of lines, perimeters and areas of polygons, and the size of a group of cells with the same class. Sample applications are: finding all forest areas larger than 200 sq km that are potentially suitable for use as a conservation area or locating airports less than 10 km apart that might be unnecessary. Where digital terrain data are used, 3-dimensional measurements are often needed for engineering applications, such as calculating the amount of cut and fill material involved in road construction. (Measurements along a network, such as along a road system, are a special case discussed in the section on network functions.)

In a vector-based GIS, spatial elements can vary in size and shape. New attributes can be calculated for the spatial elements

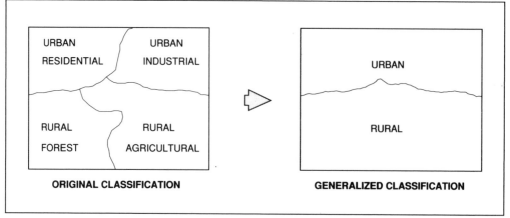

Figure 7.13   Generalization. Generalization is the process of combining classes to reduce the level of classification detail. A land use data layer with four different classes has been generalized to two classes to emphasize the boundary between the urban and rural areas.

of an overlay. The calculation of areas and centroids (the center point of a polygon) is commonly done automatically as part of the polygon creation process. Measures of shape, narrowest and broadest distance across a polygon, the length and sinuosity of a line are other useful measurement functions. In a raster-based GIS, these types of functions become neighbourhood operations because they involve the identification of connected cells. As a result, the software algorithms are different and the strategy for using them is conceptually different as well.

OVERLAY OPERATIONS

Arithmetic and logical overlay operations are part of all GIS software packages. Arithmetic overlay includes such operations as addition, subtraction, division, and multiplication of each value in a data layer by the value in the corresponding location in a second data layer. A logical overlay involves finding those area where a specified set of conditions occur (or do not occur) together.

For example, desirable areas for cottages might be defined as those areas that have a forest vegetation cover, have well-drained soils, and have a south-facing exposure. If vegetation, soils, and exposure are represented as separate data layers in the GIS, then a logical overlay operation could be used to identify the locations where these conditions occur together.

The flexibility provided to the operator and the level of performance of overlay operations vary widely among GISes. One of the major factors affecting the performance of these functions is the data model being used. Raster and vector models differ significantly in the way arithmetic and logical operations are implemented. Overlay operations are usually performed more efficiently in raster-based systems. Because this has been a critical issue, the difference in approach will be illustrated by the following examples.

Figure 7.14 illustrates an arithmetic function, multiplication, applied to rain guage data. The example shows a raster and a vector implementation. The procedure is being used to convert the data from units of inches to units in millimeters by multiplying each rain guage value by 25.4 mm/inch. In the raster case, the rain gauge data for the five locations are entered directly into a data layer (the input data layer in the Figure). The cells with no rain gauge data are shown to be blank in the figure for clarity,

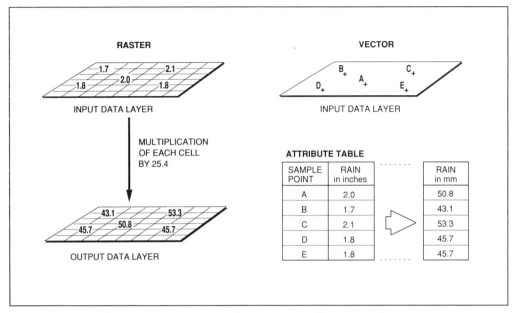

Figure 7.14 An Arithmetic Operation on a Single Data Layer in the Raster and Vector Domains.

however each cell would be assigned a value (commonly these blanks would be zeroes). The multiplication operation is performed on every cell of the input data layer and the result is written to the corresponding cells of the output data layer. The operation is performed as many times as there are cells, regardless of the fact that most cells are blank.

In the vector case, the rain gauge locations are represented as points in the input data layer. The attributes of the points are stored separately in an attribute table, as illustrated. The attribute data consist of only the five rain gauge values and the multiplication procedure can be done using the attribute data alone. In the Figure, a second column containing the calculated millimeter values is shown to have been added to the attribute table. Only five multiplications were needed. For sparse data, this operation on one data layer is much more efficiently processed in the vector domain than in the raster domain.

Figure 7.15 illustrates an arithmetic function of two data layers as implemented in the raster domain. The value at each location in Input Data Layer A is to be added

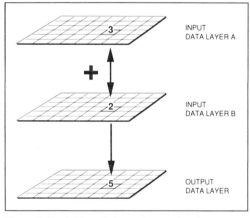

Figure 7.15 Arithmetic Operations on Two Data Layers in the Raster Domain. An addition operation is applied to cells with corresponding positions in the two input data layers and the result is written to the corresponding cell positions in the output data layer. Since each data layer is organized using the same grid of cells there is no need to create any new boundaries between spatial entities.

to the value in the corresponding location in Input Data Layer B. The result is to be written to the corresponding cell of the output data layer. Because the map area is regularly subdivided into cells and the cells of each data layer are in perfect registration, there is no uncertainty in the location of boundaries since they are the cell boundaries and always coincide. The value assigned to a cell represents the condition for all points within the cell, i.e. the cell is the smallest unit of division.

Figure 7.16 illustrates the same operation in the vector domain. Here areas are represented as polygons. Each polygon can differ in size and shape and so the boundaries of the polygons in one data layer usually will not coincide with the boundaries of polygons in the other data layer. The polygon values are stored separately in attribute tables, as shown. The value assigned to a polygon represents the condition for every location within the polygon, as for a raster cell. However, the size of the polygon is variable. In order to add the two polygon data layers, new polygons must first be created in the output data layer. The area where polygons A and D overlap must be defined as a separate new polygon and then assigned the new attribute value. This process of subdividing polygons is termed **clipping**. Although there will be fewer arithmetic operations, because there are fewer polygons than cells, clipping makes overlay operations considerably more complex in the vector domain than in the raster domain. When a large number of irregularly shaped polygons are involved, the vector overlay procedure generally requires significantly more processing time than raster overlay.

In the raster domain, the data file consists of an ordered list of values. The spatial locations to which these values refer (i.e. the row and column positions) are determined by

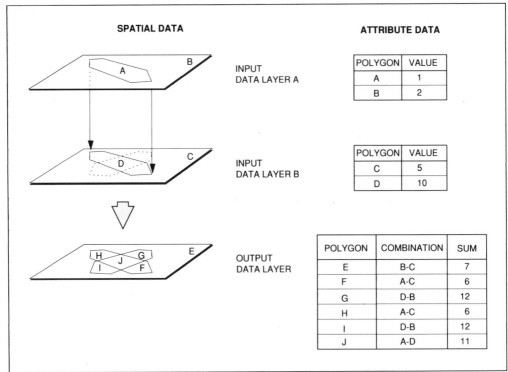

Figure 7.16 Arithmetic Operations on Two Data Layers in the Vector Domain. To perform an overlay operation in the vector domain, polygons A and D in the input data layers must be subdivided or clipped to create the new boundaries for the output polygons (F,G,H,I,J). Then the addition operation is performed using the values from the attribute table. The clipping operation is complex and can require considerable processing time when there are large numbers of irregularly shaped polygons.

the position of the value within the file. As a result, overlay processing involves retrieving and comparing the data in corresponding positions in the data files. There is no need to calculate intersections of boundaries or make any modifications to the feature boundaries because each spatial element is a single cell of standard size.

The regular subdivision of space makes overlay operations easy to implement in the raster domain. However, sparse data sets require as much processing as densely populated ones because the operation is performed on every cell regardless of the data it contains. In the vector domain only the data of interest are processed. The more sparse the data set, the faster the processing. However, the overlay operation is much more complex to begin with. Where operations are performed on the attribute data for a single data layer, the storage of attributes separate from the spatial data is advantagous. A single data layer can have an attribute table assigning multiple attributes to each spatial element. Operations on these attributes are effectively independent of the spatial data.

In many GISes a hybrid approach is used that takes advantage of the capabilities of both data models. A vector-based system may implement some functions in the raster domain by performing a vector-to-raster conversion on the input data, doing the processing as a raster operation, and converting the raster result back to a vector file. In a raster-based GIS, data compression techniques can be used in effect to create elements that represent contiguous areas (a connected group of cells) that have the same value. (See the discussion of data compression in Chapter 6.) There is a trend to introduce more integrated raster-vector processing in GIS because the two data models offer different advantages and because many digital data sets are available in only one of the two formats.

NEIGHBOURHOOD OPERATIONS

Neighbourhood operations evaluate the characteristics of the area surrounding a specified location. Counting the number of residential dwellings within a 5 km radius of a fire station (Figure 7.17) is an example of a neighbourhood operation. Every neighbourhood function requires the specification of at least three basic parameters: one or more target locations, a specification of the neighbourhood around each target, and a function to be performed on the elements within the neighbourhood. In this example, the target is the fire station, the specified neighbourhood is the area within a 5 km radius, and the function is to count the number of elements that are residential buildings.

Virtually all GIS software packages provide some form of neighbourhood operations. They vary in the flexibility and sophistication with which the three basic parameters can be specified and in the specialized operations provided. The most common types are the search function (which includes the generation of summary statistics), topographic functions, and interpolation.

Neighbourhood operations are particularly valuable in evaluating the character of a local area, for example to find all residential areas that are in the vicinity of a school,

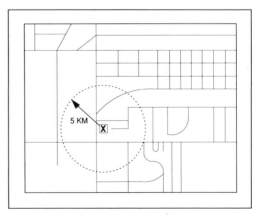

Figure 7.17 Defining a Search Area With a 5 km Radius.

a park, and a shopping area. Another example is the assessment of the quality of an area as wildlife habitat. Wildlife require a specific combination of vegetation and terrain types with access to water, food, and shelter within a limited size of neighbourhood.

Many neighbourhood operations must be implemented using some regular division of the study area. For this reason, a raster model is commonly used. In some vector-based systems, the geographic data is converted into a raster form for the analysis and then converted back to vector format. Other vector systems generate the neighbourhoods directly and intersect them with the geographic data as an overlay operation.

Search

The search function is one of the most common of the neighbourhood operations. This function assigns a value to each target feature (such as the fire stations in the previous example) according to some characteristic of its neighbourhood. (The search function, as presented here, is defined in a general way. GISes will differ in the names used to refer to this function, and several individual procedures may be needed to provide some of the search operations discussed.)

The three basic parameters to be defined in a neighbourhood search are the targets, the neighbourhood, and the function to be applied to the neighbourhood to generate the neighbourhood value. The target elements and the elements of the neighbourhood are commonly stored in one or more data layers. In the previous example, the locations of the fire stations and residential buildings could be in a single data layer of the buildings in the municipality or in separate data layers, one containing locations of all emergency facilities and the other containing the data on housing.

The search area is most commonly square, rectangular, or circular with a size selected by the operator (e.g. a circular area with a

5 km radius). The search functions are usually predefined and the operator selects one of those offered. Search functions are of two types: those that operate on numerical data and those that operate on thematic data. Typical numerical functions are the total, average, maximum, minimum, and measures of diversity (see Table 7.1). In each case, the function is applied to the corresponding neighbourhood for each target.

Table 7.1   Neighbourhood Functions.

FUNCTION	DESCRIPTION
AVERAGE	The average of the values in the neighbourhood.
DIVERSITY	A measure of diversity of the values in the neighbourhood is calculated such as the variance or standard deviation.
MAJORITY	The number of occurances of each value in the neighbourhood is evaluated. The value occurring most frequently is the calculated result.
MAXIMUM, MINIMUM	The values of each element in the neighbourhood are evaluated and the maximum or minimum value is returned.
TOTAL	The total of the values in the neighbourhood.

The data layer used to define the neighbourhood may be different from the data layer to which the function is applied. For example, consider an analysis of the average dollar value of residential buildings within 5 km of the fire stations in a metropolitan area. The target locations (fire stations) may be identified on a data layer of emergency service facilities, the residential buildings identified on a second data layer, and the values of each residential building stored in a separate attribute file. The neighbourhood would be defined using the first data layer but would be applied to the data of the second data layer — to identify the residential buildings within the 5 km search radius. The values of these residences would then be retrieved from the attribute file and averaged. The result of this analysis could be provided in the form of a map showing the fire stations and their 5 km surrounding neighbourhoods shaded according to the average value of the residences in their neighbourhoods. The results could also be output as a table listing each fire station and the average value of the residences in each neighbourhood.

Neighbourhood functions using thematic data commonly use the following operators: the majority (also termed the mode), the maximum, the minimum, and diversity measures (such as the number of different classes in the neighbourhood).

In more sophisticated implementations of the search function, more flexibility in the specification of the three basic control parameters is provided. Instead of using only predefined functions, the operator may be able to enter an equation. The selection of the target points or target areas might be defined using a query such as "select hospitals that have more than 500 beds and are less than 5 years old". This type of query might use data from both the attribute and spatial data bases to generate a list of locations that would then be used by the search routine.

In many applications, it is useful to specify a neighbourhood area that may be different for each target and may not be a regular shape. The neighbourhood might be a political region like a county or state. The neighbourhood might also be generated by other GIS functions, such as an overlay operation. The application of a function to a user-defined neighbourhood is termed **region of interest** processing, the interactive definition of a search area is often termed **windowing**, and the search area itself is termed a **window**.

Consider the example in Figure 7.18 in which the boundaries from the county data layer are used to define a processing window for analyzing land uses in County 25. Note that the land use polygons and county polygons do not have the same boundaries. In effect, the county polygon has been used like a "cookie-cutter" to extract the corresponding portions of the land use polygons. Then the area for each type of land use in

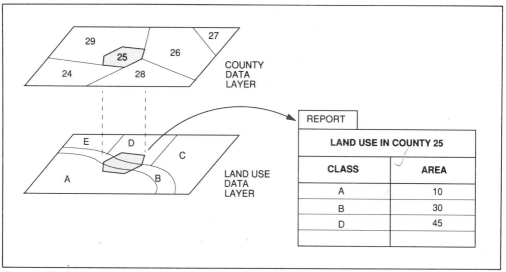

Figure 7.18 An Overlay Operation Used to Define a Region of Interest. A land use report for County 25 is generated by overlaying the *County* and *Land Use* data layers.

the county has been totalled and reported in tabular form.

The polygons used as windows can be quite complex. A network function might be used to define areas within a 20 minute ambulance ride from a hospital, as shown in Figure 7.19. In this case, the window might be quite irregular because it would depend on the traffic capacity and speed of travel along the surrounding streets. Once the area has been defined, it could be used to further analyze that service area, such as to retrieve information about the residents

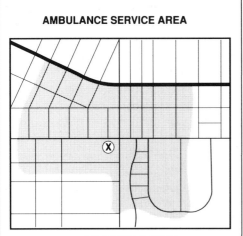

AMBULANCE SERVICE AREA

(X) INDICATES HOSPITAL LOCATION

AGE DISTRIBUTION OF RESIDENTS
WITHIN 20 MINUTE SERVICE AREA

YEARS	POPULATION	PERCENT
0-5	200	16
6-10	100	8
11-15	50	4
16-20	25	2
21-30	300	23
31-40	250	19
41-50	100	8
51-60	175	13
60+	90	7
TOTAL	**1290**	**100**

Figure 7.19 Twenty Minute Ambulance Service Area. To define the areas that can be reached by ambulance within twenty minutes, a network function is used to evaluate travel times along the street network.

of the area. In this example, a report is shown of the population age structure within the 20 minute service area.

In some cases, particularly in a raster-based GIS, search functions are applied to every location (i.e. every cell is a target). The processing is then in effect like moving a window the size of the neighbourhood through the data layer, cell-by-cell. At each step the neighbourhood function is evaluated and the result assigned to the corresponding position in the output data layer, as illustrated in Figure 7.20. Here the objective was to find all locations surrounded by at least a 3 km by 3 km area of forested land. Since each cell represents a 1 km by 1 km area, a 3 cell by 3 cell window was used. This window was applied at each cell location in the input data layer. (No values are calculated for the cells at the edge of the data layer because some of the cells of the neighbourhood would be off the map.) The function applied to each neighbourhood is to count the number of non-forested cells (shown here in black) and assign that value to the cell in the ouput data layer that has the same position as the center cell of the window.

Line-in-Polygon and Point-in-Polygon Operations

In a vector-based GIS, the identification of points and lines contained within a polygon area is a specialized search function. In a raster-based GIS, it is essentially an overlay operation, with the polygons in one data layer and the points and/or lines in a second data layer (see the section on overlay operations earlier in this chapter). Quadtree-based systems also deal with this type of function in a specialized way. Using a quadtree data structure, this type of search operation can be implemented very efficiently, usually with performance levels significantly higher than vector- or raster-based systems can provide.

Figure 7.21 is a map showing the major highways (lines) crossing a metropolitan area

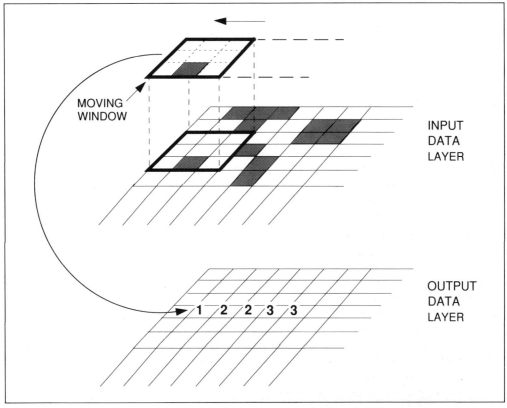

Figure 7.20 A Neighbourhood Search Using a Moving Window. A 3-cell by 3-cell neighbourhood is used as a window that is, in effect, moved through the input data layer. At each location, the number of non-forested (black) cells are counted and the value entered in the output data layer in the cell location corresponding to the center of the window. The window is then moved one step, and the evaluation process is repeated.

and the location of three repair depots (points P1, P2, and P3). A simple point-in-polygon operation might be to find all repair depots in the metropolitan area. This type of search would require that the coordinates of all repair depots be evaluated and those that fall within the polygon representing the city limit be reported.

A search for all highways that cross the city is an example of a line-in-polygon operation. At first glance, it would appear that this search could be done as a point-in-polygon operation by searching for points that comprise the line and fall inside the boundary of the polygon. However, in a vector-based GIS, lines or arcs are stored as a series of points connected by straight-line segments.

The line itself is only created when needed, such as when a map is produced. If a straight line crosses a polygon it may be represented by a single line segment with both endpoints outside the polygon.

In Figure 7.21, the black dots indicate the points in the line for which coordinates are actually recorded. Notice that Highway 80 has no coordinate points within the city limit, which is the search polygon. To identify all highways crossing the city, a search for coordinates that fall within the polygon would miss Highway 80. For this reason more sophisticated algorithms are used that can properly handle these cases.

A more challenging example is the map enquiry system used by the US Geological

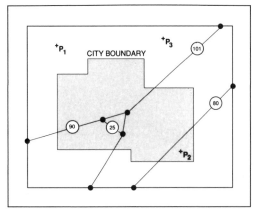

Figure 7.21 Line-in-Polygon and Point-in-Polygon Operations. The map shows the boundary of a metropolitan area and the location of three vehicle repair depots, labelled P_1, P_2, and P_3.

Survey. This system is used to find the map that contains any specified point in the United States. In effect, the operation is to find the polygon (the area covered by a particular map) in which the specified point, anywhere in the country, is located. Though there are thousands of map areas, the system response time is less than a minute using a quadtree data structure.

The details of the algorithms used to optimize point-in-polygon and line-in-polygon functions are beyond the scope of this discussion. Software packages vary widely in the levels of performance they achieve. If these operations are frequently used, the speed and flexibility of this function should be carefully evaluated.

Topographic Functions

Topography refers to the surface characteristics, i.e. the relief, of an area. The topography of a land area refers to the hills, valleys, and plains of which it is comprised. The topography is thus defined by the elevation of each location within the area. The topography of a land surface can be represented in a GIS by digital elevation data. This data set consists of the elevation of a large number of sample points distributed throughout the area being represented. These sample points are commonly organized as a grid of points, essentially a raster form of organization.

An alternative form of representation is the Triangulated Irregular Network or TIN used in vector-based systems. In a TIN, a network of triangular facets is generated by the GIS from a set of elevation sample points that can be irregularly distributed. These facets can then be manipulated as polygons and the elevation, slope, aspect, and other parameters can be assigned to the facets as polygon attributes. Raster and vector data models and TINs are discussed in Chapter 6.

Topography can be used to study data other than elevation. Any characteristic that has a continuously changing value over an area can be represented as a surface. In geology, aeromagnetic data and geochemical data are often represented as a surface. The noise levels in the vicinity of an airport, the income levels of neighbourhoods in a city, or the levels of pollution in a lake can also be represented in this way.

Topographic functions are used to calculate values that describe the topography at a specific geographic location (e.g. the elevation at the location) or in the vicinity of the location (e.g. the slope of the area immediately surrounding the location). Most topographic functions use a neighbourhood to characterize the local terrain. The two most commonly used terrain parameters are the slope and aspect, which are calculated using the elevation data of the neighbouring points.

Slope is defined as the rate of change of elevation. **Aspect** is the direction that a surface faces. Conceptually, the calculation of the slope and aspect at a point can be thought of as fitting a plane to the elevation values of the neighbouring points. The slope of this plane and the direction it faces are the slope and aspect of the point. As shown in Figure 7.22, the slope may be calculated in the X-direction, the Y-direction, or in the direction of maximum slope. The maximum slope is termed the **gradient**.

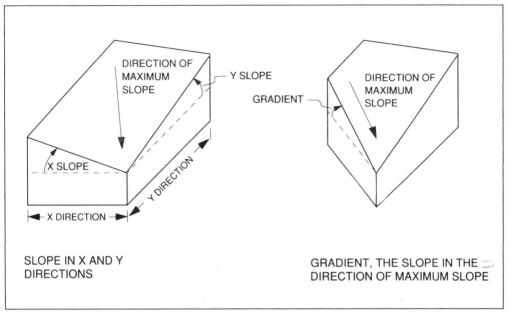

Figure 7.22 Measurement of Slope. The slope at a point is the angle measured from the horizontal to a plane tangent to the surface at that point. The value of the slope will depend on the direction in which it is measured. Slope is commonly measured in the direction of the coordinate axes e.g. in the X-direction and Y-directions. The slope measured in the direction at which it is a maximum is termed the **gradient**.

Slope is usually measured in degrees of arc or as a percentage (the change in elevation divided by the corresponding horizontal distance). Aspect is defined by the horizontal and vertical angles that the surface faces. The horizontal angle is usually measured in degrees of azimuth, the angle formed by moving clockwise from north, as shown in Figure 7.23. The vertical angle or angle of elevation, the positive angle measured from the horizontal to a line drawn perpendicular to the surface, is sometimes used as well. This angle is equal to 90° minus the gradient.

Although commonly used in the analysis of elevation data, slope and aspect can be usefully applied to other data sets as well. Slope measurements are commonly used in the analysis of gravity and aeromagnetic data in geology. In an urban setting, slope values could be calculated for land costs. High values of slope would then indicate areas where land costs change abruptly with distance. Such

areas might represent zones of potential social conflict, or they might also indicate areas with good investment potential.

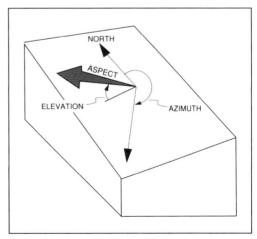

Figure 7.23 Measurement of Aspect. There are two components in the measurement of aspect. The horizontal angle or azimuth is the aspect direction. It is the angle formed by moving clockwise from north to the direction of maximum slope. The vertical angle or elevation angle is measured from the horizontal to a line drawn perpendicular to the surface.

Other important functions used in topographic analyses are illumination, viewshed modelling, and perspective view generation. These are discussed in subsequent sections.

Thiessen Polygons

Thiessen or voronoi polygons define individual areas of influence around each of a set of points. Data from rain gauges are commonly analyzed in this way. It is an approach to extending point information which assumes that the "best" information for locations with no observations is the value at the closest point with a known value. Thiessen polygons are commonly used in the analysis of climatic data, such as rain gauge data. In the absence of a local observation, the data from the nearest weather station are used.

Thiessen polygons are constructed around a set of points in such a way that the polygon boundaries are equidistant from the neighbouring points. In other words, each location within a polygon is closer to its contained point than to any other point.

Figure 7.24 illustrates the use of Thiessen polygons to analyze rain gauge data. The rain gauge locations are represented in the GIS as points. Thiessen polygons are then generated around each point and the rainfall value for the rain gauge is assigned to its surrounding polygon. The rainfall at all locations within each Thiessen polygon is considered to be that of the contained rain gauge station. The amount of rain falling on each polygon can then be calculated as the amount recorded by the rain gauge multiplied by the area of the polygon. For the entire study area the total rainfall would be estimated by totalling the rainfall calculated for each Thiessen polygon.

Thiessen polygons, in effect, are used to predict the values at surrounding points from a single point observation. The method has a number of limitations. The division of a region into Thiessen polygons is completely dependent on the location of the observation points. This can produce polygons with shapes that are quite unrelated to the phenomenon being mapped. The position of rain gauges may produce long thin polygons, a pattern in which rainfall would not normally occur. The value assigned to each polygon is estimated from a sample of one, the observation point. Estimates of error cannot be calculated from a single sample. Finally, Thiessen polygons do not assume that points close together are more similar than points far apart, an assumption that is usually appropriate in geographic analyses (Burrough 1986).

Interpolation

Interpolation is the procedure of predicting unknown values using the known values at neighbouring locations. The neighbouring points may be regularly or irregularly spaced. Figure 7.25 is a simple example of this function, in this case presented using raster data layers. (The function can also be implemented using TINs in a vector-based system.) The stippled cells contain the known values. A simple linear function, derived by analyzing the known points, has been used to generate the missing values. Interpolation programs employ a range of methods to predict unknown values, including polynomial regression, fourier series, splines, moving averages, and kriging.

The quality of the interpolation results depends on the accuracy, number, and distribution of the known points used in the calculation and on how well the mathematical function correctly models the phenomenon. Interpolation assumes that the phenomenon being predicted (e.g. terrain elevation) is closely approximated by the mathematical function used. The unknown values are then calculated according to this function. The choice of an appropriate model is therefore essential in order to obtain reasonable results.

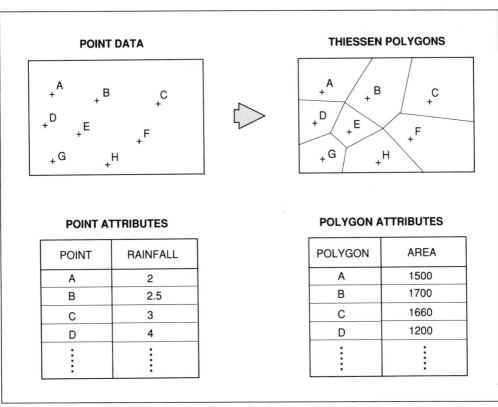

Figure 7.24 Thiessen Polygons Used to Analyze Rain Gauge Data. A Thiessen polygon is constructed around each rain gauge location. The rainfall value for every location within the polygon is considered to be that of the contained rain gauge station.

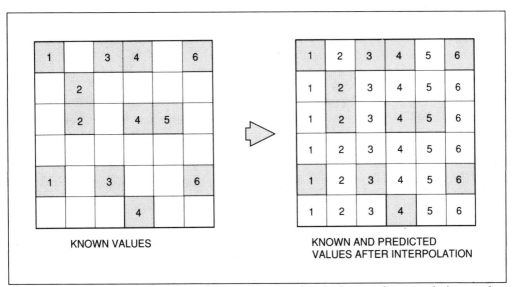

Figure 7.25 Interpolation. Interpolation is the procedure of predicting unknown values using the known values at neighbouring locations, usually by means of a mathematical function.

The best results are obtained when the mathematical function behaves in a manner similar to the phenomenon. For example, air temperatures over a uniform landscape would be expected to have gradual changes. An algorithm that interpolates temperature values by averaging the changes smoothly over distance might be appropriate. However, for elevation data in rough terrain, such an algorithm would tend to smooth abrupt elevation changes, such as steep cliffs and ridges, features that should be retained. Alternative functions that preserve abrupt changes can be used to overcome these types of problems. Of course, more complex algorithms tend to exact higher processing costs.

Many different mathematical models are used for interpolation. Some of the more common ones used for elevation data are the average of the neighbouring values or a weighted average where the weights are inversely proportional to the distance. More complex operations, such as **kriging**, take into account the trend of the known values in the vicinity of each point to be predicted.

A detailed discussion of interpolation is beyond the scope of this work. It is important to realize that in calculating missing values from neighbouring points, the data are assumed to behave in a spatially predictable manner over the map area. As the assumptions of the model are more severely violated, the interpolation results will become less accurate. For this reason, the quality of the interpolation result should be reported as well as the results.

Contour Generation

Contour lines are used to portray surface relief as a set of lines that connect points of the same value. In a topographic map, perhaps the most familiar application, contour lines connect points with the same elevation value. Contours are routinely used to portray a wide range of spatial data sets that can be represented as a surface: crime rates, housing values, geochemistry, wildlife population counts, and climatic data are a few examples. Although contour generation is used to produce output products, it has been included here because of the importance of interpolation in its implementation.

The process of generating contours is a more involved procedure than connecting data points with the same value. It also requires that predictions be made for missing values. Contouring functions make use of interpolation routines to generate those missing data points. Figure 7.26 shows how a single set of data points can produce two different patterns of contour lines. Although both results are reasonable, they differ in the way the elevation values have been interpolated at key locations to generate the contours. The topography that can be inferred from the 30 m data points is ambiguous. As a result, there is more than one reasonable solution. In Result A, no additional 30 m data points were produced. As shown in the perspective view, the contours portray a surface with a north-south ridge. In Result B, additional 30 m points were generated, and the resulting contours portray an area with two hills separated by an east-west gap.

Software packages will differ in the way they handle ambiguous conditions such as this. Often the results are judged by comparison with the way a cartographer would have drawn the contours, with the one that more closely matches the hand drawn version being deemed the ''better'' solution. The evaluation of contouring software must, therefore, take into account the expectations of the users. Users that have become accustomed to a particular style of contouring may judge a different, though equally correct style, to be unacceptable.

CONNECTIVITY FUNCTIONS

The distinguishing feature of connectivity operations is that they use functions that accumulate values over the area being traversed. That is, they require that one or more attributes be evaluated and a running

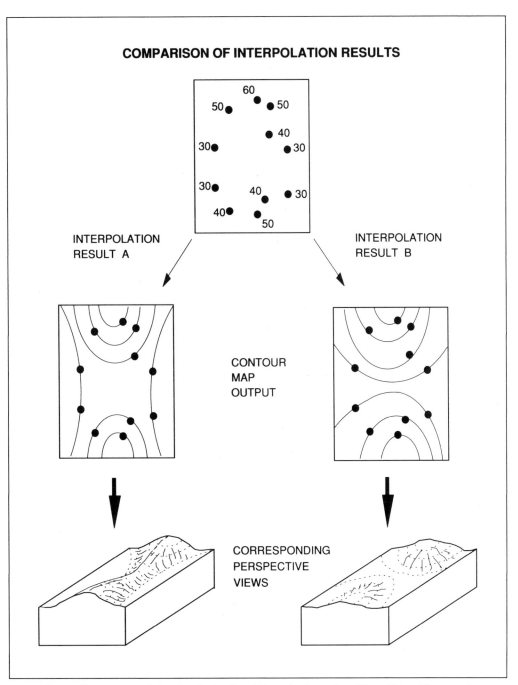

Figure 7.26 Different Interpolation Results Produced From the Same Set of Data Points. There can be more than one solution to the interpolation of a set of values. The figure illustrates two different results that could reasonably be obtained from the same set of data points. Each result is shown as a contour map and in perspective view. Though both solutions are technically correct, only one solution correctly depicts the actual landscape which these data describe.

total of the results be retained in a step-by-step fashion. Each step represents a movement in space, such as a 100 m segment along a street. The running total may be quantitative, such as the accumulated distance travelled or the accumulated travel time. The running total can also be qualitative, such as whether a point is or is not still visible. Every connectivity function must include the following:

1. a specification of the way spatial elements (such as roads) are interconnected;

2. a set of rules that specify the allowed movement along these interconnections;

3. a unit of measurement.

Consider, for example, a connectivity function for travelling along city streets. A street map could be used to define the way the elements (the streets) are interconnected. The rules for movement might include observing one-way streets and speed limits. The measurement unit might be distance, e.g. find the travel distance between two points along the street network. To find the distance along the street network, the route must first be defined and then the measurement of each segment progressively totalled. This is a considerably more complex problem than calculating the straight-line distance between two points, which can be done as a single calculation using the coordinates of the start and destination points.

Connectivity functions provide considerable flexibility in defining units of measure. Often the unit of measurement is not distance. In the travel example, the travel time might be of greater interest. Depending on the sophistication of the software package, such factors as the traffic flow at different times of day and automated route-finding might also be provided. Then an automated route selection procedure could map out the optimum route (i.e. the route with the least travel time) to go from a given starting location to any specified destination.

Software packages vary considerably in the connectivity functions they provide and the algorithms used to implement them. Both vector- and raster-based methods are used. The approach used in a specific software package depends on the data model best suited to the problem and the format in which the data are stored in the GIS. Conversions between raster and vector data structures are sometimes used when a function is more easily implemented in a data structure other than the one used for storage. In the following discussion, connectivity functions have been grouped into the following categories: contiguity, proximity, network, spread, stream, and intervisibility.

Contiguity Measures

Contiguity measures evaluate characteristics of spatial units that are connected. A contiguous area consists of a group of spatial units that share one or more specified characteristics and form a unit, as illustrated in Figure 7.27. The contiguous areas may be specified to have unbroken adjacency, i.e. no gaps are permitted. In other cases, gaps may be allowed, e.g. when a greenspace is considered to be contiguous even though it is crossed by a road.

The definition of *unbroken* may change with the application. A corn field containing an unplanted area may still be considered one contiguous unit even though the field area would include both the planted and unplanted portions. In other cases, only the planted part might be considered the contiguous area.

Commonly used measures of contiguity are the size of the contiguous area and the shortest and longest straight-line distances across the area. A common application of these measures is to identify areas of terrain with specified size and shape characteristics. For example, a search for a land unit to be used as a park might be specified as a contiguous land unit of forest having a minimum area of 1000 square km with no section narrower than 20 km.

Proximity

Proximity is a measure of the distance between features. It is most commonly measured in units of length but can be measured in other units, such as travel time or noise levels. Four parameters must be specified to measure proximity: the target locations (e.g. a road, a hospital, a park), a unit of measure (e.g. distance in metres, travel time in minutes), a function to calculate proximity (e.g. straight-line distance, travel time), and the area to be analyzed.

Figure 7.28 illustrates one application of a proximity function. In this example, the 300 ft buffer zone drawn around the dirt roads defines the forest area where logging is not permitted. This type of proximity analysis is often called *buffer zone generation*. A **buffer zone** is an area of a specified width drawn around one or more map elements. A report can be generated of the forest stands within the buffer zone and the area unavailable for harvest by overlaying the buffer zones on a forest cover map, as illustrated in Figure 7.29. The results of this analysis can be reported in tabular form as shown in Table 7.2.

More complex proximity analyses may require values to be calculated for a large number of point locations and may also involve overlay operations with multiple data layers. For example, the noise level from an airport would be expected to decrease with distance. This could be analyzed using a proximity function to calculate the distance of each location from the sound source. A mathematical model of sound propagation would be used to calculate the reduction in sound level for each increment of distance. Then, using the mathematical model together with the proximity data, a separate data layer could be generated showing the expected sound levels at each geographic location.

Figure 7.30 illustrates the results of this type of GIS analysis for noise levels. The map shows the expected noise levels for the area surrounding a proposed airport. The outermost contour represents the 65 LDN noise level. The inner contours represent higher noise levels. This map was produced by computing the noise levels expected when each type of aircraft landed. The

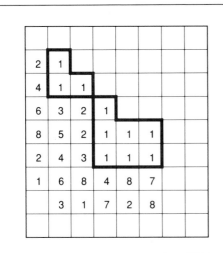

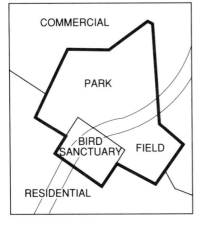

Figure 7.27 Contiguous Areas. The area of 1s would be a single contiguous area if corner-to-corner adjacencies are considered contiguous. If contiguity were limited to edge-to-edge connections, there would be two contiguous areas of 1s. The park, field, and bird sanctuary are grouped as a single contiguous greenspace area. In this case, the road crossing the area is treated as an allowable gap.

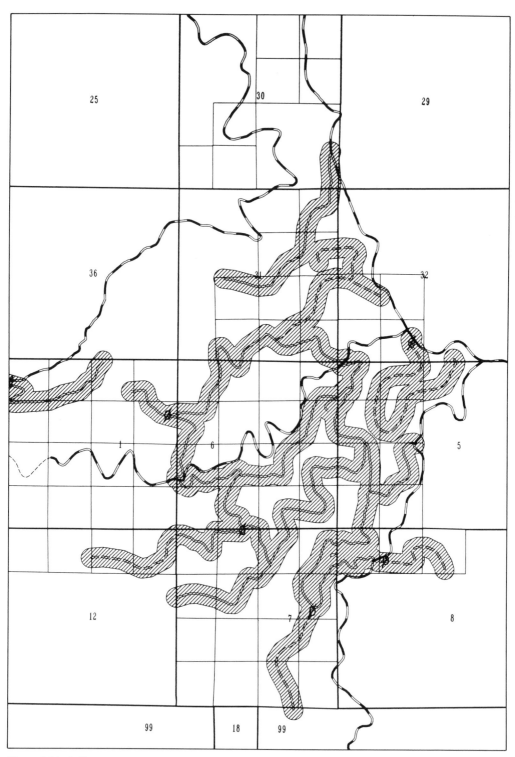

Figure 7.28 Buffer Zone Generation. A 300 ft buffer zone around the dirt roads defines the forest areas where logging is not permitted. (Courtesy of ESRI. Redlands, California.)

results of this analysis were overlayed with the land use data layer to assess the area of residential land that would be affected.

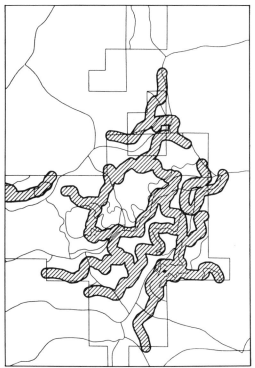

Figure 7.29 Reporting Forest Cover Area Within A Buffer Zone. A previously generated buffer zone is used as a search window to evaluate the area and types of forest that fall within the buffer zone. (Courtesy of ESRI. Redlands, California.)

By evaluating different landing patterns and aircraft types, an aircraft traffic schedule could be developed that would minimize the amount of residential land exposed to high noise levels.

Most GIS packages include some form of proximity function. They differ widely in the sophistication of the analyses that can be performed. One fundamental distinction among proximity functions is whether they can generate an *accumulation surface* (see discussion of spread functions). An accumulation surface is generated by moving outward incrementally from a target. At each step, a function (such as a travel time calculation) is analyzed for that geographic location and added to a running total. The value in the output data layer for that geographic location is set to the current total. Then processing moves to the next increment where the process is repeated. The values in the output data layer are thus totals, each of which have been accumulated over the previous incremental steps. This type of proximity function is sufficiently different that it is discussed later as the *Spread* function.

Network Functions

A network is a set of interconnected linear features that form a pattern or framework.

Table 7.2 Report of Forest Areas in Road Buffers. (Adapted from ESRI 1984).

MANAGEMENT UNIT	STAND NUMBER	STAND AREA (acres)	BUFFER AREA (acres)
1	1	327.7	84.8
1	2	92.1	18.9
1	3	42.9	2.8
	TOTAL	462.7	106.5
2	3	19.7	0.0
2	7	97.1	43.4
2	12	87.1	24.7
2	23	69.5	0.9
2	24	57.0	21.4
	TOTAL	330.4	90.4

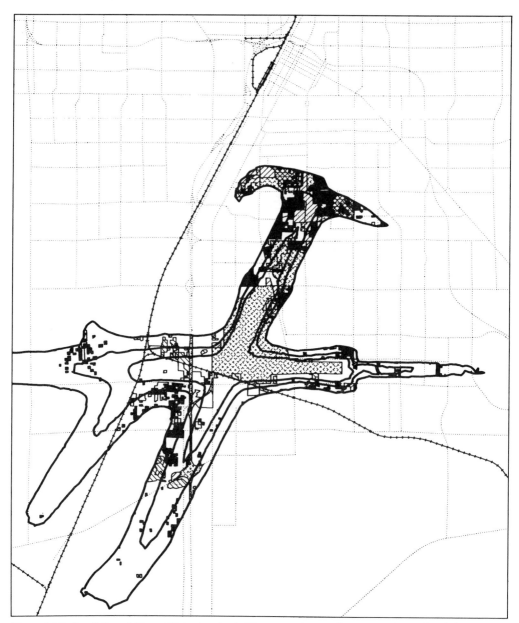

Figure 7.30 Prediction of Airport Noise Levels. A GIS was used to predict expected noise levels for the area surrounding a proposed airport. The outermost contour line represents the 65 LDN noise level, the inner contours represent progressively higher noise levels. (Courtesy of ESRI. Redlands, California.)

Networks are commonly used for moving resources from one location to another. A city's streets, a grid of power transmission lines, an airline's service routes, or the streams of a drainage basin are familiar examples of networks.

A GIS is used to perform three principal types of network analyses: prediction of network loading, route optimization, and resource allocation. The transport of water and sediment in a river system can be predicted using a network model. When several storms occur in a region the effects of the increased stream flow can be complex. By correctly predicting the flows of water through the stream network, the magnitude and location of a flood can be predicted in advance so that emergency services can be prepared. Route optimization applications range from emergency routing of ambulances, fire, and police vehicles to airline scheduling and the routing of bus services, mail delivery, and municipal garbage collection. A common resource allocation application is the division of a metropolitan area into zones that can be efficiently serviced by individual police and fire stations. For example, an area to be policed might be divided into zones that can be patrolled in equal amounts of time. In the upper diagram of Figure 7.31 the solid line defines an optimum route through a network of streets. In the lower diagram the street network has been divided into two zones each of which can be patrolled in the same amount of time.

Networks have unique properties that require special analysis functions. The resources to be transported are usually dispersed throughout the network. For example, traffic is dispersed throughout city streets, power is dispersed throughout the power grid, water and sediment are dispersed throughout the stream network. The links within a network have characteristics that determine the type of resources and the conditions under which they can be transported. For example, some city streets are one way, some are closed to trucks, and streets have different speed limits and capacities — as we are reminded when caught in rush-hour traffic.

Network analyses usually involve four components:

1. a set of resources (such as goods to be delivered);

2. one or more locations where the resources are located (such as the warehouse where the goods are stored);

3. an objective, to deliver the resources to a set of destinations (such as the location of the customers who will receive the goods) or to provide a minimum level of service to an area (such as a police patrol zone);

4. a set of constraints that places limits on how the objective can be met (such as the maximum speed that vehicles can travel).

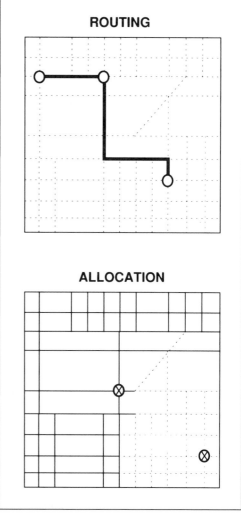

ROUTING

ALLOCATION

Figure 7.31 Network Functions for Routing and Resource Allocation. Network functions can be used to optimize vehicle routing and to divide an area into service districts to optimize the allocation of resources.

The network functions of a GIS are used to simulate the behavior of networks that would be too difficult, expensive, or impossible to measure. In a network model, the elements of the actual network are represented by a set of rules (such as the permitted direction of travel along a street) and mathematical relationships (such as the power loss along a transmission line as a function of distance). The more sophisticated the network functions and the representation of the network, the more closely the behaviour of the model can be made to mimic reality.

There is considerable variability in the network capabilities of commercially available GIS software. They differ in the size and complexity of network model that can be defined, the level of performance, and the degree of interactive control.

Spread Functions

The spread function is a very general yet powerful operation that can be used to analyze a wide range of phenomena. It can be used to evaluate transportation time or cost over a complex surface. It can also be used to define drainage basins (e.g. by spreading out from a point and allowing movement only to adjacent cells with the same or higher elevation).

The spread function has characteristics of both network and proximity functions. A spread function evaluates phenomena that accumulate with distance. Its operation can be thought of as moving step-by-step outward in all directions from one or more starting points and calculating a variable, such as travel time, at each successive step.

The distinguishing feature of a spread function is that a running total is kept of the function being evaluated. It is the value of the running total at each location that is written to the output file. (The output is sometimes termed an **accumulation surface** or **friction surface**.) In a simple case, the accumulated value may be the straight-line distance from

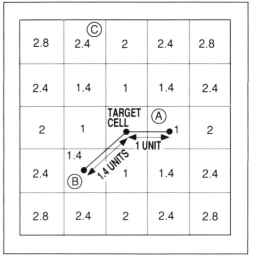

Figure 7.32 The Calculation of Distance Using a Spread Function.

the starting point. (This is the case in a proximity analysis, which is, in effect, a special case of the more general spread function.) In more complex applications, the accumulated value may represent travel time and take into account multiple constraint factors. Because values are accumulated incrementally over relatively small spatial units, constraint factors with irregular spatial distributions can be accommodated. For example, movement may be constrained by partial barriers that reduce the rate of movement or by absolute barriers that stop.

In Figure 7.32, a spread function has been used to evaluate the travel distance from a target cell. The distance was accumulated by stepping outward one unit in every direction. All the cells adjacent and in line with the target were assigned the value 1, indicating 1 unit of distance (e.g. cell **A**). Cells diagonally adjacent to the target were assigned a value of 1.4 (the diagonal distance between the centers of the two cells, e.g. cell **B**). Cell **C** can be reached by moving diagonally across one cell and up one cell for a total distance of 2.4 units. It can also be reached by moving up two cells and across one cell for a total distance of 3 units. The spread function assigns the shorter distance in the case of multiple routings. This procedure is used to calculate a value for each cell.

Rather than writing in all the values for each cell, the spread function results will be represented by contours. In Figure 7.33, the contours represent distances in km away from the starting point at **A**. The shortest distance from location **A** to location **B** is given by the dashed straight line connecting them. In this case, the function is essentially the same as the proximity function discussed previously.

Figure 7.34 shows the effect of an absolute barrier, i.e. one that does not allow any movement across it. A lake would be an absolute barrier to truck travel, for example. As before, a spread function has been used to generate the travel distances. Now the shortest travel distance is not the straight line connecting locations **A** and **B** because the route must go around the obstacle. The spread function evaluated the distances incrementally and accumulated the distance to go around the obstacle. This type of analysis involving obstacles cannot be accommodated by proximity functions.

In Figure 7.35, the effect of a partial barrier is illustrated. A partial barrier, such as rough terrain, impedes progress but does not stop travel. Instead of measuring distance, the spread function is used here to

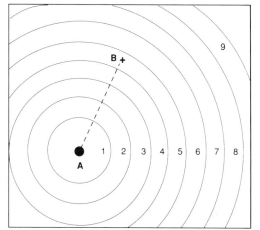

Figure 7.33 Travel Zones Defined Using a Spread Function. Equidistant travel zones in 1 km increments from the target (A) are indicated by the concentric rings. The shortest travel distance from A to B is shown by the dashed line.

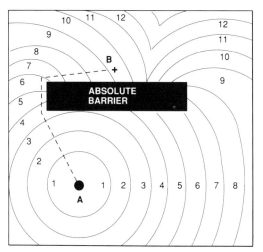

Figure 7.34 The Effect of an Absolute Barrier on Travel Zones Defined Using a Spread Function. Travel zones in 1 km increments from the target (A) are defined by the travel distance contours. The shortest travel distance from A to B must follow a route around the barrier as traced by the dashed line.

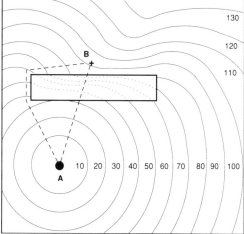

Figure 7.35 The Effect of a Partial Barrier on Travel Time Defined Using a Spread Function. The travel time from the target (A) to any location on the map is defined by the travel time contours. The label on each line indicates travel time in minutes. The rate of travel is 6 km/hr except through the partial barrier where the rate of travel is reduced.

evaluate travel time. The fastest travel time from Location **A** to **B** is 90 minutes. There are two routes with equal travel times: a longer route around the obstacle at a faster rate of travel and a shorter but slower route through the obstacle.

One of the advantages of a spread function is that irregularly distributed factors can be accommodated. In Figure 7.36, a land use map is shown on the left. The walking time is to be mapped from a designated point to any location in the map area. First a travel time data layer in raster format is generated. Based on testing, previous experience, or other information, the travel time to traverse one cell is determined for each land use type. A new data layer, shown on the right in Figure 7.36, is produced by assigning to each cell the travel time for its land use type. Roadway cells have the lowest values because they can be traversed the most quickly. Other types of terrain have higher values, indicating the slower rate of travel across them. Once the travel time data layer has been generated, the spread function can be used to evaluate the travel time from one or more target locations to any point on the map. The two input data

layers and resulting travel time contours are shown graphically in Figure 7.37.

The calculation used to generate the travel time map is illustrated in the simplified example shown in Figure 7.38. Here only two classes are used. An additional simplifying assumption is made about the calculation of distances; they are calculated from cell edges instead of the more usual cell centers. The travel time data layer defines the time to traverse each cell. The more general term **friction surface** is often used to describe this type of data layer (the value in each cell represents the degree to which movement across it is retarded). The second data layer identifies the location of the start points (in this case only one point is used). Multiple starting locations are often used, e.g. to calculate the minimum travel time from any of several emergency facilities.

The third data layer shows the cumulative travel time to any point from the target. It is produced by accumulating the travel times for each cell while moving outward from the starting location. Where a cell can have two values, the smaller value is selected. The computer implementation is somewhat different in order to improve

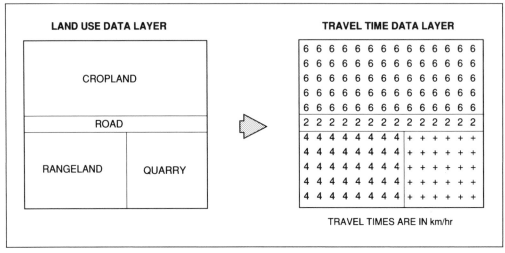

Figure 7.36 Generation of a Travel-Time Data Layer. The time to traverse cells of each land use class is assigned to a travel time data layer. Cropland is assigned a travel time of 6 minutes per cell, the roadway is 2 minutes per cell, and the rangeland is traversed at 4 minutes per cell. The quarry cannot be traversed and is assigned a +.

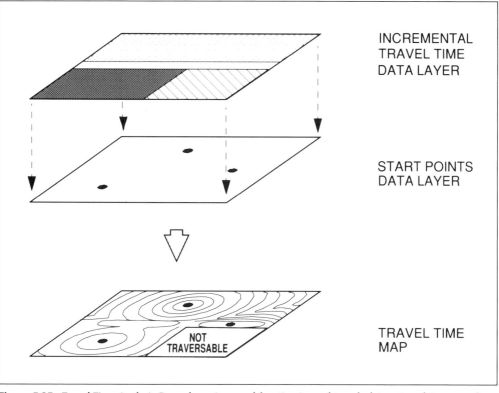

Figure 7.37 Travel-Time Analysis Procedure. A spread function is used to calculate a travel-time map from the incremental travel time values and the location of the start points.

processing efficiency, but the principle is the same.

One application of this type of analysis is in predicting the time needed to traverse terrain with variable conditions. It is an application often encountered in military operations, and has been termed **terrain trafficability**. The trafficability, or ease and speed of movement, will vary with the type of ground cover, topography, mode of transport, and season of travel. The analysis is complex because the route taken to reach a given destination is not necessarily a straight line and the travel time to cross each unit of terrain may change suddenly along a route as a result of changes in land cover or other factors. The spread function is able to incorporate these diverse and irregularly distributed constraint factors.

Spread functions have been implemented in research GIS software packages. The spread function was originally developed by Dana Tomlin and is included in the Map Analysis Package, available from the Harvard Graphics Laboratory, Harvard University. Although some features of the spread function are provided in commercial GIS software, the more general version of the spread function as described here is to the author's knowledge available only in non-commercial research software. In part, this may be a result of the difficulty of implementing the task for large numbers of cells. The function requires that the area be regularly subdivided into relatively small terrain units to provide the cells over which values are progressively accumulated. For this reason the spread function is implemented using raster-based techniques.

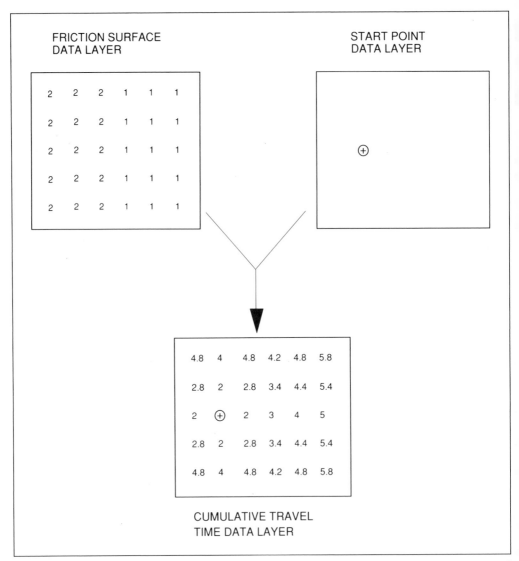

Figure 7.38 Spread Function Sample Calculations. The rate of travel across each unit of the terrain is stored in the travel time data layer or friction surface. The time required to travel to the adjacent cell is calculated by multiplying the rate of travel by the distance. The distance between cell centres is 1 unit in the horizontal or vertical direction and 1.4 units across the diagonal. The travel time to reach any cell from the starting point is found by moving outward from the start point, summing the individual travel times at each step. Where there is more than one route to reach a cell, the lower value is used.

What makes the spread function particularly important is that it can incorporate information about the variability of conditions and tally the cumulative effects of this variability. Spread functions are a valuable and very flexible GIS capability, particularly for cartographic modelling. As users become more involved with cartographic modelling and more familiar with contiguity functions, the demand for general spread function capabilities may lead to their wider availability.

Seek or Stream Functions

The seek function (also termed a *stream function*) performs a directed search outward in a step-by-step manner from a start location using a specified decision rule. The procedure is repeated until any further movement would violate the decision rule. The output from the seek operation is a trace of the one or more paths taken in moving from the start point(s) until the function stops. For example, a seek function can be applied to a digital elevation model to trace the path of water flow. The rule used might be to move from a start location to the adjacent point with the lowest elevation. This operation would be repeated until a location is reached where all the adjacent points have higher elevations (a local depression) or the edge of the study area is reached.

This procedure could be used to evaluate erosion hazard. The seek operation could be applied using a regular grid of start points. The number of times each location or cell is traversed would be a measure of the potential flow of water over each unit of terrain. This value could be used as a measure of erosion potential from surface runoff. Other factors that affect erosion potential, such as vegetation cover and soil type, could also be incorporated into the analysis.

The seek function can be used with the spread function to provide an automated route selection capability. Figure 7.39 illustrates this concept using the travel time overlay developed in the previous figure. The spread function was used to generate a data layer in which the value at each location is the shortest travel time from the start point. The fastest route from any point back to the start can be defined using a seek function with the decision rule to seek the adjacent cell having the lowest value. Since the travel time data layer was constructed outward from the start point, the seek function must end at that same point, and in the process trace the quickest route.

Intervisibility Functions

Intervisibility functions, also termed **viewshed modelling** or **viewshed mapping**, is also a cumulative type of operation. The viewshed is the area that can be "seen" (i.e. is in direct line-of-sight) from the specified target locations. Intervisibility functions can be used to map the area visible from a scenic lookout, map the area that can be detected by a radar antenna, or assess how effectively a road will be hidden from view. It is valuable for such diverse applications as landscape planning, military planning, and communications.

Intervisibility functions use digital elevation data to define the surrounding topography. Depending on the sophistication of the software, additional data can be included in the analysis such as the heights of individual features (e.g. buildings or transmission towers) or the heights of different land cover classes. These features may be represented in separate data layers or as a list of point locations and corresponding heights.

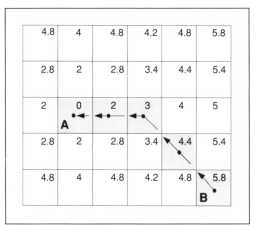

Figure 7.39 Use of a Seek Function to Optimize Route Location. The cumulative travel time data layer from Figure 7.38 can be used to provide a simple automated route selection capability. The route with the shortest travel time from point *B* to *A* can be found by starting at point *B* and selecting the adjacent cell with the lowest value. Because the data layer was generated by spreading out from cell *A*, the seek procedure will trace the fastest route back to that cell.

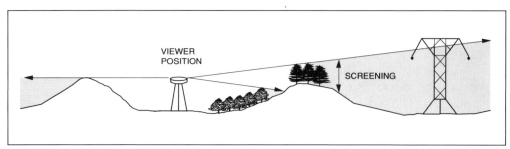

Figure 7.40 The Concept of Intervisibility. An intervisibility analysis identifies locations that are within the unobstructed line-of-sight of a viewing position. Areas that are screened from view are shaded in the diagram. (Adapted from an illustration by D. Tomlin. Ohio State University. Columbus, Ohio.)

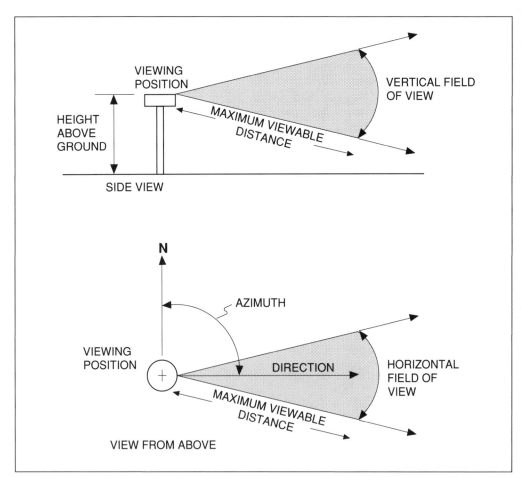

Figure 7.41 Viewing Parameters for Intervisibility Analysis. The viewing parameters used in an intervisibility analysis are: the 3-dimensional position of the viewer, the horizontal and vertical angles of view, the viewing direction, and the maximum viewable line-of-sight distance.

The concept of intervisibility analysis is illustrated in Figure 7.40. The areas that are hidden from the viewer are shown stippled. The data used in the analysis are the viewing parameters, 3-dimensional data for the landscape to be analyzed, and any specific targets to be considered. Figure 7.41 illustrates the viewing parameters needed. They are the maximum viewable line-of-sight distance, the 3-dimensional location of the viewing position within the landscape, the vertical and horizontal angle of view, and the viewing direction (usually specified as degrees of azimuth, which is the angle in degrees measured clockwise from north).

In Figure 7.42, the components of an intervisibility or viewshed analysis are shown using separate data layers for each component. The target data layer is needed if the visibility of specified targets is to be assessed. Often the purpose of the analysis is to generate a viewshed map, in which case a target data layer would not be required.

The intervisibility function is a powerful tool for planning the siting of features in a

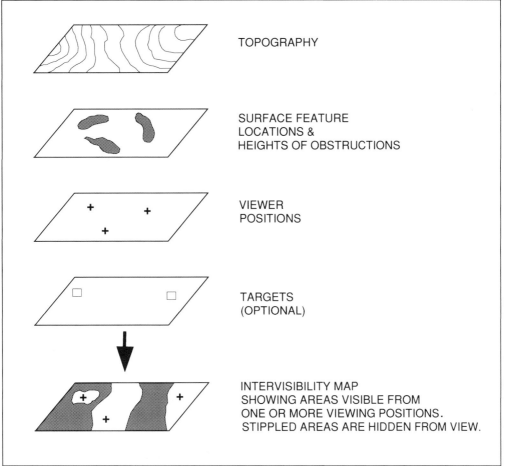

TOPOGRAPHY

SURFACE FEATURE
LOCATIONS &
HEIGHTS OF OBSTRUCTIONS

VIEWER
POSITIONS

TARGETS
(OPTIONAL)

INTERVISIBILITY MAP
SHOWING AREAS VISIBLE FROM
ONE OR MORE VIEWING POSITIONS.
STIPPLED AREAS ARE HIDDEN FROM VIEW.

Figure 7.42 Data Sets for Intervisibility Analysis. Conceptually, an intervisibility analysis requires three sets of spatial data: a surface topography and a surface features data set to define the landscape, and a set of viewing positions from which intervisibility is to be assessed. A set of targets may be included as a fourth data set, in which case the output of the analysis could be a tabulation of the targets visible from each viewing position. The more common output is a map showing the areas visible from each viewing position.

landscape. It is a tool that lends itself well to a trial-and-error analysis procedure in which the placement of objects is progressively refined by re-evaluating the viewshed as the location of objects is changed. The siting of such facilites as microwave and radar systems that require an unobstructed line-of-sight can be optimized in this way. In landscape planning, there is often the need to design facilities that are hidden from view, e.g. minimizing the roadway or cleared land that can be seen from a scenic lookout. Alternative locations for the lookout and the facilities can be evaluated by generating a viewshed map using the scenic lookouts as the viewing positions. Successive modifications to the plan could be assessed by generating revised viewshed maps.

As computer processing power becomes less expensive, interactive viewshed modelling will probably become more readily available. Two similar and more widely available functions are *perspective view generation* and *illumination mapping* (used to generate a shaded relief image). They are computationally similar to intervisibility analysis since they involve tracing rays of light, and can be implemented using the more general intervisibility function. However, they are usually discussed as separate functions, and that convention is followed here.

Illumination

Illumination functions portray the effect of shining a light onto a 3-dimensional surface. The three sets of factors that control this function are the nature and position of the illumination source, the topography and reflectance of the surface, and the position and direction from which the model is viewed. Software differ in the parameters under operator control. For example, the viewing position and surface reflectance are often fixed parameters.

Figure 7.43 Shaded Relief Image of the La Honda, California Map Area. Digital elevation data with a point spacing of 30 m was used to generate this shaded relief image with an apparent illumination from the left. The distance across the image represents approximately 11 km. (From Pike et. al. 1987, courtesy of the US Geological Survey and the American Society for Photogrammetry and Remote Sensing.)

Figure 7.43 is the output from one of the more common types of illumination functions. It is termed a **shaded relief image** or **shaded relief model** and was produced from digital elevation data. The landscape is represented as if it were composed of a material of uniform reflectance illuminated from the left of the image. The position of the illumination source was chosen to provide sufficient shadow for the relief to be easily percieved.

The human mind can perceive shape from an image much better than from data points plotted on a map. Similarly, the relationship between the magnitude of a variable and its spatial location is easily understood as a 3-dimensional surface. The surface need not even represent elevation. It could represent any data that behave as a more or less continuous surface, such as gravity or magnetic field data, which are measurements commonly used in geology. In viewing the shaded relief image of these data, the geologist can perceive the spatial distribution of a set of measurements as a more easily interpreted landscape.

Illumination functions are also used to assess natural conditions. Growing conditions are affected by the quantity and direction of sunlight. Erosion potential and vegetation regeneration rates can be influenced by solar illumination conditions. By modelling illumination conditions, this factor can be included in planning activities.

Shaded relief images provide a rendering of the landscape that can add the surface information to thematic maps or digital images like satellite imagery. The process of applying another data set over a shaded relief image is termed **draping**. It is as if the map were placed or draped over a 3-dimensional model of the terrain. Draping functions are available in several commercial GIS packages. These models are usually portrayed as if viewed from the side and are termed **perspective views**, as discussed in the subsequent section.

The results of an illumination analysis are usually presented in the form of an image, although tabulations are sometimes used (e.g. the average illumination for each forest stand may be used as a stand attribute). A photographic image can best reproduce the subtle grey tones of a shaded relief image so that it appears 3-dimensional. While some types of plotters can produce a range of grey tones, they cannot provide the fine gradations of a photographic image. Instead, plotter outputs are usually presented as mesh diagrams viewed from an oblique position, i.e. a perspective view.

Perspective View

A surface portrayed from a viewing position other than vertical is termed a perspective view. Perspective views are primarily a presentation tool. They are useful in showing the 3-dimensional context of features on a surface, such as a natural landscape. Whereas the vertical view tends to flatten the perceived relief, in a perspective view the relief can be exaggerated to emphasize surface features. Perspective views are commonly generated as photographic outputs or plotted as mesh diagrams. In a mesh diagram, the topography is represented as if a grid of regularly spaced lines had been draped over it. A line-shaded image uses parallel lines with variable width to provide a perspective rendering, as shown in Figure 7.44. In a similar way, thematic maps or satellite imagery can be draped over a shaded relief model to give a 3-dimensional perspective view of the landscape. Plate 13 is a perspective view generated in this way from a satellite image and digital elevation data. By generating a series of perspective views like this one, a motion picture flying sequence can be produced. Perhaps the most sophisticated use of computer-generated perspective views has been in the production of flying sequences for the commercial film industry.

OUTPUT FORMATTING

Output formatting is the preparation of analysis results for output. In the case of tabular data summaries, the preparation is generally incorporated into the analysis function itself and the output file need only be sent to the printer. Outputs in the form of maps are generated in hardcopy formats by such devices as pen plotters, electrostatic plotters, and photographic devices. Map-like outputs are also displayed as electronic images (also termed **softcopy**) on monochrome or colour monitors.

The software functions provided to create these types of output vary widely in flexibility and ease of use. The simplest approach has been to provide one or more standard presentation formats. The operator may be restricted in the placement of titles, legend blocks, and

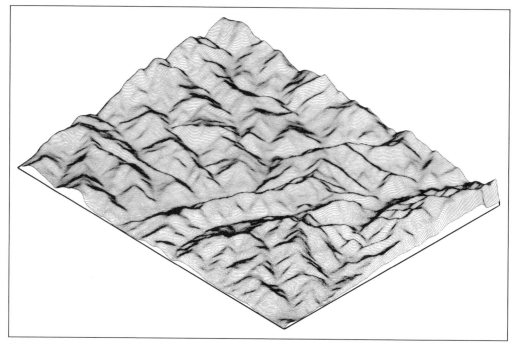

Figure 7.44 Line-Shaded Image of the La Honda, California Map Area. This image was produced from 30 m digital elevation data with a vertical exaggeration of 1.5. It shows the same area as in Figure 7.43 viewed from the southwest. The distance in the east-west direction is approximately 11 km. (From Pike et. al. 1987, courtesy of the US Geological Survey and the American Society for Photogrammetry and Remote Sensing.)

other annotation. More sophisticated systems provide a range of digital cartographic functions, such as the generation of coordinate grids, a wide selection of text fonts, line weights and colours, the definition of symbols, and even the automated placement of text labels within the map. Some of the more common types of output formatting functions are discussed in the following sections.

MAP ANNOTATION

Titles, legends, scale bars, and north arrows are perhaps the most common form of annotation. In its simplest form the title block and legend blocks have fixed positions on the map and the operator can only enter the text and legend symbols. More flexible implementations allow the operator to select the placement and size of these map features. These types of map annotation are placed either outside the map boundary or

they overwrite a portion of the map information. They are not generally interspersed with the map information itself.

TEXT LABELS

Text labels (also termed name labels) are placed within the map area and interspersed with the map information. They may be placed next to a point location (city names), along a linear feature (the name of a river placed along the curvilinear river edge), or within a polygon (the name of a country). Text labels form an important part of a map. Not only do they provide the name of the geographic feature, they can also be used to show the orientation of the feature, its relative size, and even its class. This is done through the font, size, spacing, and placement of the label. Labels also affect the appearance of the map — whether it looks cluttered or is clearly legible. Some of the general principles used in map label design and placement are as follows (Imhof 1975):

1. The names should be legible and located close to the feature they describe.

2. The association between the name and the object it identifies should be easily recognized.

3. Labels should not overlap and the covering or concealing of map information should be minimized.

4. The format and positioning of a name label should directly assist in showing relative importance, territorial extent, connections, and in distinguishing among groups of map features. For example, the name of an area feature should span the entire area and conform to the general shape of the element.

The basic requirement of any system of labelling is that the name label must unmistakably refer to the feature it designates and must not overlap point data. To achieve this in a systematic way, name labels with a smaller degree of freedom are positioned before those with a greater degree of freedom. Area names have the smallest degree of freedom. They must be spread from one end of the feature to the other, conforming to the general shape of the feature without overlapping any point features. There is a greater degree of freedom in the placement of point feature labels. They must be placed near the point described, usually with a preferred placement (such as above and to the right) to give the map visual consistency. Line feature labels have the greatest degree of freedom and are positioned last. The name can be placed almost anywhere along the line, although tight curves and the endpoints of the line are best avoided. As the number of labels and restrictions on overlaps increase, there are very few degrees of freedom and the label placement task becomes quite complex.

Most GIS software has some text labelling capability. More limited implementations severely restrict the size and orientation of the labels. More comprehensive text labelling software can allow the operator to position labels interactively while viewing an image of the output map. Interactive scaling of text size, the automated retrieval of labels from the data base, and even automated label placement may also be provided. Systems designed originally as digital cartographic systems to which GIS functions have been added have tended to provide more sophisticated capabilities, such as automated label placement. However, over the past few years the cartographic quality of GIS systems has improved considerably.

TEXTURE PATTERNS AND LINE STYLES

The selection of line widths and colours is dependent on the output device. Most devices can generate texture patterns. Line widths and colours are used to portray attributes of the line. Lines that represent such features as highways, railways, or political boundaries are commonly distinguished in this way. Line types, such as dashed lines or dotted lines, are also used to distinguish elements. Some systems provide for user-specified dash patterns. In a similar way, patterns (including solid colours) can be used to distinguish different types of areas. The patterns generally include different patterns of cross-hatching, shading, and colours. Software differ in the amount of effort needed to select these drawing parameters. In some cases, standard drawing parameters can be saved and applied to other maps containing the same types of elements. Otherwise, the definition of the drawing parameters may have to be done separately for each map.

GRAPHIC SYMBOLS

Graphic symbols are used to represent map objects. The symbols used to designate a city, a mountain peak, a bridge are common examples. Some systems provide a standard set of symbols but do not allow the operator to create symbols. Others provide the capability to create symbols and store them within the GIS so they can be recalled as needed, termed a **symbol library**. Some systems enable symbols to be assigned according to a user-specified attribute. In this way the appropriate symbol can be automatically plotted.

CARTOGRAPHIC MODELLING: A GIS ANALYSIS PROCEDURE

The previous sections presented an overview of the analysis functions available in geographic information systems. The key to using these functions effectively is to use a systematic approach in defining the information needed and in designing the analysis procedure to meet them. Cartographic modelling is one procedure that has been used for predictive modelling using a GIS.

The term **cartographic modelling** was coined by Tomlin (1983) to mean the use of basic GIS manipulation functions in a logical sequence to solve complex spatial problems. It was developed to model land use planning alternatives, an application that requires the integrated analysis of multiple geographically distributed factors.

The cartographic modelling concept is illustrated first using a contrived example and then an actual application. Figure 7.45 diagrams a land use planning application of cartographic modelling to site a roadway in a hypothetical National Park. The design of a cartographic modelling procedure is best approached by working backwards from the required final result. In the Figure, the final result, a map of the final route location, is generated by Procedure 6, at the extreme right. It is an output formatting procedure that plots the route selected in the previous step.

Procedure 5, the immediately preceding step, is the selection of the final route from several proposed alternatives. The selection is guided by an evaluation of the quantifiable costs and benefits of the competing routes, as well as by the consideration of qualitative factors, such as aesthetics, public sentiment, and so on. In this example, the competing concerns are to minimize construction costs, to minimize the visibility of the road from scenic lookouts, and to minimize the loss of those land use types most critical to wildlife.

Trade-offs must often be made between qualitative and quantitative objectives. Is it worth an additional $10,000 to minimize the visibility of a roadway from scenic lookouts? Is the value of preserving an additional 15 sq km of wetland worth the cost of constructing an additional kilometer of roadway? The cartographic modelling process provides a systematic means to explicitly identify these issues and provide information to support the decision. However, it does not automatically provide the decision; subjective value judgments must still be made.

Procedure 4 is the process of generating the alternative routes. The inputs used are a map showing the relative cost of road construction in the study area, a map of the areas visible from the scenic lookouts, a map of the land cover types, the location of the start and end points for the road, and judgments that reflect the design objectives. The ideal solution might be a road with the minimum construction cost, that is not visible from the lookouts, and does not disturb any critical wildlife habitat. Usually the ideal solution is unattainable, in which case trade-offs must be made among the competing objectives. An iterative process is used to develop route locations that satisfy these objectives to different degrees. The relative importance of these objectives is a judgment made by the participants in the planning process to guide the search for alternative routes. Of the many alternatives considered in this step those considered to be the "best" would be passed to the final route selection process (Procedure 5).

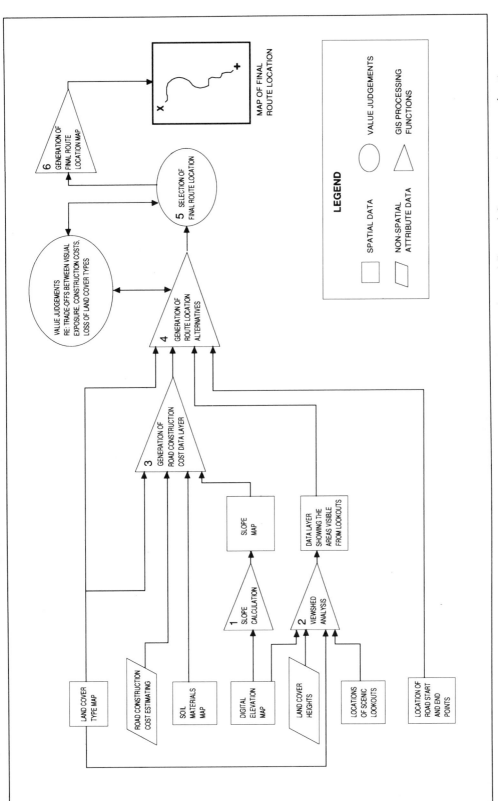

Figure 7.45 Cartographic Modelling for Route Selection. The flow chart illustrates the use of cartographic modelling to define an optimum route location.

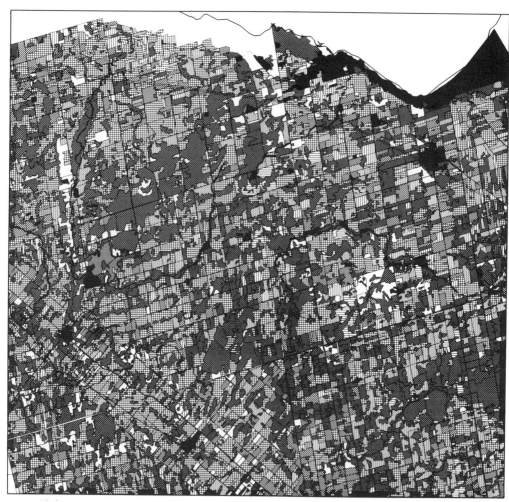

Figure 7.46 Agricultural Land Use Systems Map. This map was produced from one GIS data layer. Some 23 classes of land use are shown. During the planning process, a GIS was used to group these classes into constraint categories relevant to transmission line route selection. In the figure, the darker areas have more severe constraints for transmission line routing. (Courtesy of the Environmental Planning Branch, Ontario Hydro. Toronto, Ontario.)

Procedure 3 is the generation of the road construction cost data layer that shows the cost of roadway construction at each location in the study area. The factors taken into account are the slope of the terrain, the type of soil material, and the type of land cover. Rules are used to calculate the construction costs for each terrain condition. These values are then used to produce the construction cost data layer used as one of the inputs to Procedure 4.

Procedure 2 is a viewshed analysis to identify the park areas that are visible from the scenic lookouts. The inputs to this procedure are the lookout locations, the average heights of the different land cover types, the land cover data layer, and digital elevation data for the study area. The output from the viewshed analysis is a data layer showing those areas visible from the lookouts and those that are hidden.

Procedure 1 is the generation of a slope data layer that is calculated from the digital elevation data.

The process of working backwards through the analysis ensures that all data that will be needed are identified, data that

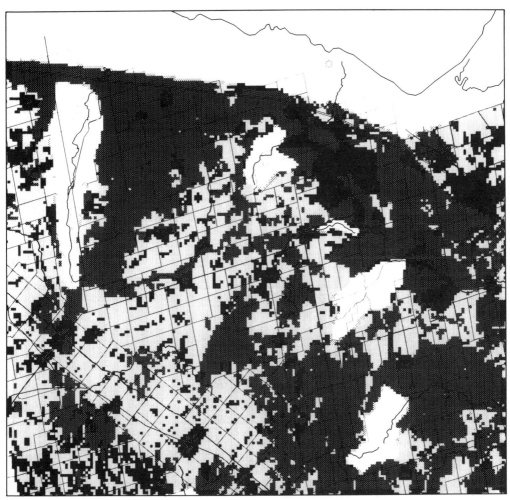

Figure 7.47 Environmental Constraint Map for Transmission Line Routing. Environmental constraint categories were produced using the data from ten resource data sets weighted according to priorities expressed by regulatory agencies and the public. (Courtesy of the Environmental Planning Branch, Ontario Hydro. Toronto, Ontario.)

will not be used are not collected, and that steps where value judgments must be made are explicitly identified. It is a systematic approach that can be applied to a wide range of planning activities. This approach also facilitates the documentation of how design decisions were reached, making it easier to refine the analysis and to scrutinize the process.

A TRANSMISSION LINE PLANNING EXAMPLE

Figures 7.46 to 7.49 illustrate several steps in the cartographic modelling procedure used by Ontario Hydro to select a right-of-way for a power transmission line. The planning process required that engineering, environmental, and social factors be considered and that the design trade-offs be justified before internal reviews as well as before government regulatory agencies and public hearings. It is required that the criteria used to select a route location be documented and that outside concerns be addressed.

To support this type of planning activity, Ontario Hydro used an in-house GIS. The source data consist of existing geographic

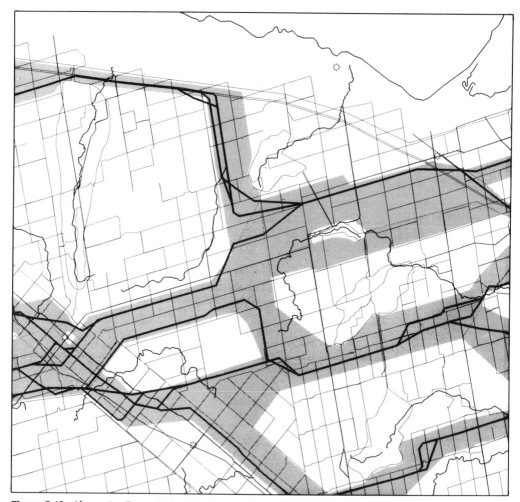

Figure 7.48 Alternative Transmission Line Routes. Maps of individual resource categories as well as derived maps produced by GIS analyses were used to develop several transmission line routing alternatives. (Courtesy of the Environmental Planning Branch, Ontario Hydro. Toronto, Ontario.)

information in map and tabular form, as well as original data collected by field surveys and analysis of aerial photography and satellite imagery. Digitizing facilities are available to convert non-digital data sets for input to the GIS. An image analysis system is used for reconnaisance level land cover and land use mapping. The remote sensing results are output in a digital form that can be input directly to the GIS data base.

Figure 7.46 is a map produced from one of the GIS data layers used in the route location study. It is a map of agricultural land use

systems produced by the Ontario Ministry of Food and Agriculture. Some 23 classes are shown in this map. The GIS was used to group these detailed land cover and land use classes into constraint categories appropriate for the route selection analysis. Other data sets used in the study were agricultural capability, current land use, forestry capability, mineral potential, recreation, hydrology, heritage features, and human settlements.

Figure 7.47 is a constraint map that takes into account all the environmental concerns. It is not a simple addition of the constraint

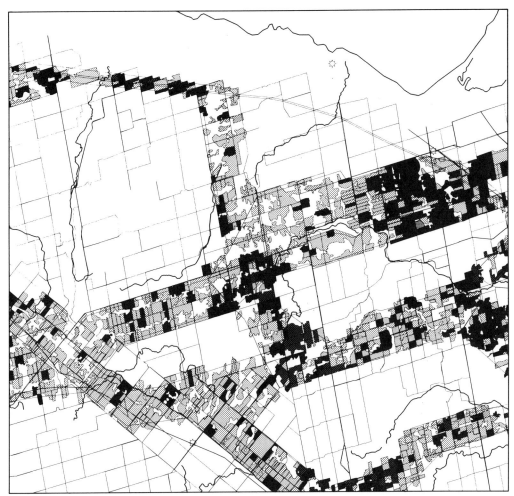

Figure 7.49 Map of Agricultural Resources Along the Proposed Route Locations. This map illustrates the process of extracting resource information for areas defined by a buffer zone around the proposed transmission line routes. (Courtesy of the Environmental Planning Branch, Ontario Hydro. Toronto, Ontario.)

ratings for the individual resources. The constraints were weighted according to the priorities expressed by the participating advisory and regulatory groups. A total of nine constraint categories were defined. These were combined into five categories to produce a more legible map for illustration. However, the nine categories were retained in the data base. In the Figure, the more severe constraint classes are darker.

Figure 7.48 shows the alternative routes proposed for the transmission line. They were developed using maps of the individual resources and derived maps produced by integrated GIS analyses, such as the constraint map. The routes were actually drawn on an airphoto mosaic and then digitized for input to the GIS. These proposed routes were then analyzed to determine the quantity of various resources that would be affected by each right-of-way. This was done by generating buffer zones (the shaded areas in the figure) around the transmission line routes (shown as heavy black lines).

Table 7.3 Tabulation of Constraints for A Proposed Transmission Line Route. (Courtesy of the Environmental Planning Branch, Ontario Hydro. Toronto, Ontario.)

AFFECTED AREA (ha)	AFFECTED LENGTH (km)	% OF ROUTE AREA	CONSTRAINTS
2	.32	.2	major non-urban settlement
6	.79	.5	military base, communication towers
26	3.47	2.2	proposed settlement or airport
2	.32	.2	wetland, waterfowl nesting/staging
84	11.06	7.0	deer yards
6	.79	0.5	classes 1,2,3 wetland habitat
43	5.69	3.6	sensitive woodland wildlife habitat
6	.79	.5	class 1-4 productive wildlife habitat
461	60.67	38.4	class 1-3 forest land
127	16.91	10.7	class 4-5 forest land
63	8.37	5.3	principal fruit & vegetable producing areas
687	90.53	57.3	class 1 agricultural soil
201	26.54	16.8	class 2-3 agricultural soil
22	3.00	1.9	existing surface and mineral extraction
480	63.20	40.0	potential surface and mineral extraction
22	3.00	1.9	class 4 wetlands
109	14.38	9.1	cold water fish, areas sensitive to erosion effects
2	.32	.2	warm water fish, areas sensitive to erosion effects
383	50.40	31.9	cold water fish, areas with low sensitivity to erosion effects
20	2.69	1.7	areas with risk of erosion effects for which no fish information is available

An overlay operation was then used to generate quantitative information for the areas that would be affected by each of the routes. Figure 7.49 illustrates this procedure graphically using one of the resource maps. The agricultural resource types are shown only in the buffer zone areas. The detailed constraint data were analyzed in a similar manner. Table 7.3 is a tabulation of these constraints for areas within one of the proposed transmission line routes. In fact, it was the tabular data summaries more than the buffer zone maps that were used to compare the alternative routes.

By using the map and tabular information produced in this way, alternative route locations could be developed through a process that was documented, could be scrutinized by internal and outside reviewers, and was defensible. By means of this systematic analysis procedure, the logic used to reach the design alternatives could be reviewed and refined throughout the process, and effective use was made of the analytical capabilities of the GIS.

REFERENCES

Ahn, J. and H. Freeman. 1983. A Program for Automatic Name Placement. In *Proceedings of the Sixth International Symposium on Automated Cartography*. University of Ottawa. Ottawa, Ontario. pp.444–453.

Beard, M.K. and N.R. Chrisman. 1986. Zipping: New Software for Merging Map Sheets. In *Proceedings of the 1986 ACSM-ASPRS Annual Convention*. Amercian Society of Photogrammetry and Remote Sensing. Falls Church, Virginia. Volume 1:153–161.

Berry, Joseph K. 1987. Fundamental Operations in Computer-Assisted Map Analysis. *International Journal of Geographical Information Systems* Vol 1. (2): 119–136.

Burrough, P.A. 1986. *Principles of Geographical Information Systems for Land Resources Assessment.* Clarendon Press. Oxford, U.K.

Congalton, R.G., R.G. Oderwald, and R.A. Mead. 1983. Assessing Landsat Classification Accuracy Using Discrete Multivariate Analysis Statistical Techniques. *Journal of Photogrammetric Engineering and Remote Sensing* 49(12):1671–1678.

Dangermond, J. 1983. *Software Components Commonly Used in Geographic Information Systems.* Environmental Systems Research Institute. Redlands, California.

ESRI. 1984. *Annual Map Book.* Environmental Systems Research Institute. Redlands, California.

Jackson, M.J., and D.C. Mason. 1986. The Development of Integrated Geo-information Systems. *International Journal of Remote Sensing* 7:723–740.

Imhof, E. 1975. Positioning Names on Maps. *The American Cartographer* 2(2):128–144.

Lam, N.S., P.J. Grim, and F. Jones. 1987. Data Integration in Geographic Information Systems: An Experiment. In *Proceedings of the ASPRS-ACSM 1987 Annual Convention, Volume 5.* Amercian Society of Photogrammetry and Remote Sensing. Falls Church, Virginia. pp.53–62.

Lynch, M.P. and A.J. Saalfeld. 1987. Conflation: Automated Map Compilation — A Video Game Approach. In *Proceedings of AutoCarto 7.* American Society of Photogrammetry and Remote Sensing. Falls Church, Virginia. pp.343–352.

Lupien, A.E. and W.H. Moreland. 1987. A General Approach to Map Conflation. In *Proceedings of AutoCarto 8.* American Society of Photogrammetry and Remote Sensing. Falls Church, Virginia. pp.630–639.

McKeown, D.M., and R.C.T. Lai. 1987. Integrating Multiple Data Representations for Spatial Databases. In *Proceedings of AutoCarto 8.* American Society of Photogrammetry and Remote Sensing. Falls Church, Virginia. pp.754–763.

Newcomer, J.A. and J. Szajgin. 1984. Accumulation of Thematic Map Error in Digital Overlay Analysis. *The American Cartographer* 11(1):58–62.

Pike, R.J., G.P. Thelin, and W. Acevado. 1987. A Topographic Base for GIS from Automated TINs and Image Processed DEMs. In *Proceedings of the GIS'87 Symposium.* American Society for Photogrammetry and Remote Sensing. Falls Church, Virginia. pp.340–351.

Tomlin, C.D. 1983. *Digital Cartographic Modelling Techniques in Environmental Management.* Unpublished Doctoral Dissertation. School of Forestry and Environmental Studies, Yale University. New Haven, Connecticut.

Tomlinson, R.F. and A.R. Boyle. 1981. The State of Development of Systems for Handling Natural Resources Inventory Data. *Cartographica* 18(4):65–95.

Walsh, S.J., D.R. Lightfoot, D.R. Butler. 1987. Recognition and Assessment of Error in Geographic Information Systems. *Photogrammetric Engineering and Remote Sensing* 53(10):1423–1430.

8. IMPLEMENTING A GIS

INTRODUCTION

The design and implementation of a GIS is a major, long-term undertaking. The entire process, from when an organization first becomes aware of the technology through to when a system is finally operational commonly takes one or more years. To the uninitiated, the acquisition of a GIS centers on technical issues of computer system hardware and software, functional requirements, and performance standards. But experience has shown that, as important as these issues may be, they are not the ones that in the end determine whether a GIS implementation will succeed or fail.

The issues responsible for implementation failures are almost always *people problems*, not technology problems. Upon reflection, this is to be expected for it is the technology that has a concrete existence. In principle, the capabilities of the technology can be rationally assessed by means of a physical test. The results can be expected to be repeatable. If you can produce a certain map output once, the same procedure will probably produce it again. Even unpredictable system failures can be anticipated and contingencies can be planned. However, the decisions that people make are not nearly as predictable. People are significantly influenced not only by the facts pertaining to the technology but also by the context of the situation. A personality clash or a power struggle can reverse a seemingly logical decision. It is not that the decisions made by people lack a logic, it is that the people involved have multiple objectives. The parties to a decision are commonly operating with several diverse agendas that can change abruptly. Wherever people interact, there are politics.

The implementation of a GIS is where technology and people meet. One of the reasons for the complexity of the implementation process is that it is, necessarily, political. It is the people in an organization that adopt and learn to use a new technology. In so doing, the organization must evolve, for in adopting any new technology, especially one with an influence as pervasive as a GIS, the organization itself is changed. Information flows are shifted, and different people exert different degrees of control over information, its distribution, and use. In so doing, they exercise power. Information is power, and the power of the information goes to the organization whose budget generates and controls it. Computer technology is political in that control over centralized information systems tends to increase the power of the administrators and technical experts who control them at the expense of those who lack the expertise to use them effectively. For this reason, there is a strong tendency for each agency to want to control its own information.

When the public works department of a municipality has the only up-to-date facilities maps, they can exercise control over access and use of that information. They are able to control the information, even if they are required to provide access to other departments. However, if those maps are available on a GIS, then any user with access privileges can use the information directly. Previous informal control and review of information requests may be lost when the information must no longer be requested from a single source. It may also be more difficult to deny access or hide preferential treatment.

Computer-generated data is a potent political tool in its power to influence. Policy-makers and the public generally consider computer-generated information to be more accurate, credible, and objective simply because it was computer-generated. The technical language surrounding the preparation, use, and evaluation of computerized data tends to be neutral, expressing no particular value or political bias. This tends to bias the listener to view the computer-generated data as more authoritative than comparable information generated in a non-computerized manner. Of course, there are also those who distrust information if it is computer-generated.

Computers and associated analytical techniques are often assumed to be unbiased and objective tools in the hands of neutral, unbiased, non-political technical experts. In fact, computerized data and analysis techniques are subject to the same types of political bias and inaccuracies as other data. The bias enters in the selection of the data to be included, the analytical methods to be used, and the way the results are presented. These choices are inherently political because they influence the analysis of the results, the perception of issues, and the range of potential solutions. An astute use of computers can effectively hide political choices beneath a cloak of mystifying technical analyses (Klosterman 1987).

The mere conversion of information to digital form can have dramatic implications. Property information that is publicly available in the form of hand-written entries can be used in an entirely different way than the same data in digital form. Behrens (1985) cites several examples. One became a 1973 New Hampshire Supreme Court ruling that a university economics professor had the right to acquire a computer tape of property record information for 35,000 properties in the city. The data were to be used in a tax study. The city argued that providing the data would constitute an invasion of privacy for every citizen on the file. The court ruled that the information could not be regarded as confidential because it could be acquired by a visual inspection of the property.

Yet the digital data could be used to generate information about individuals that would be impractical to determine otherwise. Finding the names of the one hundred largest property holders would be impractical using a manual search. By computer the analysis is fast, simple, and inexpensive.

The introduction of a GIS will change the existing management information systems. In most cases, by the time an organization considers a GIS, it already has some form of computer-based information system in place. The system may not adequately handle spatial information, but it does support some operational needs. The existing system is also the one with which personnel are familiar and it probably reflects the structure, values, and management philosophy of the organization. Unless the existing system is totally inadequate, the introduction of a GIS must be integrated with the existing information system. Some functions that are duplicated will eventually be relegated to one system or the other. However it is important to coordinate the integration process so that the operation of the organization during the transition is not jeopardized.

The introduction of new technology will change an organization in ways that cannot be entirely predicted. Attention is usually focussed on the desired changes and their ramifications, but there will be numerous subtle but significant organizational changes as well. For example, if additional staff are hired to operate the GIS equipment, their salaries may be significantly higher than the managers under whom they will work. These types of changes are stressors within the organization that can easily result in non-cooperation. Similarly, jealousies and rivalries can develop between departments in an organization when responsibility for collecting and maintaining data is "rationalized" and one department loses control over "its" data.

Stresses such as these will occur at a time when the organization is already stressed. The organization often does not have sufficient staff to cope with the extra work of introducing a GIS. There is typically insufficient skill in handling the new technology, less than optimal funding, and what funding is provided may be planned for too short a duration. In addition, the GIS is often being implemented in response to external pressures for improved information services at a lower cost.

The organizations may be as large as a national agency or one person in a rural community. The specific considerations that become critical will differ, but the range of issues faced by each are basically the same. The list of issues that have been identified in the literature and discussed among practitioners is long indeed. Aside from the impracticality of recounting the causes of past successes and failures, such a list would probably not be very useful. Instead, this chapter presents a framework for implementation. GIS implementation is considered here as the entire technology transfer process, from when an organization becomes aware of GIS technology through to when it adopts it. "Adoption" is used here to mean that an organization has incorporated a GIS into its operations and regularly uses it where appropriate in its day-to-day activities. Implementation can be seen as a six phase process:

1. *Awareness:* People within the organization become aware of GIS technology and the potential benefits to their organization. Potential uses and users of the GIS are postulated.

2. *Development of System Requirements:* The idea that a GIS could benefit the organization is formally acknowledged and a more systematic and formal process is instituted to collect information about the technology and to identify potential users and their needs. A for-

mal needs analysis is often done at this stage.

3. *System Evaluation:* Alternative systems are proposed and evaluated. The evaluation process takes into account the needs analysis of the previous phase. At the end of this phase, a formal decision must be made whether or not to proceed with acquisition of a GIS.

4. *Development of an Implementation Plan:* Having made the decision to proceed with acquisition of a system, a plan is developed to acquire the necessary equipment and staff, make organizational changes, and fund the process. The plan may be a formally accepted document or a more or less informal series of actions.

5. *System Acquisition and Start-Up:* The system is purchased, installed, staff are trained, creation of the data base is begun, and operating procedures begin to be established. Creation of the data base is usually the most expensive part of the implementation process. Considerable attention is needed to establish appropriate data quality controls to ensure that the data entered meet the required standards and that suitable updating procedures are implemented to maintain the currency and integrity of the data base.

6. *Operational Phase:* By this stage the initial automation of the data base is complete and operating procedures have been developed to maintain the data base and provide the information services that the organization requires. In this phase procedures are developed to maintain the GIS facility and upgrade services so that the GIS continues to support the changing information needs of the organization. Operational issues concerning the responsibilities of the GIS facility to provide needed services and

to guarantee performance standards become more prominent.

The issues that arise at each of these stages have both a technical and organizational component. The remainder of the chapter discusses each of these phases, the issues that are addressed, and some of the lessons that have been learned.

PHASE 1: AWARENESS

Information about new technology can enter an organization at several levels. Ideally, new ideas would be examined in the same way regardless of the messenger. In fact, new ideas are treated differently when presented by senior management or junior staff. New ideas, such as GIS, have to be "sold" within an organization. The selling may be from the bottom up, from the top down, or from an independent third party.

APPROACHES TO INTRODUCING THE NEW TECHNOLOGY

Top-Down. Top-down promotion occurs when ideas for doing things differently are pushed from the management level to the production level. The advantage is that the power to provide funding and organizational support either resides at this management level or is more readily obtained by this level. The management level may also be capable of taking a broader view of potential benefits — a view that spans different types of applications. For example, a GIS may be seen as a means to organize all the spatial data of a municipality, not just the land records. Diverse departments could be beneficiaries. The system could be used by police and fire services for emergency dispatching, by public works for facility maintenance, by the school board for bus route allocation, and so on.

The disadvantage of the top-down approach is that the management level may not be fully aware of how information actually enters and is processed by the organization. They may not be technically competent to evaluate the feasibility of actually obtaining the benefits promoted for a GIS. As a result, their proposals may lead to inflated expectations that are easily dismissed. There is also the problem of resistance from staff. Changing the organization will require the re-training of some staff and perhaps the elimination of some traditional jobs. For example, weather forecasting once depended on large numbers of technicians to draft maps. Now most of this work has been automated.

Bottom-Up. Bottom-up promotion occurs when those doing the work become aware of improvements that could be made to the current procedures. The individuals at this level in the organization are probably the most capable of recognizing the limitations of the current system and most keenly aware of the bottlenecks that might be removed through the introduction of a GIS. This approach is usually easier to implement because it has the support of the working level.

One of the difficulties with bottom-up promotion is the way that management justifies expenditures for a GIS. It is rare that solid cost-benefit cases can be made for computer systems, whether they be as complex as a GIS or as commonplace as the office word processing system. When examined in detail, cost-benefit justifications tend to include some quantification of benefits, but still are more or less a leap-of-faith. Many of the changes will be non-tangibles, such as conditions of work, speed of access to information, and improved levels of service. Often the decision may be stalled until the need to adopt the new technology becomes "common knowledge". Once the need to automate spatial data has been discussed frequently enough and at high enough levels, it becomes accepted as the *status quo*. The need to justify automation then ceases to be the issue and attention focusses on the mechanics of implementation.

Independent Third Party. The champion for the GIS idea may be a third party, i.e. someone from outside the organization. They may be a GIS vendor, a group of GIS users in the same field, or a consultant. If they are seen as being technically competent and having no vested interest, or at least a known interest in the outcome, then their suggestions may carry considerable weight. Third party support may also come from a group within the same organization that has similar needs. However, if the affiliation is too close, the group may end up competing for control of the new resource. Data processing departments sometimes fall into the position of competing with the departments they serve for control of computer resources.

THE IMPETUS FOR ACQUIRING A GIS

Part of the impetus for introducing a new idea like a GIS is a recognition that there are problems with the current information system. Some of the more common problems that prompt a GIS investigation are as follows:

1. Spatial information is poorly maintained or is out of date. This may take the form of outdated maps, long delays in processing map revisions, or inaccurate data records and summaries. User mistrust of the quality of the information may lead to the use of alternative data sources.

2. Spatial data is not recorded or stored in a standardized way. The geographic coordinate systems may differ and map scales may vary, making it difficult to use multiple data sets together. These problems can significantly limit the usefulness of the existing data.

3. The spatial data may not be defined in a consistent manner. For example, departments may be using different definitions for their land use categories, making it necessary to capture and store more than one data set for similar information. It is often possible to develop a single classification that will satisfy the needs of all users. In some cases, the mandates of different organizations require that similar data be organized using different classification systems. However, even then, a cost reduction can often be achieved by collaborating on the collection and encoding of the data.

4. Data are not shared. This may result from fear of misuse or because potential users would not know of the existence or whereabouts of the data. As a result, users may keep their own copy of the original data base or may create and maintain their own data base. Not only can there be costly duplication of data storage and collection, duplication of data almost guarantees that there will be discrepencies. The different versions may be created in slightly different ways and will probably not be updated at the same time.

5. Data retrieval and manipulation capabilities are inadequate. The retrieval of information, such as routine reports, may be too slow and the ability to perform complex or special purpose analyses of spatial information may be limited or non-existent.

6. New demands are made of the organization that cannot be met using the current information system. The organization's mandate may be changed, or a new legal requirement may take effect that cannot be satisfied without the capabilities of a GIS. To be prepared for the unexpected, unlimited expansion of the data base is often made a system requirement.

For the second phase of implementation to be reached, the political challenges must be reconciled to the point that some group within the organization can be assigned the responsibility of analyzing the need for a GIS. There must also be some general recognition of the problems that a GIS might address. Then one or more individuals can be asked to more formally investigate how a GIS might benefit the organization.

PHASE 2: DEVELOPING SYSTEM REQUIREMENTS

Most discussions of GIS implementation begin with this phase. Yet the system requirements developed here will be influenced by which group has been given the mandate for defining what needs and whose needs the system will address. Of course the proposed systems will tend to reflect the particular orientation of this group.

Requirements are commonly developed by documenting the different ways the organization currently handles spatial information, termed a **functional definition** or **user needs analysis** of the existing system and the anticipated future needs. This is done using interviews, analyzing the information products and services handled by the organization, and the systems, and procedures used to provide them. For purposes of analysis the existing system can be subdivided into a set of processing modules. Within each module, a user, the one accessing and using the data, obtains the required input data, performs one or more procedures on that data, and generates an information product. That product may be a legal description of a land parcel, a map of critical wildlife habitat, or a tabulation of forest stands to be harvested.

The analysis of the existing information system should provide a systematic report of the information tasks the organization is mandated to provide, the input data needed, the output products required, and the procedures used to generate those output products. The form, accuracy, timeliness, and volume of those products are critical characteristics. Time sensitive data, such as weather information or emergency routing will have entirely different processing requirements from map updating or land records maintenance.

The definition of functional requirements is derived from the evaluation of the existing system and the anticipated future processing needs. The information system is then considered as a whole within the context of the organization. Given the required inputs, outputs, and processing to be done, an attempt is made to improve efficiency by reducing duplication of effort. When information systems evolve, there is a tendency for special purpose functions to be created as needed. After a while there may be several functions that perform what is fundamentally the same task. Often these functions can be replaced by a single software function.

Similarly, different data formats may have developed. By standardizing the way data are encoded, they can be made accessible to a wider range of application programs. Standardizing the way data are entered, analyzed, and displayed makes a system easier to use and to learn. It also facilitates the integrated analysis of diverse data sets, procedures that the system designer may not have anticipated. Common hardware requirements are also identified, e.g. hardcopy output devices, such as pen plotters, electrostatic printers, graphics terminals, digitizers and scanners.

Ideally, the functional requirements definition process will produce a set of data input, processing function, and output product needs, and a list of suitable hardware devices. The incorporation of the complete information processing requirements into a common system provides the opportunity to minimize redundant data collection and analysis and to maximize the sharing of information. Once the functional definition is complete, it should be considered in relation to the overall direction of the organization by the appropriate level of management.

While the definition of system requirements may appear to be straightforward at first glance, there are a number of factors that make it a difficult and not entirely systematic procedure. At one extreme, an existing system that is chaotic could simply be replicated without improving it. On the other hand, a system might be proposed that eliminates duplicated effort and redundant information but is unacceptable. What appear to be "irrational" methods in the larger view have often arisen for quite specific reasons. When viewed in the narrower frame of a specific agency, the method may be quite rational and necessary. A failure to understand the reasons for what appear to be irrational elements in an existing system can lead to the failure of the new one.

As Chrisman (1987) noted, a user-needs analysis has the potential to do too little and replicate a sub-optimal system. Or the analysis can do too much and rationalize the system to the point of overruling institutional mandates and generating political in-fighting. If done well, a needs analysis has the potential to recognize the constraints that determined the structure of the old system. Then the improved efficiency, speed, and analytical capabilties of newer technology can be used to better satisfy the needs of the mandated institution. To achieve this, the institutional factors, and in the case of a public agency the social factors, must be taken into account, as well as the technical factors.

Within the public sector, geographic information is a social commodity. Important data collection functions are not carried out for technical reasons. The creation of topographic maps, property maps, or zoning maps are not driven by a favourable cost-benefit ratio. These information products are created because an organization has been given a mandate to produce them. (The overall benefit to society may not be measurable; however, the democratic process assumes that the benefits exceed the costs.)

Legal and political systems give organizations mandates to collect geographic information in response to social need, the need for a legal description of land ownership, or the need for zoning regulation. The mandate may be given within the context of a law, an administrative regulation, or even a customary practise. A mandate implicitly or explicitly provides definitions of the objects of interest, the procedures for processing this data, and the actions that may result. Often what appear to be duplications of effort are the result of independent mandates using independent definitions having some overlapping data needs. For example, Gurda et. al. (1987) note that three participating agencies had three different and overlapping definitions of *field*. Though it would appear to be more efficient to develop a common definition for agricultural fields, the three definitions pertained to three different mandates, and so three separate field data sets had to be retained within the GIS.

Thus, the definition of system requirements is not simply a technical issue. It is an organizational one as well, with the organizational needs commonly taking precedence. What may appear technically to be redundant efforts may in fact reflect subtle yet essential distinctions.

PHASE 3: EVALUATION OF ALTERNATIVE SYSTEMS

The evaluation phase is usually conducted in two stages. First, a number of candidate systems are identified based on such information as recommendations from colleagues, vendor literature, and demonstrations at conferences and trade shows. Using the analysis from Phase 2, a set of requirements is drawn up. To do so requires that the goals or objectives of the GIS be identified. For example, system objectives might include improving the timeliness and quality of information, providing improved *ad hoc* query capabilities, and reducing data redundancy. These broad objectives must then be used to guide the specification of which functions, what data, and which products the system will be expected to provide. It is useful to divide the system requirements into those capabilities considered essential and those that would be nice to have. Selection criteria are then defined. Munro (1983) suggests dividing these criteria into three categories: hardware, software, and user-friendliness.

User-friendliness is one of the most overused terms in computer software descriptions. Though critical to the performance of a system, user-friendliness is difficult to define. In part this is because the term is relative; what is perceived as user-friendly today compared with similar software may not be considered user-friendly when more interactive, easier-to-use systems become available. User-friendliness tends to evolve as a software package is updated, bugs are fixed, "help" facilities are improved, and new hardware becomes available. The user-friendliness of a system will directly affect the time and, therefore, money needed for training, how widely the system will be used within the organization, and which levels of the organization will directly interact with the system.

If a GIS is to be used by a handful of trained technicians and programmers, user-friendliness might mean an efficient command language and good programming support. However, in most organizations, one of the objectives of implementing a GIS is to increase the accessibility of information to the non computer-specialist user. In this context the characteristics of a user-friendly system would include the following (adapted from Crosley 1985):

1. System commands that are easily understood by the non-computer specialist.

2. Efficient interaction for the expert user. A menu structure may be easy for the occasional user to work with but may be too slow for the expert. Providing a faster means of interaction, such as a command line option, may better satisfy the needs of the expert.

3. Error messages should clearly state the nature of the error, its probable cause, and give some direction to correct the problem.

4. A context-sensitive "help" facility should provide a complete explanation of what is expected of the user at a given point in the program.

Systems will differ in their approach to specific functional requirements, reflecting the trade-offs that were made in the system design. As a result, systems will differ in their ease of operation, system performance, efficiency of storage, and flexibility of processing; issues that are commonly recognized and addressed in an evaluation. What is often not considered are the potential problems associated with unexpected failures.

All computer systems have failures. Consideration should be given to the probability of failures, the potential effect on operations (some operations will be more critical than others), and the recovery procedures that would be used. The more serious the consequences of a failure, the more important it becomes to have alternative methods to generate specific information. For example, digitizing is commonly more costly than expected, software problems may be encountered when additional functions are added, the system may prove to be slower than expected, or training may be inadequate (see Table 8.1). Thus a comparison of risks should be included in the GIS system evaluations. Most GISes are complex systems. It is rare that they work perfectly when installed. The efforts of a highly trained and costly software person (either at the vendor's or customer's expense) is commonly needed to bring the system to the required operating level.

Table 8.1 Potential Problems of a GIS Product. (Adapted from Munro 1983).

1. Poor training
2. Poor documentation
3. Software does not perform as expected
4. System installation and start-up is late
5. Customer support is too slow or inadequate
6. Data entry is more costly and slower than expected
7. Price increases for the system hardware, software, or maintenance
8. Back-up or recovery systems fail and data are lost
9. Software cannot be modified to provide additional functions or handle unexpected problems

---

Implementation costs are high. Staff must be trained, and additional staff may be needed. As the system begins to serve users, user support services will be needed to deal with the day-to-day glitches that occur such as files that can't be opened, data that are lost due to a system crash, or other incidents common to any computer installation. Then, as GIS products begin to be delivered, a demand for service is created that the organization will expect to have satisfied at the same time that system start-up is continuing. If expectations are raised that cannot be met, the GIS implementation plan will be called into question, and so will the credibility of those responsible for the GIS. Finally, the initial entry of the organization's existing data into the GIS is the major bottleneck encountered in implementing a GIS. The cost of this operation varies widely, but is commonly several times the cost of the hardware and software. Faster, more accurate, and easier data entry procedures provide a direct cost advantage that may overshadow differences in system purchase price.

GISes are commonly evaluated by means of vendor demonstrations, demonstrations at existing installations meeting similar needs to the system being sought, interviews with knowledgeable users, and a review of published system specifications. There are a number of important limitations of this selection process. It is the nature of enthusiastic salesmanship to instill high expectations, especially when the potential buyer is as excited by the prospect of acquiring a new GIS as the seller is at the prospect of making a sale. Quantitative performance measures reduce but don't eliminate the excitement factor because many of the characteristics to be evaluated are non-quantitative (e.g. user-friendliness). The data sets commonly used in demonstrations are designed for that purpose. They are usually small, "well-behaved" (i.e. designed to show the system in its best light), and have been thoroughly tested. In an operational setting, a GIS usually must handle large data sets that have the uncanny habit of possessing unusual and unexpected characteristics. Where one system may crash, another may provide a simple override. These subtle nuances can make the difference between a production system on which an organization is willing to depend or a system that managers are apprehensive to use for critical tasks.

For these reasons, a second stage of evaluation is commonly used in which a few selected systems are tested more thoroughly using data sets supplied by the user and following a prescribed set of processing functions. **Benchmarking**, the administration of a standardized test procedure, provides a systematic means of comparing the performance level of competing systems. In this way the purchaser can better assess how a system will perform under the expected operating conditions.

The evaluation should start with data that the vendor has not seen before. These data should be samples of real data sets that will actually be used in the GIS. The evaluation should include every major processing function and output product that will be needed and should be performed without significant interruption such as to modify software. The system should perform as advertised without requiring software specialists to coax it along. An evaluation of this type, performed under the constant observation of the evaluators, will highlight unexpected difficulties and demonstrate the flexibility and robustness of the system. Examples of system evaluation procedures can be found in Joffe (1987), Marble and Sen (1986), and Tomlinson and Boyle (1981). Goodchild and Rizzo (1986) present an approach to evaluating GIS performance taking into account projected workloads.

The evaluation process can be taken a further and more costly step by undertaking a pilot study. A pilot study is usually designed to illustrate the effectiveness of GIS techniques to meet an organization's needs. The project is commonly structured to demonstrate the more important functional requirements using a real data set. Such a project can provide a number of important benefits. Since the organization's staff are directly involved, they become familiar with GIS technology and methods. The experience provides a realistic view of the difficulties of implementing a system. The difficulties of learning a new system, inputting data, reconciling data that disagree, assessing accuracy, satisfying user needs, developing programs to handle unusual situations, working around software limitations, etc. are all experienced first hand.

There is also the opportunity to quantify the potential benefits by carefully recording the resources used and products generated using the GIS and when the same analyses are done manually. By judicious selection of the project, it can be used to win the support of key departments or individuals who may be sceptical. It can also be used to identify potential problems such as data entry that is too slow, analyses that cannot be effectively performed, or output products that are not of sufficient quality or accuracy.

The pilot project, though a costly undertaking, is the most effective way for an organization to predict how well a GIS will meet their needs. It provides different information than a systematic evaluation. The pilot project provides direct experience; the look-and-feel of using a GIS, handling the problems, exercising the system's capabilities, and evaluating its performance and products. Brown (1986) notes that in his study of GIS sites, all successful installations had conducted pilot studies on their own or with the assistance of an outside consultant.

By the end of the system evaluation phase, a report is generally produced that includes recommendations for the system to be acquired and the financial and staff resources needed to support the GIS implementation. The system evaluation is usually presented as an objective technical analysis. However, it is important to recognize that the choice of functions that were required and the way in which the evaluation was structured **reflect the orientation and judgment of the evaluators**. For the process to have been successful, the criteria of the evaluation team must closely reflect those of the organization. More specifically, **the criteria must reflect the needs of those who will actually use the system**. The evaluation process can be easily and inadvertently biased toward the interests of the evaluators instead of the actual users. For example, an evaluation might emphasize analytical functions that are useful in planning and forecasting, yet the major need may be for operational data retrieval and updating. It is important that the organizational issues be addressed early so that the interests of the appropriate groups are adequately represented.

PHASE 4: SYSTEM JUSTIFICATION AND DEVELOPMENT OF AN IMPLEMENTATION PLAN

An implementation plan is formally or informally developed in order to proceed toward the acquisition of a GIS. It may be started during the system evaluation phase or only after the evaluations are completed. The plan is commonly part of a formal submission to proceed with acquisition. Prior to issuing a purchase order the plan should be developed and agreed upon by those who will be involved in funding, implementing, operating, and using the system. Some form of justification for acquiring a GIS may be included in the implementation plan or addressed separately. The way a GIS acquisition is justified will directly affect the steps needed to successfully implement it and the budget that can be obtained.

JUSTIFYING A GIS ACQUISITION

The arguments used to justify a GIS acquisition will define who in the organization is to benefit (i.e. whose problems will be addressed). Will it be managers who need better forecasting tools, staff who will be able to retrieve and maintain data more easily, or personnel responsible for quality assurance and financial control who will be better able to monitor the organization? A GIS can be of benefit to all of these groups. However, the way the acquisition is justified in terms of solutions in these problem areas will define how the success of the GIS will be recognized and assessed by the organization. It will also define who are expected to be the direct beneficiaries and where political support for the project is likely to be found.

The implementation of a GIS is a costly, long-term undertaking. Deciding whether or not to proceed can be greatly assisted by a systematic and quantitative analysis of the expected costs and benefits. Some of the more commonly recognized benefits of a GIS are:

1. better storage and updating of data;
2. more efficient retrieval of information;
3. more efficient production of information products;
4. rapid analysis of alternatives; and
5. the value of better decisions.

These benefits are predicated on selecting an appropriate system and entering suitable data. If the right data is not in the system, the best decision will not surface.

It is important to recognize that no matter how rigorous and extensive the analysis, in the end the decision to acquire a GIS will be a judgment call. While that judgment can be substantially aided by cost-benefit analyses, there does not exist an objective, unbiased analysis procedure that can, by itself, provide the correct "go/no-go" decision. There are two principal reasons for this. First, cost-benefit analyses, feasibility studies, workload projections and similar analyses are not as objective and rigorous as we would like to believe. Second, and perhaps more important, is that the decision to provide geographic information is rarely justified on the grounds of costs and benefits in the first place.

Cost-Benefit Analyses

In principal, a comparison of the costs and benefits of the current system and the new GIS should provide solid data to support the acquisition decision. The problem is that a good deal of judgment must be exercised in deciding how to define and measure those costs and benefits. The items selected for measurement and the way they are quantified directly affect the result.

First an assessment is done of the costs associated with the current system. There are many costs that can be compared in a fairly straightforward manner. For example, the time needed to generate a given product, such as a map, using the current method and using the proposed system is fairly easy to quantify. Time estimates for labour and estimates of material costs can be developed for the existing system by analyzing existing records or by collecting the data directly. The same can be done for the proposed system by producing the same products on a test system. However, there will always remain some judgment needed to estimate such items as the cost of operating the new system, the custom programming required, the rate of depreciation, and the way the cost of shared resources (such as facilities and person-years) should be apportioned.

At first glance, estimating the costs of the new system should be easier to quantify than the costs of the current one. However, there are many fixed price items, such as hardware, software, system maintenance, and training courses. The costs to create the data base are less easily predicted because they depend on the expected but unknown level of performance of the staff that have yet to be trained. If the data base creation task is contracted out, the costs may be more predictable, depending on the contractual arrangements. There are also the unexpected difficulties, such as analytical capabilities for which special software must be written, software that does not perform adequately, poor documentation or training, hardware failures, and other glitches which inevitably occur in any major computer installation. These difficulties have a dollar cost that can be estimated based on past experience and good judgment.

Staffing estimates are not as objective and quantifiable as they at first seem. The new system will require staff with additional skills. To some extent, existing staff might be trained. If new staff are hired, portions of other jobs may become redundant, but it may not be possible to reduce staffing levels to capture those portions of person-years. Also, the salaries needed to attract new staff may require that the salaries of existing job categories be adjusted. The organization may not be able to tolerate managers earning less than their staff.

In cost-benefit analyses, the benefits are the more difficult to quantify. Five types of benefits need to be considered:

1. *The benefit of increased efficiency.* The more efficient system will require less resources to perform an operation. This type of benefit is generally measured for existing tasks, such as producing a map or generating a report. It is assumed that the time saved will be used to benefit the organization. Hence the dollar value assigned to the time saving is the pro-rated salary of the personnel whose time is saved. If the time-savings occur in such a way that they can be used for other activities, then this method is quite reasonable. However, there are many improvements that will save time in quantities too small to be of use. Saving ten or twenty minutes here and there may be of no benefit if there is no additional work that can be scheduled into those time periods.

2. *The benefits of new non-marketable services.* A new GIS has the potential to provide useful products and services that were previously unavailable. Some of these will have been anticipated, in fact, they may have been a major justification for the system. Others will be unexpected benefits. How valuable is it to produce more impressively formatted reports or to produce better graphics? Quantifying these benefits is not straightforward.

3. *The benefits of new marketable services.* If the organization sells its products and/or its services, then improved production efficiency may be turned into increased revenues. The organization might also be able to sell the GIS expertise that it develops, either to outside organizations or to other departments.

4. *The benefits of better decisions.* More accurate information and faster and more flexible analysis capabilities can improve the decision-making process itself. This is recognized by users as a very important benefit of a GIS. However, predicting how decisions would be changed and the value of those better decisions is an uncertain exercise at best.

5. *Intangible benefits.* These benefits may include better communication within the organization, improved morale, and a better public image. Though not directly quantifiable these benefits can also have a direct and important effect on the efficiency of the organization.

There is extensive literature on cost-benefit analysis in the business administration field. Within the GIS literature, good discussions are provided by Kevany (1986) and Goodchild and Rizzo (1986). Examples of costing the production of geographic information can be found in Green and Moyer (1985), Kenney and Hamilton (1985,1986) and Laroche and Hamilton (1986).

The limitations of cost-benefit studies noted above are not meant as a criticism of systematic, quantitative evaluation. These studies provide valuable input to the decision making process. They must, however, be viewed as a part of the story, not as the whole story. There are many important factors that will not be included. It is not unusual for other considerations, completely separate from cost-benefits, to determine whether a system is acquired. Aside from undesirable influences, such as political in-fighting or poor decision-making, there are valid reasons why other considerations should take precedence. These arise from the fact that the decision to collect and provide geographic information is rarely based on cost-benefit grounds.

The Justification for Collecting Geographic Information

Within the public sector, geographic information is a social commodity. Legal and political systems mandate organizations to collect geographic information for such purposes as social need, the need for a legal description of land ownership, the need for an inventory of forest and mineral resources, the need to preserve wetlands, and so on. If an organization can only satisfy its mandate by using a GIS, then the cost-benefit arguments may not in fact enter the acquisition decision at all. For example, when Florida's 1985 Planning and Development Act came into effect, local governments were required to provide much more detailed information in the form of maps and map overlays. To satisfy these new legislated information requirements, local governments had to use GIS technology (Gilbrook and Sheldon 1987).

There is a trend for government agencies to be required to provide more accurate, more detailed, and more current geographic information. Increased environmental concerns are driving many of these requirements. In some cases, agencies are simply directed to automate their data. Under these kinds of conditions, the justification for acquiring a GIS may not be an issue. Instead the issues become when, how, and at what cost the GIS will be implemented. The British Columbia Forest Service justified its GIS acquisition largely on the basis of its expanded responsibilities as legislated by a new provincial forestry act (Hegyi and Sallaway 1986). Raines (1987) presents a similar situation where the changing mandate of the US Forest Service made it essential for GIS technology to be used and Hansen (1987) describes a similar scenario from the perspective of a municipality.

Within private organizations, cost-benefit considerations tend to be of greater importance. Clearly, if a GIS will be more costly and cannot provide improved information services to the organization, it will not be viewed as a cost-effective step. However, it is not simply a matter of more cost-effective ways to do the current work. Changes in analysis methods, new legal requirements, or pressure from competitors may require faster or more detailed levels of information processing. For example, more detailed information may be needed for more intensive forest management, to update forest harvest plans more frequently, or to assess the value of prospective land purchases more quickly than competitors (e.g. see Wakeley 1987). Although some form of cost-benefit analysis is generally used to help justify a GIS acquisition, a leap-of-faith is commonly needed as well.

The Data Base as an Asset

The largest expenditure in implementing a GIS is in the creation of the data base. It involves organizing existing data and converting it into a suitable digital form. This has commonly been considered an expense. However, the data base is in a very real sense a valuable asset. Not only is it valuable for the data it contains, but conversion should also have made it a more valuable asset because the data would be more accessible, usable for a wider range of applications, and have improved accuracy (e.g. by referencing the data to a more accurate base map). In addition, the data base has the potential to increase in value more quickly because it can be more easily maintained. Sety and Chang (1987) discuss the rationale for considering a GIS data base an asset and suggest an approach to its evaluation. The asset value of the data base may be an important factor in cost-benefit assessments of a GIS.

One important consideration in data base maintenance that is often overlooked is the value of historical data. Usually when the data in a GIS are updated, the old information is lost. Yet this information can prove extremely valuable. Past conditions can be important in deciding future land uses. For example, the type of forest that was removed can provide a valuable indication of which species might best be re-planted. Historical records in the form of old maps, airphotos, and satellite images have proven invaluable for assessing characteristics of landscapes that have since been changed.

THE IMPLEMENTATION PLAN

The implementation plan describes how technology, information, and people will be molded into an operating information system. The fundamental challenge is that all three of these factors must work together. To do so requires a significant investment of time and money to bring each one into operation. For the implementation of a GIS to be successful, the various development activities must be appropriately sequenced and the necessary financial and political support must be maintained.

An organization can acquire an operational GIS capability in several ways. These range from contracting for all services and purchasing virtually none of the GIS hardware and software, to purchasing a complete GIS system, to developing the entire set of hardware and software components in-house. Table 8.2 illustrates some of the considerations of these alternative approaches.

Technology

Computer hardware and software are developing at an ever increasing rate. Over a one-year period, hardware prices change significantly, new hardware becomes available, and new software versions are released. The development of a data base, training of personnel, and development of in-house expertise to reach full operation usually requires several years. Depending on an

Table 8.2 Alternative Approaches to Implementing a GIS.
(Adapted from Dangermond and Smith 1980).

CONSIDERATIONS	IMPLEMENTATION ALTERNATIVES				
	User Creates System	Buy Some Software	Buy Complete Software Package	Buy Complete Software and Hardware Package	Purchase GIS Services
Dependence on supplier	Very Low	Low	High	Very High	Nearly Complete
Time until system functions	Long	Long to Moderate	Short	Very Short	Not a Problem
Initial cost	Low	Moderate	Moderate	High	High
Labour costs paid by user	High	Lower	Moderate	Moderate	Very Low
Risk and uncertainty	High	Lower	Low	Low	Low
Customizing	Complete	Complete	Moderate	Moderate	Varies
Technical skill required of user	Extremely High	High	Moderate	Moderate	Quite Low
Use of existing resources	High	High	Moderate	Low	Very Low

organization's resources and budgeting needs, the hardware and software may be purchased before, after, or during the data base development.

Purchasing the Technology First. If the hardware and software are purchased early in the implementation process, then the data base can be constructed using the GIS in which it is to operate. It also increases the visibility of the project by the presence of tangible evidence (the equipment) and a forced allocation of funds and space to purchase and accommodate it. Staff can begin gaining experience on the GIS they will actually use and the organization can directly control the availability of system time. The disadvantages of this approach are that the initial investment will appear to be under-utilized until staff are trained and data are entered. During this start-up period, new hardware and software will become available and by the time the full capabilities of the system are actually needed, a more advanced system might be purchased for the same price. Because software and hardware change so rapidly, it is usually best to delay system purchases as long as possible.

Develop the Data Base Before System Purchase. Another alternative is to first develop the data base and then purchase the technology. In this way, the latest technology is acquired. Also, attention is focussed on the most costly component of GIS implementation, the data base. Construction of the data base commonly costs 5 to 10 times that of the hardware and software. However, the data base is an investment, and becomes more valuable over time, whereas the hardware and software depreciate.

The trade-offs in this approach are that the data base must be developed by an outside contractor using a GIS other than the one in which it will operate. There may be difficulties in converting the data base to run on the GIS eventually selected. Also, in-house staff will not gain experience from the data base development process. Data base maintenance, which must begin soon after the data base is created, will have to be postponed or done by the contractor. As a result, there will be additional expenses for creating the data base in advance of purchasing the system.

Commit to Purchase a System for Delivery After Data Base Development is Underway. A variation of the previous approach is to select the software and hardware first, negotiate a firm contract with the vendor, but not actually purchase the system until the data base has been partially or completely constructed. In this way, a minimum performance level and a fixed price can be locked-in with provisions made to take advantage of price reductions or system improvements that occur during the interim. The resistance of vendors to this approach makes the contract a difficult one to negotiate. The advantage to the customer is that the implementation effort can still be focussed on data base development. Yet much of the risk of incompatiblity is removed because the system has already been selected. The installation can be timed so that the complete data base or a portion of it is ready for use. System acceptance can then be made conditional on the proper functioning of the GIS using the actual data base. Using this implementation approach, staff can develop expertise using their own data, and maintenance activities will not be delayed.

Single versus Multiple Vendors. The data base, software, and hardware could be acquired simultaneously, either through a single vendor or multiple vendors. While the single vendor approach can greatly simplify project management, it limits the range of systems available. Many major system vendors do not offer data base creation services, thus reducing the system options. Contracting data base construction separately makes project management more difficult as two or more simultaneous contracts must be coordinated (Antenucci 1986).

Data Base Development

Data base development involves converting existing data into a digital format and entering the data into the GIS. It is the single most expensive part of the implementation process, representing 75% or more of the total expenditure. Data base development requires experienced personnel. A new GIS site will either have to hire or contract for this expertise. A complete discussion of data base development is beyond the scope of this section however some of the more important management considerations are presented here.

The implementation plan should consider the priority of data conversion that will be most effective in bringing the GIS into operation early. It should explicitly state when the organization will begin to see returns from its GIS investment. By planning some early paybacks, it will be easier to maintain the organization's support through a lengthy data base development effort. However, the pressure to demonstrate benefits early conflicts with demands to address the needs of all departments at once and to satisfy a higher level of accuracy than may be immediately needed. It may be sufficient to demonstrate system capabilities using existing digital geographic data sets such as Digital Line Graph or TIGER files (see Chapter 4).

The progress of the GIS implementation will be judged by the first products. These first products should be important but non-essential information. This ensures that unexpected delays are not catastrophic while the results are seen to be valuable. It should be expected that once these first products are shown, there will be a steady and increasing demand for them. Sufficient resources should be planned to satisfy this demand without compromising the data base development and maintenance effort.

The timing of the data base development effort will determine when the first products can be generated. The data base specifications and analysis procedures will determine their quality. Procedures should be defined to coordinate among users how data types will be defined and the types of output products required. Written procedures are needed for source data collection, interpretation, accuracy verification, and the preparation of data for input. It is essential that high standards of data preparation and quality control be maintained. The output of the GIS will not be trusted if the data are unreliable, and early impressions tend to be long remembered. Also, data that are inaccurate or incorrectly entered can be difficult and very costly to correct.

People

Ultimately, the success of a GIS site will depend on the people who implement the GIS. It is their enthusiasm and commitment that will see the project through the inevitable stumbles and set-backs. Managing the process is critical. User groups must be coordinated, detailed data base design must be completed, equipment purchases, training, and contractor services must be managed. The implementation plan should define the group or groups within the organization who will be responsible for the implementation and operation of the GIS.

Often, the most attractive option is to assign this responsibility to an existing unit within the organization. This approach limits the creation of additional administration. However, it is usually necessary to provide additional staff. The number of additional staff needed would depend on the skills of the existing unit, their familiarity with GIS, and their workload. The staff functions to be provided are a project coordinator, a GIS system manager, a data base manager, systems analysts/programmers, and data entry personnel.

The assignment of the GIS facility within the organization will have a direct effect on the level of service that will be provided to the users. If user services are inadequate, the overall benefit of the GIS facility will be significantly reduced. Resolving competition for the GIS facility and reaching agreement on its placement within the organization may be difficult. The Data Processing unit, often the first unit suggested to house the GIS, is probably not the best place. Geographic data and processing operations are fundamentally different from those used in Management Information Systems (MIS). Also, by placing the system within the Data Processing unit, a non-user department, the GIS may not be implemented in a way that adequately addresses the needs of the users. Another disadvantage of this approach is that existing workloads and responsibilities may divert attention away from the GIS project.

To avoid this dilution of effort, a separate organizational unit could be created. This group would then be able to focus all its efforts on GIS implementation and operation. The disadvantage here is that expanded administrative and management support may be needed both to support the group and to provide for its effective interaction with the rest of the organization. It may also be necessary to adjust the responsibilities and mandates of existing organizational units. Increasingly, organizations are cooperating to share the GIS data base creation expense or the entire GIS facility. For example, Antenucci (1988) discusses partnership arrangements between municipalities and utilities to share the cost of data capture. Though there tends to be an increase in management effort, sharing the high data conversion costs can be quite worthwhile.

Finally, the implementation plan should include a budget that provides sufficient funding to complete the job. The budget should allow for unexpected set-backs, such as data that must be re-entered and maps that need to be re-drafted, as well as the development of new software to handle unexpected processing needs and production bottlenecks. Running out of funds leads to going back to ask for more money and the associated questioning of the credibility of the project team. A considerable marketing effort is needed to garner support and maintain it. Pilot projects should be planned that develop the organization's expertise and also provide a steady stream of system demonstrations and output materials to attract users and get their active involvement. The importance of the continued selling of the benefits of the GIS facility should not be underestimated. Maintaining political support is key to maintaining the financial support and cooperation needed to complete the project.

PHASE 5: SYSTEM ACQUISITION AND START-UP

Having developed the implementation plan, won the organization's support, and received a formal commitment to go ahead, the next step is to contract for equipment and services.

CONTRACTING

The buyer will be better protected by carefully reviewing and negotiating the terms of a contract rather than accepting "standard" agreements. Unless vendor obligations are clearly defined, the buyer can be faced with an expensive, outdated, and non-functioning system. System purchases generally include hardware, software, training, documentation, installation, and maintenance. Pricing, though often the focus of negotiation efforts, is usually not where the most important benefits are to be gained. In the case of smaller orders (relative to the vendor's sales volume), significant discounting usually cannot be supported. However, terms and conditions, such as guarantees, warranties, upgrade options, delivery schedules and penalty clauses, can ensure that the system delivered is appropriate for

the application, will meet the buyer's specifications and the vendor's claims, and clearly assigns responsibilities in the event that there are failures to meet obligations.

In the course of marketing and selling a GIS, many claims are made about the capabilities of the system. There are verbal presentations, casual telephone discussions, sales literature, written specifications, and demonstrations. All these exchanges will affect the buyer's choice of system. Yet contracts commonly contain clauses that negate all claims not explicitly stated in the contract documents. For the buyer to be legally protected, specific guarantees must be written into the contract. Those claims considered by the buyer to be critical to the intended use of the system should be specifically written into or referenced in the contract. There should also be guarantees of the completeness of the hardware and software, i.e. all hardware and software needed for the system to function as claimed should be included in the purchase price except for stated exceptions. Guarantees of system performance, adherence to schedules, and penalty clauses for non-compliance can also be included. Ideally, the purchase contract should be reviewed by a lawyer who is knowledgeable in this area.

System integration includes all those steps necessary to make the individual system components perform together as specified. A GIS is a complex system. Hardware and software components from several manufacturers must be correctly installed and initialized for them to operate correctly. It is essential that the contract clearly define who is responsible for proper integration, define how the acceptability of the integration will be determined, and what is the buyer's recourse should the integration be unacceptable.

An acceptance test is normally conducted after the system has been installed and integrated. Warranties, payment milestones, and other vendor responsibilities are triggered upon successful completion of the test. Vendors usually supply a "standard" test procedure. However, it is in the interest of the buyer to ensure that the proposed test procedure is exhaustive and includes all the critical functions needed for the system to operate as claimed. In some cases, an acceptance test at the site of manufacture is done before shipment as well as on-site testing of the system after installation. Testing should be performed using a reasonably large data base provided by the buyer. The entire test should be observed by a knowledgeble representative of the buyer. Partial payment will normally be required upon successful completion of the test. Final payment is often postponed until satisfactory performance can be verified with the full data base over a specified period of time.

Maintenance contracts for hardware and software should ensure that the system will provide an acceptable level of service. The vendor should provide regularly scheduled preventive maintenance and should respond promptly to reported problems. Annual maintenance contracts cost on the order of 5–15 % of the purchase price. The frequency of preventive maintenance, required response time, and location of the support personnel will all affect the level of service provided. In the event that response time is inadequate, equipment fails repeatedly, or software bugs are not corrected, adequate remedies should be identified in the contract. This should include conditions under which the vendor will be required to replace equipment that fails repeatedly and provide credits against maintenance payments for poor service.

Software upgrades may be included as part of the maintenance agreement or may be sold separately. Software upgrades are usually desirable and often required for continuance of the maintenance contract. The buyer should consider the software upgrade policy when comparing system operating expenses and ensure that the stated upgrade policy, discounts, and other software protections are included in the purchase agreement. Vendors generally try to provide for a smooth transition between software versions. However, major changes in the vendor's products can create sudden obsolescence of hardware or software. Some protection can be gained by stipulating a period of time, such as one year, during which software can be exchanged for full credit against purchase of the replacement software package (Antenucci and Roitman 1987).

START-UP

The issues surrounding the start-up of a GIS installation were discussed as part of Phase 4 where the implementation plan was developed. Unexpected problems will always occur. The methods used to deal with them will differ with the organization and the situation. Bad news usually travels more quickly than good. So it is important to continually keep users and managers aware of the progress being made on the GIS start-up and to demonstrate useful applications and products. In this way, unexpected problems when they occur will not be the only progress reports.

Pilot studies are a valuable means of assessing progress and identifying problems early, before significant resources have been wasted. A complete data set should be developed for a small study area if this was not previously done in the system evaluation phase. Representative application projects should be developed that are formally planned and executed as a full scale application and then carefully evaluated. The evaluation should document what went right, what went wrong, and include suggestions for improvements the next time around. The participation of all users in these early post-mortems is important so that the full training benefits are derived from these first projects. In this way, the pilot project experience can reduce the repetition of mistakes and also provide a series of tangible products, such as completed projects.

Consultants and contractors can provide valuable assistance during the start-up phase. The GIS staff working with a consultant can learn successful approaches and techniques directly instead of by trial-and-error. Visits to similar installations and contacts with similar organizations can also be a valuable source of useful ideas and experience. Contractors can be used to shorten the initial data conversion time and bring the full data base into operation sooner.

However, in using outside help, there is a trade-off. It is the organization's own staff that must in the end be able to operate the GIS. Outside help can become a crutch if every time a problem arises the staff turn to a consultant. To be able to confidently work through difficult problems, it is necessary to have had the experience of working through difficult problems before. Outside help can provide short term efficiency by getting the job done sooner — finding out how similar problems have been successfully resolved is much more efficient than re-inventing the wheel. If judicious use is made of outside assistance, the start-up phase can be shortened and the learning period reduced without compromising the development of in-house expertise.

The successful development of in-house expertise will require that at least some staff be permanently assigned to the GIS facility and given the opportunity to become expert with the system. This expertise will also make them more marketable. So, in order to keep these individuals, it may be necessary to re-write their job descriptions, raise their status, and probably raise their salaries. GIS experience is a highly sought and valuable commodity.

PHASE 6: THE OPERATIONAL SYSTEM

The GIS facility can be considered to have reached the operational phase when end-users are making effective use of the system. The data conversion may not have been completed but a standardized procedure would be in-place to complete the task. In the operational phase, the organization would have developed sufficient expertise to handle routine tasks and special projects in a systematic and effective manner.

Once the facility is operational, the organization must develop procedures to keep staff current with new developments in the GIS field. Software and hardware are under constant development and require periodic upgrading. Perhaps more important than the technology related issues are those that will arise as the effects of the information provided by the new facility are recognized. In order to maintain financial and political support for the GIS facility, it is necessary to actively promote the benefits of the system. The completion of projects can be publicized, the benefits to specific users can be highlighted, project reports can be distributed, and tours of the facility can be given to make people aware of the benefits of the GIS. The increased visibility also demands that the GIS be operated in a manner that is seen to be responsive to the needs of the users. Having developed an organization that satisfies the operational demands of users, some of the more contentious issues of operating a GIS facility will come to the fore. One of these issues, especially critical for publicly-funded organizations, is the issue of accountability for the consequences of providing information. This is the subject of the next section.

WHO IS RESPONSIBLE?

A man, somewhat shabbily dressed, appeared at the information counter of a municipality. He claimed to be doing some research and needed a list of the names and addresses of all single women living in a specific area of the city. The municipality's land records were in digital form and the clerk easily used the analysis capabilities of the GIS to run a search and generate the report. As the clerk was about to hand the listing to the customer, his supervisor came by interested to know how the system was being used. After looking at the report, and the customer, the supervisor decided to veto release of the information. The supervisor felt there was a significant risk that the customer would use this information inappropriately — to target these women.

In another case, a manager in a municipality decided to produce a map showing the distribution of potentially vicious breeds of dogs. His objective was to demonstrate the capabilities of the municipality's GIS using an interesting but non-controversial data set. The digital files for dog licenses were imported into the municipal GIS. Then an analysis was done to plot the distribution of these dogs on the city map by the addresses registered with the dog licences. The map showed a few areas of the city to have an unusually high density of these breeds of dogs. While the results were interesting and potentially useful in analyzing crime and public hazards, the manager recognized that the map of this information was potentially damaging to the landowners in these areas. If this information were made public, the desirability of these areas might be reduced, causing a fall in property values. As a result, the information was deemed to be confidential.

A final example is that of a municipality that had been asked by the real estate board to provide a digital file of landowner's names, addresses, and tax arrears. The board already had received the digital file of the city assessor's data base. This data base provided such information as the property address, size, land use status, and valuation. But without the owner's name and address, it was still necessary to go to the land registry office and manually search each property for the ownership information.

The city council denied access to arrears information but could see no harm in providing a digital file of owner's names and addresses, since this information is publicly available at the land registry office. The only difference was that the search could be done much faster by computer. To their surprise, there was a vehement public outcry against releasing these files. Despite guarantees by the real estate board that the data would only be searched one property at a time, the public believed the data would be used for computer screening to target individuals for sales campaigns, mailings, or the data might be used with other files to find out information that individuals consider private. As a result of the public outcry, the real estate board withdrew its request.

These examples are not fictitous. They are recent events that happened in North American municipalities. These examples highlight what is probably the most serious challenge to the operation of geographic information systems: who takes responsibility for the consequences of distributing the information produced using a GIS. In the course of researching this book, I have found the issue of responsibility and liability to be one of the most serious issues facing GIS managers, particularly at the local level. Yet it is an issue that has received relatively little attention in the literature.

Computer security systems are well-developed. Systems of passwords, records of operator activities, and data encryption are routinely used to secure data from unauthorized access. However, the responsibilities that arise from the voluntary distribution of data are not well-defined and tend to be ignored.

Information is power. Where the consequences of exercising that power are significant, society assigns responsibility. In such professions as engineering, architecture, and medicine the professional is held legally responsible for the quality of the information he or she provides. There are professional standards that must be met. Even when the standards cannot be precisely defined, the professional can be held responsible for damages. For example, a doctor may use one of several approaches to diagnose and treat a particular illness. Yet, even though a standard performance level cannot be precisely defined, a doctor can be held liable for malpractice if the procedures he chose to use did not conform with generally accepted medical practice.

The information in a GIS is for the most part information produced by professionals. Survey engineers, soil scientists, geologists, foresters, cartographers, and many other professionals generate the information that comprise a GIS data base. One of the great advantages of a GIS is that it is designed to make the integration of diverse data sets easy. Where, in the past, a skilled technician or professional might have been needed, today staff with relatively little formal understanding of the data can modify maps or change data records. In many cases, the GIS does not even keep track of which individual makes these changes. However, an individual with the ability to skillfully operate a GIS may not have the background to assess whether an analysis procedure is valid.

Perhaps more importantly in the context of a GIS facility, critical information policy decisions may inadvertently be left to the discretion of the GIS operator or researcher. Such decisions as what information should be generated, how its validity should be expressed, and to whom the information should be released should ultimately be made by a professional who has been explicitly assigned this responsibility. Perhaps the job title would be "GIS Information Officer". It would be the responsibility of this individual to obtain whatever expert advise he or she may require to take into account

the technical, legal, economic, and political factors in deciding what information should be generated and released. These decisions will often be judgment calls. There may not be an obvious correct answer. However, it is important that the judgment call be made by someone with an appropriate level of training, authority, perspective, and responsibility.

This formal assignment of responsibility should be an arrangement analogous to that of an engineering office. The engineer who stamps a set of engineering drawings accepts responsibility for that information. He probably did not draft every line or check every calculation, yet the engineer's stamp on the drawings indicates that a qualified professional has reviewed and approved the work and accepts the legal liability if the information is found not to meet accepted standards.

The managers of GIS facilities need to be aware of the responsibilities they knowingly or unknowingly may assume. The objective of this section on responsibility and liability is to present a framework of issues and to suggest some approaches to managing the risks. The subject is a difficult one because each case is unique. Such factors as the level of detail of the information, the potential for harm to individuals, and the public right of access to information will influence how information can and should be distributed. However, there are precedents. Organizations responsible for national statistical reporting, such as Statistics Canada and the US Bureau of the Census, have well-developed methods to assess and protect the confidentiality of information.

In the following section, a few references are made to court cases. They are presented as indicators of the responsibilities recognized by society. The outcome of a particular case may differ considerably depending on the jurisdiction. From a management perspective, the point being stressed is that the distribution of information can have significant consequences for which someone must accept responsibility. As the consequences become potentially more serious, greater care is needed in selecting the analysis procedures and the form in which the information will be distributed. To manage the risks associated with inaccurate information, or the inappropriate use of information, requires that the management structure of the GIS facility be organized so that the flow of information to the outside world can be effectively managed.

FOUR ISSUES OF RESPONSIBILITY

Four issues of responsibility are discussed here. While they are not an exhaustive list, they introduce major issues that the manager of a GIS facility should consider.

Accuracy of Content

Accuracy of content is the degree to which the data represent the condition they describe. The accuracy with which the geographic position or the attributes of a feature are recorded are examples of accuracy of content. Errors may be introduced during data collection or when the data are digitized, classified (i.e. when new groupings are created), or processed. These sources of error were discussed in Chapter 5 on Data Quality.

The provider of information can be held liable for inaccuracies of content. For example, the United States federal government was held legally responsible for inaccurately and negligently showing the location of a broadcasting tower on an aeronautical chart. This inaccuracy was found to have directly contributed to a fatal airplane crash (Epstein 1987, Reminga v. United States 1978).

Issues related to accuracy of content are usually addressed in the context of standards. Standards are an important issue and considerable effort has been focussed on establishing suitable data standards, not only for accuracy but also for such characteristics as format and the definition of categories, such as land use classes. Standards make it easier for data to be shared by outside users because they know what to expect. It is in the interests of both producers and consumers to have standardized information products. Information provided in a standard form, widely distributed and commonly used, make it possible to presume standard levels of information quality and standard levels of knowledge among users. It improves the predictability of the entire process of generating, disseminating, and using information. In so doing it makes the assignment of responsibilities much clearer.

There are two sides to standards: the producer side and the consumer side. Most of the effort has been focussed on the producer side, i.e. the perspective of the organization generating the information product. The consumer side is the perspective of the user of the information product. It is not possible for information to be 100% accurate and it is usually not cost-effective to demand the highest obtainable level of accuracy. At some point the cost of potential errors and probability of their occurrence is not worth the cost of imposing a higher accuracy standard. The level of error that remains represents the risk in using the data. Should the remaining level of error result in damages when the information is used, then someone will absorb the cost of these damages, i.e. the cost of this residual error. Bédard (1987) has refered to this as **uncertainty absorption.** Technical means can be used to reduce uncertainty, but there will always be some risk of error that could result in damages for which someone will have to pay.

The concept of risk assessment, as used in evaluating alternative policy decisions, can be used to evaluate decisions based on imperfect information. Decision analysis is a problem solving procedure used to guide decision-making under conditions of uncertain information (see for example Stokey and Zeckhauser, 1978). It involves estimating the costs and benefits of each possible outcome and then discounting these amounts by the probability of their occurrence.

Aronoff (1983) proposes using decision analysis theory and map accuracy assessment methods to compare alternative decision-making procedures that use geographic information. Both the risk and the cost of potential errors are taken into account in assessing construction hazards in a route location application. An index termed the **minimum accuracy value** is used to discount the value of the map information according to the expected accuracy of the map data. In effect, the data from the map accuracy test is used to calculate the expected "worst case" map accuracy. (Tables of minimum accuracy values are provided in Aronoff, 1985.) A similar approach is demonstrated for labelling land cover classes (see Aronoff, 1984). In this case the minimum accuracy value is used to minimize the cost of classification errors by defining land-cover classes such that the *consequences* of incorrect class assignment are minimized.

For example, consider forest fire hazard mapping. For large areas it is too costly to produce detailed vegetation type maps. Instead, digital analysis of satellite imagery can be used to identify land-cover classes. Each class will contain several vegetation types, some of which may constitute a high fire risk and others a low risk. The cost of a classification error will be very different depending on the consequences. If an area is misclassified into a high risk class, then in an emergency there may be an over-response to a fire in that area. However, misclassifying a high risk area into a low risk category could have more serious results, such as the outbreak of an uncontrolled wildfire. By taking into account the probabilities of such errors and their costs, the designations of the fire risk for each land-cover class can be adjusted so that the chance of a costly error

is minimized. Errors cannot be eliminated but a choice can be made about the frequency with which they will occur. For example, a land-cover class might be designated as high risk even though there are only minor occurrences of high risk vegetation types. As a result there will be a higher risk of over-response to a fire involving this class, however the chance of an inadequate response and resulting wildfire will be reduced.

These approaches attempt to minimize the cost of the residual error in geographic information by evaluating the consumer and producer risks and incorporating these residual uncertainties into the decision-making process. In so doing, the accuracy of content is explicitly recognized and taken into account.

Accuracy of Context

When information is shown in the context of a map, an assertion is being made about the interrelationship of all the map elements. Even though the content of each data set may be correct, when the data from different sources are presented together as a map, the information conveyed can be quite misleading.

Consider the following hypothetical example. The locations of three storage sites for environmentally hazardous PCB wastes are recorded in the form of a table of map coordinates. To communicate this information, the locations are represented in the form of a simple map (Figure 8.1). There is virtually no information on this map other than the symbols representing the three PCB sites. About all that can be learned from this map is the relative positions of the sites (assuming that a rectangular coordinate system has been used to plot the locations). The information conveyed is fairly innocuous.

Now suppose that the PCB sites were plotted on an existing map, a map that showed the locations of the schools in the area (Figure 8.2). The information conveyed now

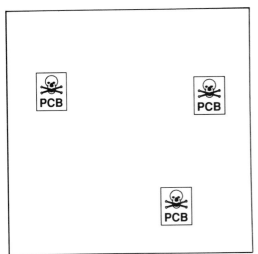

Figure 8.1 Map of PCB Disposal Sites.

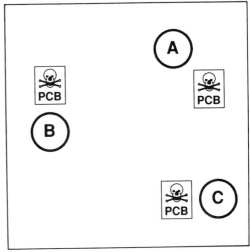

Figure 8.2 Map of PCB Disposal Sites Shown with School Locations A, B, and C.

is far from innocuous. It suggests that the schools are located next to the PCB sites. You could just imagine the headlines in the local newspaper! Figure 8.3 presents some of this information at a larger scale and with more information about the scale of mapping. It is clear that the PCB site is located a considerable distance from the school, some 70 km.

The point of this example is not that geographic information must be presented at an appropriate scale, or that one cannot use a map without knowing the basic map information, both of which are true. The point is that if the information is presented in such a way that the user could reasonably draw an incorrect conclusion, then the information product should be seriously re-evaluated. A different presentation might communicate the information in a more accurate context.

Consider now a more subtle example, that of a company producing aeronautical navigation aids. In a liability case it was shown that the aircraft navigational charts produced by Jeppeson Company were technically correct. However, the company was held to be at fault because the diagrams of the airport approach used a perspective different from that usually provided in charts

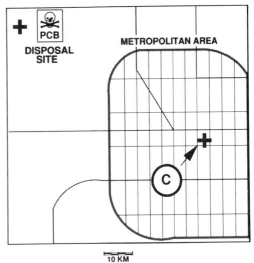

Figure 8.3 Large Scale Map of a PCB Site Relative to School C.

of this type. The specifications for the diagram were clearly and correctly indicated. However, it was judged that the use of a non-standard perspective was confusing to the pilot and contributed to the pilot error and fatal crash (Epstein and Roitman 1987, Aetna Casualty and Surety v. Jeppeson and Co. 1981). In other words, the content was sufficiently accurate, but the context produced an insufficiently accurate product for the intended use.

The following is another illustration of the responsibility problems that can arise from inaccuracy of context. At a regulatory hearing in Wisconsin evidence of the location of the ordinary high water mark (OHWM) of a lake was presented. This evidence was in the form of observations and measurements by botanists and surveyors indicating the OHWM to be at the 990 ft elevation contour. The current lake level was 10 ft lower, at 980 ft. The hearing examiner decided to depict this evidence in map form in his final report. He used a standard 1:24,000 US Geological Survey topographic map and highlighted the 990 ft contour.

A short time later, a landowner with property adjacent to the lake sued the state. In Wisconsin, land below the OHWM is deemed to belong to the state. This landowner had not been involved in the hearing, but the information in the hearing examiner's report suggested that a significant portion of her land was below the OHWM and therefore the property of the state. Even though the hearing examiner's report was withdrawn, the Wisconsin Supreme Court awarded damages to the landowner because the state had, for a period of time, called into question her title to the land.

In this case the hearing examiner had not recognized the implications of placing the survey and botanical observations in the context of a particular map. The statement that "evidence was presented that the OHWM is at an elevation of 990 ft" is not the same as highlighting the 990 ft contour on a topographic map. The statement by itself requires that other means be used to establish the relation between the OHWM and other ground features. The map presentation asserts that the particular spatial relation between the highlighted contour and all the other features on the map are true (to the level of accuracy of the map). The hearing examiner, by highlighting the contour line on the map, had inadvertently rendered

a judgment of the true position of the OHWM (Epstein 1987, Zinn v. State 1983).

Data Format

A more difficult issue arises when a change in the format of existing data changes the way they can be used. Changing the format can significantly alter the consequences of distributing the information.

Consider land ownership information for example. It is legally required that certain land ownership information be made publicly available. These data are usually provided in the form of printed land registry books that must be used in a reading room. Although all of the information contained in these documents is in the public domain, the fact that the data is in a printed format severely limits the way they can be used. It would be impractical to find every property owned by a specified individual or to identify all owners who transferred properties totalling more than $10 million within the last year. This information could be produced by manually searching every property record. However, for a metropolitan area a manual search would be prohibitively expensive. By providing practical limitations to the way the data can be used, the format of the data provides a degree of protection to the public.

However, if the property data were made available in a digital format, the data could be easily and quickly searched by computer and could be matched with data files obtained from other sources. (**File matching** is the process of combining data sets that have common data fields, such as a name, address, or ID number. The procedure can be used to relate data from files that were not intended to be used together. In this way information about individuals that would not normally be released might be derived.)

Computer analyses can be used to screen large data files and identify individuals who match a certain profile: who own a specified amount of property, whose land taxes are in arrears and might be pressured into selling, or to identify individuals who have any number of other characteristics. These analyses can be used to target individuals for special attention they may not want, ranging from determined real estate agents to the tax collector. The additional uses that can be made of the data once they are in digital form may be unintended by those providing the information and may compromise the right to privacy of the individual.

When the system of property data distribution was first set up, these data analysis capabilities did not exist. The balance between the right of the public to have information and the right of the individual to privacy was taken into account in the context of a manually searched record system. However, when the data are made available to the public in an electronic form, this balance of rights is significantly changed even though there has been no change in the content of the data. So, even though digital versions of data sets exist, there may be undesirable consequences of distributing it in a digital format even though the same data are publicly available in non-digital form.

Problems in Combining Data Sets

One of the great benefits attributed to a GIS is the ease with which diverse spatial data sets can be combined. Yet it is often forgotten that the data being combined are essentially data produced by experts, and so the validity of the input data and the analysis procedure should be verified by qualified personnel. The more serious the consequences of error, the greater the effort that should be invested in checking the information.

One could take engineering drawings and combine sections from different buildings to create a new set of building plans. But for a non-engineer to even consider this would be viewed as absurd. Although one could

perform this procedure, the results would not be considered valid — unless they were checked by a suitably qualified engineer.

Overlay analysis is one type of GIS procedure where results with unacceptable levels of error can easily be produced. When two spatial data sets are overlayed, the accuracy of the resulting information will generally be less than the least accurate input data set. This is intuitively obvious. Every location that is in error in the least accurate input data set will still be in error in the output. In addition, those points in error in one data set probably will not all occur at the same location in the other, so there will be some additional points in error.

Newcomer and Szajgin (1984) conclude that the error rate in the information produced from an overlay analysis will be no lower than that of the least accurate input data set and could be as high as the sum of the error rates of all the input data sets. Walsh et. al. (1987) provide a further discussion of error propogation in GIS analysis procedures.

When data sets are combined, both explicit and implicit inaccuracies can be created or increased. Explicit inaccuracies include errors in the geographic location or the attribute assigned to a feature. Implicit inaccuracies may result when the combined data product is incorrectly assumed to have the accuracy level of the data sets from which it was produced. For example, when new information is plotted on an existing base map, such as a topographic map, the user will assume that the accuracy of the new information meets the standards of the topographic map. There is an implied positional accuracy and attribute accuracy.

When data sets are combined, new relationships are established among all the features not previously represented in the original data sets. Some of these relationships may not be intended or meet the required level of accuracy. Finally, there is the implied accuracy that may result from similarity to other information products. The Jeppeson case discussed previously is an example involving a single data set. A similar problem could easily arise when more than one data set is involved.

The accuracy of combined data sets is a complex and controversial one. It requires the skills of discipline experts. However, the accuracy of combined data sets is often not assessed at all. The responsibility often falls on the manager to ensure that appropriate levels of accuracy assessment are undertaken.

PREVENTIVE MEASURES

GIS technology provides the capability for personnel, acting entirely within their designated level of authorization, to generate information that is not valid. Sophisticated output capabilities give all reports and maps the authoritative look that high quality printers and plotters provide. However, the quality of the data used may not have been suitable for the analyses performed, thereby invalidating the results. Many of these difficulties can be avoided by recognizing key decision points in the data selection, analysis, and distribution process and incorporating review procedures at these stages.

Verification of Data Quality by the Original Producer

An organization can be protected from inappropriate use of data from outside sources by consulting directly with the provider of that data. For example, a County in Southern Ontario needed to enter the floodplain boundaries into their GIS. They obtained the data from the local Conservation Authority (the organization legally responsible for defining the floodplain boundaries in the area). The County recognized that the procedure of digitizing the boundary data from paper maps could introduce errors in the boundary positions. So an arrangement was made for the Conservation Authority's personnel to digitize the data and verify its accuracy. In this way the organization with the mandated responsibility and the appropriate personnel was asked to verify that

1. the data were correct *as represented within the GIS*, and

2. the data were sufficiently accurate for the scale of mapping *at which they would be used within the GIS*.

By working directly with the mandated producer of the data, the user ensured that the data quality was suitable for the application. Such an arrangement can be attractive to the producer of the data as well as to the user. If the data have not yet been digitized, there is the opportunity to share the digitizing costs. Also, in preparing the data for digitizing, there may be opportunities to enhance the data quality, e.g. by improving the geodetic control of the mapping base. Where data are produced within the same organization, this consultative process can be made quite routine.

Verification of Data Quality by Discipline Experts

In cases where the organization producing the data cannot be directly involved, procedures can be implemented to ensure that qualified personnel assess the suitability of data for an analysis. By explicitly recognizing the need for a specialist to review analysis procedures, the chance of producing and distributing invalid results can be reduced. Also, with the growing awareness of expert system methods, the GIS can be used to monitor and disallow analyses for which the data are inappropriate or of insufficient quality.

Quality assurance procedures are used routinely in the mapping profession. The quality assurance approach can be extended to the operation of a GIS facility by documenting routine analysis procedures and instituting compulsory quality assessments of results. In the case of non-standard analyses, a mandatory review can be instituted. This review should be done by experts *in the subject being analyzed* (not just GIS techniques) to ensure that the manipulation of the data is appropriate to that discipline. Only after the quality of the results have been assessed and accepted should the data be released for internal or external use. Also, by documenting the procedures used to generate each type of information, the procedure is more easily analyzed and amended when deficiencies in the information are recognized.

Setting and Maintaining Data Quality Standards

The quality of geographic data is often examined only after incorrect decisions have been made and financial losses or personal injury have occurred. Increasingly, producers of geographic information are being held liable when their information products are found to contain errors, are poorly designed, or are used in ways and for purposes unintended by the producers. Data quality standards, appropriately defined, tested, and reported, can protect both the producer and user of geographic information. When data are provided in a standard format and at a well-defined and accepted level of quality, the producer is protected from liability if the data are inappropriately used. Such standards also protect the user from relying on inappropriate information.

Introducing Changes in Data Availability

Technological changes can be introduced far more quickly than the legal, political, and social systems of our society can absorb them. It takes time for these systems to become aware of the consequences of new technology like GIS and to develop appropriate safeguards. It takes time for agencies to identify potential abuses of data files, such as file matching, to derive information about individuals from grouped data. A conser-

vative approach is to begin by allowing the new technology to provide the same types of information products as were previously available. For example, in the case of land records, the public might be able to ask an information clerk for a specific property record and the clerk might query the GIS to generate the report. But individuals would not be allowed unrestricted access to the GIS to generate reports on multiple properties.

CONCLUSION

A GIS provides the means for geographic information to be used for a broad range of applications and by users with a wide range of skills. In order for these data to be used in decision-making, their quality must be reliably known. Public organizations charged with producing and disseminating geographic information must be aware of the responsibility issues that may arise when their data are used. These issues arise from the stated or implied accuracy of the data. They also arise from the unanticipated uses of the data that may compromise individual rights of privacy or cause injury. It is in the interests of the provider of information to clarify those responsibilities in advance, rather than by discovering them in a court of law or in the midst of a political controversy.

The organizational structure of the GIS facility should reflect these concerns, not for fear of lawsuits but to treat seriously the responsibilities that come with the power of controlling information. Ultimately, it is the management of a GIS facility that will determine the quality of information and the extent of its distribution. If foresight and public service are not sufficient incentives to recognize and deal with these responsibilities, then the threat of litigation may provide the necessary encouragement. Rather than waiting for responsibility to be enforced from outside, organizations should take steps to anticipate and deal with these issues when implementing a GIS.

REFERENCES

Aetna Casualty and Surety v. Jeppeson and Co. 1981. United States 9th Circuit. 642 F 2d 339.

Antenucci, J.C. 1988. Organizational Structures and Examples: Municipal and Utility Partnerships. In *Proceedings of the 1988 ACSM-ASPRS Annual Convention*. American Society of Photogrammetry and Remote Sensing. Falls Church, Virginia. Volume 2:30–38.

Antenucci, J.C. 1986. Timing the Acquisition and Implementation of a GIS Computer System and Its Database. In *Proceedings of the URISA '86 Conference*. Urban and Regional Information Systems Association. Washington, D.C. Volume 2:13–21.

Antenucci, J.C. and H. Roitman. 1987. Negotiating for System Acquisition and Maintenance — Some Legal and Practical Guidelines. In *Proceedings of the URISA '87 Conference*. Urban and Regional Information Systems Association. Washington, D.C. Volume 2:25–36.

Aronoff, S. 1985. The Minimum Accuracy Value as an Index of Classification Accuracy. *Photogrammetric Engineering and Remote Sensing* 51(1):99–111.

Aronoff, S. 1984. An Approach to Optimized Labelling of Image Classes. *Photogrammetric Engineering and Remote Sensing* 50(6):719–727.

Aronoff, S. 1983. Evaluating the Effectiveness of Remote Sensing Derived Data for Environmental Planning. *Journal of Environmental Management* 17:277–290.

Bédard, Y. 1987. Uncertainties in Land Information Systems Databases. In *Proceedings of the Eighth International Symposium on Computer-Assisted Cartography*. American Society of Photogrammetry and Remote Sensing. Falls Church, Virginia. pp. 175–184.

Behrens, J.O. 1985. Accessibility of Public and Private Land Information — New Departures for Old Realities. In *Proceedings of the URISA '85 Conference*. Urban and Regional Information Systems Association. Washington, D.C. Volume 1:12–28.

Beidler, A.L. and R.E. Williams. 1986. A Local Government Geographic Information System Evaluation Process. In *Proceedings of the URISA '86 Conference*. Urban and Regional Information Systems Association. Washington, D.C. Volume 2:168–176.

Brown, C. 1986. Implementing a Geographic Information System — What Makes a New Site a Success? In *Proceedings of a Geographic Information Systems Workshop*. American Society of Photogrammetry and Remote Sensing. Falls Church, Virginia. pp.12–19.

Chrisman, N.R. 1987. Fundamental Principles of Geographic Information Systems. In *Proceedings of the Eighth International Symposium on Computer-Assisted Cartography*. American Society of Photogrammetry and Remote Sensing. Falls Church, Virginia. pp. 32–41.

Chrisman, N.R. and B.J. Niemann. 1985. Alternative Routes to a Multipurpose Cadastre: Merging Institutional and Technical Reasoning. In *Proceedings of AutoCarto 7*. American Society of Photogrammetry and Remote Sensing. Falls Church, Virginia. pp.84–94.

Crosley, P. 1985. Creating User Friendly Geographic Information Systems Through User Friendly Supports. In *Proceedings of AutoCarto 7*. American Society of Photogrammetry and Remote Sensing. Falls Church, Virginia. pp.133–140.

Croswell, P.L. 1987. Map Accuracy: What Is It, Who Needs It, and How Much Is Enough. In *Proceedings of the URISA '87 Conference*. Urban and Regional Information Systems Association. Washington, D.C. Volume 2:48–62.

Dangermond, J. and J. Harrison. 1987. Urban Geographic Information Systems: The San Diego Design Experience. In *Proceedings of the GIS '87 Symposium*. American Society of Photogrammetry and Remote Sensing. Falls Church, Virginia. pp. 387–395.

Dangermond, J. and L. Smith. 1980. Alternative Approaches for Applying GIS Technology. In *Proceedings of the ASCE Specialty Conference on the Planning and Engineering Interface with a Modernized Land Data System*. Denver, Colorado.

Epstein, E.F. and H. Roitman. 1987. Liability for Information. In *Proceedings of the URISA '87 Conference*. Urban and Regional Information Systems Association. Washington, D.C. Volume 4:115–125.

Epstein, E.F. 1987. Litigation Over Information: The Use and Misuse of Maps. In *Proceedings of the International Geographic Information Systems Symposium: The Research Agenda*. Association of American Geographers. Washington, D.C.

Gentles, M.E. 1987. What Are The Secrets to a Successful Conversion Effort? In *Proceedings of the URISA '87 Conference*. Urban and Regional Information Systems Association. Washington, D.C. Volume 2:37–47.

Gilbrook, M.J. and P.K. Sheldon. 1987. Coping with Florida's New Growth Management Legislation: A First Time Application of a Geographic Information System. In *Proceedings of the URISA '87 Conference*. Urban and Regional Information Systems Association. Washington, D.C. Volume 2:254–265.

Goodchild, M.F. and B.R. Rizzo. 1986. Performance Evaluation and Workload Estimation for Geographic Information Systems. In *Proceedings of the Second International Symposium on Spatial Data Handling*. International Geographical Union. Williamsville, New York. pp.497–509.

Green, J. and D.D. Moyer. 1985. Implementation Costs of a Multipurpose County Land Information System. In *Proceedings of the URISA '85 Conference*. Urban and Regional Information Systems Association. Washington, D.C. Volume 1:145–151.

Gurda, R.F., B.J. Niemann Jr., S.J. Ventura, D.D. Moyer, J. Amundson, and H. Braunschweig. 1988. Developing Data Management Models for Multi-Agency Land Information Systems. In *Proceedings of the 1988 ACSM-ASPRS Annual Convention*. American Society of Photogrammetry and Remote Sensing. Falls Church, Virginia. Volume 2:39–46.

Hansen, H. 1987. Justification of a Management Information System. In *Proceedings of the GIS '87 Symposium*. American Society of Photogrammetry and Remote Sensing. Falls Church, Virginia. pp.19–28.

Hegyi, F. and P. Sallaway. 1986. *The Integrated Three-Dimensional Forest Land Information System of British Columbia, Canada*. Presented at the 18th IUFRO World Congress held September 1986 in Ljubljana, Yugoslavia.

Joffe, B.A. 1987. Evaluating and Selecting a GIS System. In Proceedings of the *GIS '87 Symposium*. American Society of Photogrammetry and Remote Sensing. Falls Church, Virginia. pp.138–147.

Kenney, H. and A. Hamilton. 1986. Unit Costs for Parcel Indexing and Related Activities in Northern New Brunswick. In *Proceedings of the URISA '85 Conference*. Urban and Regional Information Systems Association. Washington, D.C. Volume 1:141–149.

Kenney, H. and A. Hamilton. 1985. Unit Costs for Property Mapping in Northern New Brunswick. In *Proceedings of the URISA '85 Conference*. Urban and Regional Information Systems Association. Washington, D.C. Volume 1:132–144.

Kevany, M.J. 1986. Assessing Productivity Gains in Advance: Feasibility Studies. In *Proceedings of the URISA '86 Conference*. Urban and Regional Information Systems Association. Washington, D.C. Volume 2:40–46.

Klosterman, R.E. 1987. Guidelines for Computer-Aided Planning Models. In *Proceedings of the URISA '87 Conference*. Urban and Regional Information Systems Association. Washington, D.C. Volume 4:1–14.

Laroche, S. and A.C. Hamilton. 1986. Unit Costs for Topographic Mapping. In *Proceedings of the URISA '86 Conference*. Urban and Regional Information Systems Association. Washington, D.C. Volume 1:150–158.

Marble, D.F. and L. Sen. 1986. The Development of Standardized Benchmarks for Spatial Database Systems. In *Proceedings of the Second International Symposium on Spatial Data Handling*. International Geographical Union. Williamsville, New York. pp.488–496.

Munro, P.A. 1983. Selection Criteria for a Geographic Information System — In Retrospect. In *Proceedings of the International Conference on Renewable Resource Inventories for Monitoring Changes and Trends*. College of Forestry. Oregon State University. Corvallis, Oregon. pp.273–275.

Newcomer, J.A. and J. Szajgin. 1984. Accumulation of Thematic Map Errors in Digital Overlay Analysis. *The American Cartographer* 11(1):58–62.

Ottaway, E. 1988. Personal communication. Deputy Planning Commissioner. Oxford County, Ontario.

Rains, M.T. 1987. The Role of GIS in Spatial Resource Information Management — A Forest Service Perspective. In *Proceedings of the GIS '87 Symposium*. American Society of Photogrammetry and Remote Sensing. Falls Church, Virginia. pp.111–121.

Reminga v. United States. 1978. 448 F Supp. 45 (1978, W D Michigan).

Sety, M.L. and K. Chang. 1987. A Rationale for Considering the Geographic Information System Data Base an Asset. In *Proceedings of the GIS '87 Symposium*. American Society of Photogrammetry and Remote Sensing. Falls Church, Virginia. pp.122–127.

Stokey, E. and Zeckhauser, R. 1978. *A Primer for Policy Analysis*. N.W. Norton & Company. New York, New York.

Tomlinson, R.F. and A.R. Boyle. 1981. The State of Development of Systems for Handling Natural Resources Inventory Data. *Cartographica* 18(4):65–95.

Wakeley, R.R. GIS and Weyerhauser — 20 Years Experience. In *Proceedings of the GIS '87 Symposium*. American Society of Photogrammetry and Remote Sensing. Falls Church, Virginia. pp.446–455.

Walsh, S.J., D.R. Lightfoot, and D.R. Butler. 1987. Recognition and Assessment of Error in Geographic Information Systems. *Photogrammetric Engineering and Remote Sensing* 53(10):1423–1430.

Zinn v. State. 1983. 112 Wisconsin 2d 417, 334 N.W.2d 67.

9. CONCLUSION

Human societies have become increasingly dependent for their well-being on the ability to collect and analyze geographic information. The world is becoming more crowded and resources are becoming more scarce. A new urban subdivision, a mine, a power plant, or a waste disposal site are projects now scrutinized by diverse regulatory agencies and frequently subject to public opposition. At the international scale, nuclear fallout, acid rain, desertification, toxic chemicals, and deforestation have become widely recognized problems that directly affect the economic and social well-being of the global human population.

Geographic information systems are a powerful resource for analyzing the inter-related systems involved in these types of problems. They provide flexible methods for exploring relationships among geographic data and assisting experts from diverse fields in pooling their knowledge to solve complex problems. Indeed, our success in dealing with many global environmental issues will depend on this type of multidisciplinary effort.

Using GIS technology, geographic information can be assembled and applied in new ways. The GIS offers a practical means to manage large and diverse spatial data bases and provides effective tools to understand the relationships among diverse phenomena. An increasing number of decision-makers and managers have recognized that GIS technology will be essential if they are to address the expanded mandates and complex decisions they now face.

GIS technology has developed at a remarkable pace over the past two decades. Yet, the technological changes have occurred much more rapidly than the institutional ones. We are only beginning to grapple with the many managerial, legal, and social issues that are accompanying the wide-spread use of geographic information systems.

A NEW WAY OF LOOKING AT GEOGRAPHIC DATA

The printing press not only made the production of written material faster and less expensive, it revolutionized the way information flowed within the society. In a similar way, GIS technology has dramatically changed the rate at which geographic information can be produced, updated, and disseminated. Map updating tasks that required months of manual effort are now done in hours, and spatial analysis capabilities that were unavailable a decade ago are now common place.

The GIS has not only made the production and analysis of geographic information more efficient, it is changing the way geographic information is perceived and used. It is a technology that makes geographic data more malleable, more easily shaped by the user into the form best-suited to the application at hand. It makes geographic information more easily customized. In the past, the production of a special purpose map was a costly undertaking. Today a GIS can make the special purpose presentation of data quick and inexpensive.

The traditional map is like a "snapshot" of the geographic data from which it was compiled. It represents a set of geographic information, usually at a single point in time. The map is updated at infrequent intervals, owing to the cost and time to produce an updated version. For this reason, the map tends to be a standardized, general purpose product designed to serve as wide a range of users as possible.

In a GIS, the storage of the data is independent of the mode of presentation.

The physical map becomes a relatively inexpensive output product that can be generated quickly and customized for a single application. In addition, the geographic data base used to produce the map can be continuously updated. As a result, the physical map becomes a customized "snapshot" of a continuously changing geographic data base. It is a view of the data, selected and organized by the user to best serve the application at hand.

The processing power of the GIS has also enabled geographic information to be used in a qualitatively different way. Complex analyses can be iteratively refined towards an optimum solution, an approach that would be prohibitively expensive using manual methods.

GIS TECHNOLOGY

GIS technology has continued to develop at a rapid pace. An ever increasing demand has created a competitive environment for producers of hardware and software. GIS technology has become less expensive and more reliable. Hardware prices continue to fall. However, software prices have tended to remain steady, reflecting the increasing cost of developing and maintaining ever more complex software.

The increasingly knowledgeable user community has been a significant factor in the direction of technology development. GIS users are demanding that systems accept data in diverse formats from existing digital data bases, that they be easier to use, and achieve ever higher levels of performance. As a result, GISes have incorporated advanced graphics, more powerful computers, more competent data exchange and data base functions, and, in some cases, expert system techniques to reduce the operator's workload. Although important technology issues still remain, particularly in the area of data capture, current market forces are providing the necessary incentive to continue the fast pace of development.

DATA BASE CREATION

Perhaps the greatest technical impediment to implementing a GIS is the creation of the data base. It is a difficult and costly undertaking to develop a suitably accurate data base from diverse error prone source materials. It is a task that commonly represents 75% or more of the total cost of implementing a GIS. Once the data base has been developed, there is also the substantial on-going maintenance cost to keep the data base current.

Much of the source material used to create the geographic data base is in the form of paper maps. Manual digitizing remains a very versatile and important means of encoding paper maps. However, in many situations, scanning systems are providing a more cost-effective means of data entry. The continuing improvements in scanning technology, particularly the introduction of expert system techniques, will further extend its use. There is also a steady increase in the use of digital methods in the data collection process itself, thereby eliminating the digitizing step entirely. Although the data entry problem is acute, it is one that will eventually lessen.

Unfortunately, the rush to digitize often compromises the need for a careful data base design. The creation of a large data base does not in itself guarantee that the data needed will be included or will be of adequate quality. Data cannot be simply dumped into storage; they must be in a form that can be efficiently retrieved and analyzed. Also, the priorities used in creating the data base must take into account the user needs to be supported. This requires considerable planning by experienced and highly trained personnel — which are currently in very short supply.

One of the perplexing problems in data base development has been justifying the considerable expense to create and maintain it. It represents a significant investment with an undefined asset value. The value of the data being collected and the information being produced is largely unknown. To develop a more rational means of setting data capture and analysis priorities, a realistic measure is needed of the value of GIS data and their actual effect on final decisions.

OVERCOMING INSTITUTIONAL BARRIERS

It is the political and institutional issues more than the technical ones that have been the major obstacles to the introduction of GIS technology. A GIS is not "bought", it evolves and becomes part of the information system of an organization. Its introduction fundamentally changes the way an organization can and will use data. It affects the political power structure of the organization as much as it affects the mechanics of work. Managers need to understand not only the principles of the GIS technology they use, but also the associated social, economic, and political issues. For example, the digitizing of maps can involve inter-agency negotiations to share data costs, issues of data ownership and copyright, and consideration of legal liability and individual privacy rights.

The implementation of a GIS with its many diverse data sets, all accurately registered, and with the analytical power to serve multiple organizations is a very costly and long-term undertaking. It is a project with large up-front costs and future benefits that are difficult to quantify. Tomlinson (1989) has noted that there is commonly a 4 to 5 year period before successful installations achieve a positive cost-benefit.

In many cases cooperative agreements among government agencies and with the private sector have enabled conversion costs to be shared and quality standards to be set. However, the considerable administrative overhead to initiate and manage data sharing agreements has proved to be a significant obstacle. As a result, there is an incentive for individual public organizations to minimize their digitizing efforts and maximize their use of publicly provided data. Cooperation among government agencies has been more successful in the sharing of GIS expertise and existing digital data.

The demand by local levels of government, such as municipalities, for geographic data in digital form has put pressure on mapping agencies to speed up the digitizing of geographic data. It has also led some users to generate their own data sets. While this may solve their immediate short-term information needs, it compromises longer-term objectives.

These short-term solutions have tended to undermine efforts to share data conversion costs and to standardize data formats and quality. Instead, the data collection effort is duplicated and the data quality and compatibility issues are handled on an *ad hoc* basis. This tendency has been particularly evident in land information system applications at the municipal and county levels. For example, utility companies and departments of public works have proceeded quickly in converting land records into digital form and implementing a GIS to manage them. They have tended to proceed more quickly than the agencies actually responsible for the land records (Duecker 1987). Strong institutional incentives are needed to encourage cooperative, longer-term perspectives in the creation and maintenance of GISes.

Traditional concepts of data ownership have also been an impediment to the introduction of GIS technology. For example, there is often a tendency for individual departments or organizations to think of data as "their data", when, in fact, the data were gathered at the expense and for the benefit of a larger constituency. It is sometimes difficult to shift this perception to a view of the GIS data base as a "corporate resource" to be managed primarily for the benefit of the entire user constituency it was created to serve, not for the convenience of the data base custodian. Controversy over control of data has compromised many GIS implementation efforts.

NEED FOR TRAINED INDIVIDUALS

Perhaps the greatest challenge to making GIS technology more widely available will be to find competent personnel. There is a critical shortage of people to satisfy current research, development, education, and operational needs, and the demand for personnel is accelerating. There are not enough educational institutions to provide training, and it is not even clear what educational background is needed.

Most GIS-related training tends to focus on technology training, that is, training individuals to operate a GIS. But users must develop expertise in methods of applying the technology. By its very nature a GIS is a data integration machine. To use a GIS effectively, the analyst requires sufficient background to understand the nature of the problems being addressed and the characteristics of the data being analyzed. Yet, one cannot be expert in all fields. It is this competing need for breadth across disciplines and depth within a discipline, and for theory as well as practical experience, that makes the definition of training needs so difficult. In addition, experience is in short supply and better means are needed to disseminate that hardwon knowledge.

The need for trained personnel is even more severe in less-developed countries. In these countries, the stature accorded individuals with specialized technical training often brings them immediate advancement to managerial positions in which they do not use those technical skills. Alternatively, technically competent individuals are attracted to more developed countries that offer higher standards of living and more opportunity for career advancement. As a result, the few technically trained personnel may quickly become unavailable.

LIABILITY

The legal and political liability of distributing geographic information is potentially the most explosive issue to accompany the introduction of GIS technology. A GIS facility is created for the purpose of generating and disseminating information. In so doing, the individuals who control that information resource exercise substantial power. Where there is power, there should be accountability. In the operation of a GIS, the issue is who is responsible?

Those given access to a GIS may have at their disposal a wealth of valuable information about individuals, their activities, and their property. The personnel of a GIS facility may be well trained to maintain high technical standards, but what are their ethical standards? Their knowledge will define what they are able to do, but what checks will limit what these individuals may choose to do? Who will provide the legal and ethical guidelines for their work, and who will have final authority and assume final responsibility for their actions? Is there a single individual who is given the authority and responsibility to manage the GIS, or is responsibility jointly held, with no single individual held accountable? The power of a GIS is a double-edged sword. In providing instant access to vast amounts of data, it provides the opportunity to abuse, to misinform, and to invade the privacy of individuals on a greater scale than ever before.

There is public concern about coordinated gathering and analysis of digital information about individuals and their property. There is a growing awareness of the ease with which digital data files can be transferred, accessed, and combined in ways unforeseen by the provider and without the knowledge or consent of the individuals affected. Unauthorized access and copying of data can be done instantaneously, without leaving a trace. In West Germany, citizen concern over privacy forced a federal census to be cancelled, and, in Sweden, stringent laws forbid file matching of digital data on individuals. In the United States, some 70 public laws regarding information disclosure, confidentiality, and rights of privacy have been passed in recent years (Abler 1987).

The power of a GIS to rapidly analyze large data files has the potential to compromise individual rights. Many records, such as those of municipalities, contain information about the holdings and activities of individual citizens. Though many of these records are already publicly available, the time and expense to manually access them makes it impractical to search the entire data set. The inefficiency of retrieval serves to protect the individual's privacy. However, the computer makes selective search and retrieval fast and efficient. Data files that were not intended to be integrated can be easily combined or matched using any common data field, such as a street address or a name. Large data sets can be efficiently searched to target individuals by the value of their home, their tax arrears, the schools their children attend, perhaps their ethnic background, and so on. The courts are only beginning to deal with the difficult trade-offs of protecting private interests while making available the information that public agencies require to fulfill their mandates.

There are serious unresolved concerns about the liability of the producer for the quality of its data products. In the United States, many court decisions have found the producer liable for fatal injuries caused, in part, by incorrect or non-standard data. The introduction and adoption of data standards can not only protect the interests of the producer, but can also provide the user with a more dependable and predictable information source.

There is also an uncomfortable political dimension to GIS management. To what extent can those who control the budget of a GIS facility influence the information that is produced? How immune are the individuals who operate that facility from requests to suppress or modify potentially embarassing information. Whether it is a local city councillor or high-ranking officials at the regional or national levels, there is a potential conflict of interest if the GIS they fund can generate information contrary to their interests.

THE POLITICAL NATURE OF A GIS

Computers in general, and GISes in particular, are not objective decision-making systems. As Klosterman (1987) has noted, computers are inherently political. They can be used to hide or downplay the importance of the assumptions on which their analyses are based. Centralized control of computer data bases, such as a GIS, tends to increase the power of the bureaucrats, administrators, technical experts, and computer literate groups who use them at the expense of those who lack the expertise or access to these systems. Policy-makers and the general public tend to view information as more accurate, more objective, and more current simply because it was computer-generated. Not only do the data appear more authoritative, the politically neutral language used to describe the data preparation, analysis, and results reinforces the air of objectivity. Computer modelling and analysis does, however, require that numerous choices be made in the selection of data, choice of analysis methods, and the presentation and interpretation of results. These choices are inherently political since they directly affect the results that will be obtained.

Computer-based analyses can be used to mystify as easily as they can be used to clarify. Political choices can easily be hidden within procedures made too complex for the uninitiated to decipher. For this reason, it is important that policy choices and assumptions be made explicit and that analytical procedures be fully documented, available for scrutiny, and made clear to those who will use the information.

The GIS has provided the opportunity to greatly improve the speed and accuracy with which geographic data can be handled and the diverse issues to which they can be applied. To reap the benefits requires not only technically qualified users. It requires managers who understand GIS technology and the issues that surround its use. The challenge ahead is to provide the management perspective and direction to ensure that the power of GIS technology is channeled so as to advance our values and objectives not compromise them.

REFERENCES

Duecker, K.J. 1987. Multi-Purpose Land Information Systems: Technical, Economic, and Institutional Issues. In *Proceedings of the Eighth International Symposium on Automated Cartography*. American Society for Photogrammetry and Remote Sensing. Falls Church, Virginia. pp.1–11.

Klosterman, R.E. 1987. Guidelines for Computer-Aided Planning Models. In *Proceedings of the URISA '87 Conference*. Urban and Regional Information Systems Association. Washington, D.C. Volume 4:1–14.

Tomlinson, R.G. 1989. *GIS Challenges for the 1990's*. Presentation at the National Conference on Geographic Information Systems — Challenge for the 1990's. Held February 27–March 3, 1989 in Ottawa, Canada.

APPENDIX A:

Abbreviations for Units of Measure

Abbreviation	Unit	Equivalence	
m	metre		
km	kilometre	1000	m
cm	centimetre	.01	m
mm	millimetre	.001	m
μm	micron	.000001	m
ha	hectare	10,000	sq m
in	inch	2.54	cm
ft	foot	0.305	m
acre	acre	0.405	ha

APPENDIX B:

Data Sources

NATIONAL SCALE SOURCES OF DIGITAL DATA IN THE UNITED STATES

Landsat Data

EOSAT
4300 Forbes Blvd.
Lanham, Maryland 20706
(301) 552-0500
(800) 344-9933 (US and Canada)

SPOT Data

SPOT Image Corporation
1897 Preston White Drive
Reston, Virginia 22091
(703) 620-2200

AVHRR, Coastal Zone Colour Scanner, GOES, and Seasat Data

US Department of Commerce
National Oceanic and Atmospheric
  Administration (NOAA)
National Environmental Satellite Data
  and Information Service (NESDIS)
User Services
Code E/OC21
1825 Connecticut Ave N.W.
Washington, D.C. 20235
(202) 606-4549

DLG Files, Digital Elevation Data, and Land Use/Land Cover Data

Earth Science Information Center
US Geological Survey
507 National Center
Reston, Virginia 22092
(703) 860-6045
(800) 872-6277 (US only)

TIGER File Data and Census Attribute Data Sets

Customer Services Branch
Data User Services Division
Bureau of the Census
Washington, D.C. 20233
(301) 763-4100

NATIONAL SCALE SOURCES OF DIGITAL DATA IN CANADA

National Digital Topographic Data

Database Information Services
Product and Services Division
Energy, Mines, and Resources Canada
615 Booth Street, Room 408
Ottawa, Ontario K1A 0E9
(613) 995-0314

Data from the Canada Soils Information System

CanSIS Project Leader
Land Resource Division
Centre for Land and Biological
  Resource Research
Agriculture Canada, Research Branch
K.W. Neatby Building
Ottawa, Ontario K1A 0C6
(613) 995-5011

Data from the Canada Land Data System (includes the Canada Geographic Information System)

Environmental Information Systems Division
State of the Environment Reporting Branch
Environment Canada
Ottawa, Ontario K1A 0H3
(613) 997-2510

Data from the National Atlas of Canada

Product Development and Client Services
National Atlas Information Service
Canada Centre for Mapping
Energy, Mines, and Resources Canada
615 Booth Street, Room 650
Ottawa, Ontario K1A 0E9
(613) 992-4252

Landsat and SPOT Satellite Data

Data Centre
Radarsat International Inc.
13800 Commerce Parkway
Richmond, B.C. V6V 2J3
(604) 244-0400

NOAA/AVHRR Satellite Data

Order Desk
Satellite Operations Centre
Canada Centre for Remote Sensing
2464 Sheffield Road
Ottawa, Ontario K1A 0Y7
(613) 990-8033

DIGITAL DATA FROM STATISTICS CANADA

Census Data

Electronic Data Dissemination Division
Statistics Canada
9th Floor, R.H. Coates Building
Ottawa, Ontario K1A 0T6
(613) 951-8200
(800) 465-1222 (Canada only)

Area Master File and CARTLIB Data

Client Liason
Geography Division
Statistics Canada
Jean Talon Building, 3rd floor
Ottawa, Ontario K1A 0T6
(613) 951-3889

INDEX